Exploring Elementary Mathematics

A Small-Group Approach for Teaching

A Series of Books in the Mathematical Sciences
 Editor: Victor Klee

Exploring Elementary Mathematics

A Small-Group Approach for Teaching

Julian Weissglass
University of California, Santa Barbara

W. H. Freeman and Company
San Francisco

AMS (MOS) Classification: 98B10; 98B15
  Discovery method; Laboratory method

Sponsoring Editor: Peter Renz
Project Editor: Pearl C. Vapnek
Copyeditor: Douglas Bullis
Designer: Perry Smith
Production Coordinator: William Murdock
Illustration Coordinator: Batyah Janowski
Artist: J & R Technical Services
Cartoonist: John Johnson
Compositor: Advanced Typesetting Services of California
Printer and Binder: Webcrafters, Inc.

Library of Congress Cataloging in Publication Data

Weissglass, Julian.
    Exploring elementary mathematics.

    (A Series of books in the mathematical sciences)
    Bibliography: p.
    Includes index.
    1. Mathematics—Study and teaching (Elementary)
I. Title.
QA135.5.W39   372.7   79-14931
ISBN 0-7167-1027-7

Copyright © 1979 by W. H. Freeman and Company

No part of this book may be reproduced by any mechanical,
photographic, or electronic process, or in the form of a
phonographic recording, nor may it be stored in a retrieval
system, transmitted, or otherwise copied for public or
private use, without written permission from the publisher.

Printed in the United States of America

9 8 7 6 5 4 3 2 1

To Peri, Jonathan, and young people everywhere,
in the hope that their learning will be enjoyable

Contents

Preface *xi*
To the Student *xiii*
The Equipment *xvii*

1
Set Theory *1*

Lab Exercises: Set 1 *1*
Comment: Operations on Sets *3*
Lab Exercises: Set 2 *6*

2
Counting and Numerals *12*

Lab Exercises: Set 1 *12*
Comment: The Development of Counting *14*
Lab Exercises: Set 2 *17*
Comment: Numerals *22*
Lab Exercises: Set 3 *23*
Comment: Positional (Place-Value) Numeration Systems *25*

3
Addition and Subtraction *28*

Lab Exercises: Set 1 *28*
Comment: Properties of Addition and Subtraction *33*

4
Algorithms for Addition and Subtraction *38*

Lab Exercises: Set 1 *38*
Comment: Rules for the Abacus *41*
Lab Exercises: Set 2 *43*
Comment: Algorithms for Addition and Subtraction *45*

5
Multiplication *33*

Lab Exercises: Set 1 *53*
Comment: Properties of Multiplication *55*

6
Algorithms for Multiplication 59

Lab Exercises: Set 1 59
Comment: Algorithms for Multiplication 62

7
Division 67

Lab Exercises: Set 1 67
Comment: Understanding and Doing Division 70

8
Number Theory 76

Lab Exercises: Set 1 76
Comment: Prime Numbers 79
Lab Exercises: Set 2 82
Comment: The Greatest Common Factor; Odd and Even Numbers 85

9
Topology 90

Lab Exercises: Set 1 90
Comment: The Königsberg Bridge Problem 93
Lab Exercises: Set 2 94
Comment: Euler's Formula 98

10
Rational Numbers 100

Lab Exercises: Set 1 100
Comment: Fractions and Rational Numbers 104
Lab Exercises: Set 2 109
Comment: Addition and Subtraction of Rational Numbers 112
Lab Exercises: Set 3 113
Comment: Multiplication and Division of Rational Numbers 116

11
Positive and Negative Numbers 122

Lab Exercises: Set 1 122
Comment: Addition and Subtraction of Positive and Negative Numbers 125
Lab Exercises: Set 2 126
Comment: Multiplication and Division of Positive and Negative Numbers 131

12
Introduction to Geometric Concepts 136

Lab Exercises: Set 1 136
Comment: Points, Lines, and Planes 142
Lab Exercises: Set 2 144
Comment: Curves 148

13
Lines, Angles, and Triangles 153

Lab Exercises: Set 1 153
Comment: Angles 156
Lab Exercises: Set 2 157
Comment: Triangles—Congruence Axioms, Centroid, Orthocenter 161
Lab Exercises: Set 3 163
Comment: Triangles—Circumcenter and Incenter 165

14
Area and the Geoboard 169

Lab Exercises: Set 1 169
Comment: The Concept of Area and Its Properties 173
Lab Exercises: Set 2 179
Comment: The Pythagorean Theorem 184

Contents

15
Decimals *188*

Lab Exercises: Set 1 *188*
Comment: Computing with Decimals *191*
Lab Exercises: Set 2 *194*
Comment: Terminating, Repeating, and Nonrepeating Decimals *196*

16
Measurement: The Metric System *199*

Lab Exercises: Set 1 *199*
Comment: The Metric System *203*
Lab Exercises: Set 2 *205*
Comment: Errors in Measurement *207*

17
Tessellations: Mathematics in Art and Floor Tilings *213*

Lab Exercises: Set 1 *213*
Comment: Regular Polygons *218*
Lab Exercises: Set 2 *219*
Comment: Tessellations in Mathematics and Art *220*

18
Transformations *226*

Lab Exercises: Set 1 *226*
Comment: Translations and Reflections *234*
Lab Exercises: Set 2 *237*
Comment: Products of Reflections *239*

19
Symmetry *243*

Lab Exercises: Set 1 *243*
Comment: The Concept of Symmetry *247*
Lab Exercises: Set 2 *250*
Comment: The Algebra of Symmetry *253*

20
Probability *257*

Lab Exercises: Set 1 *257*
Comment: Basic Concepts of Probability *260*

21
Statistics *267*

Lab Exercises: Set 1 *267*
Comment: Basic Statistical Concepts *270*

22
Mathematical Explorations *274*

Exploration 1 *274*
Exploration 2 *276*
Exploration 3 *277*

Bibliography *281*
Index of Quotations *285*
Index *287*

Preface

This book serves as a main text or as a supplement in mathematics courses for prospective elementary teachers. As a supplement, it covers the use of manipulative materials and small-group learning. Since this approach to mathematics is more inviting and less intimidating than others, this text may be used in courses designed to overcome fear or avoidance of mathematics, as well as in courses for future teachers. Also, since this book presents basic mathematical concepts in an experiential way, it can be used effectively with students who need remedial work. As a text for education students, this book covers all the standard mathematical material, but with a fresh approach—one that will increase prospective teachers' appreciation and understanding of the mathematics and of the hands-on learning that is so important in elementary school teaching.

Knowing how this book originated will help you understand its aims. More than six years ago, when lecturing to a large class of prospective elementary school teachers, I stated that mathematics is best taught by having students *do* mathematics rather than by having them listen to someone talking about it. After the lecture, Louis Berneman, a student, came forward to suggest that I "practice what I preach." Since then, while working with my classes at the University of California at Santa Barbara, I developed the approach presented in this book. Both my students and I have found this approach rewarding. I thank Louis for his outspokenness and hope that his example will encourage other students to speak up and take a more active role in their education.

We learn best by doing and discovering. Lectures and reading can guide us, but active involvement provides a different quality of interest and understanding. Using this text allows students—preferably working in small groups—to discover mathematics for themselves with a minimum of guidance from the instructor. Generally, students using this text will be working with the same materials available in elementary schools, but they will be using them in a far more complex way. This has the benefit of allowing future teachers to work with equipment that they can use in their classrooms and to experience the excitement and involvement of active learning—an important and often neglected aspect of mathematics education.

Since this book is designed for a small-group laboratory approach, it is different from most textbooks. Each chapter has one or more sets of lab exercises followed by a section titled "Comment." The lab exercises always come first and are designed to be done before the reading. Although the laboratory experience will leave some questions unanswered, it will provide the exploratory experiences that prepare the student for the explanation.

Since different people work at different rates and since all mathematical topics do not require the same amount of laboratory work, it was impossible to make each set of lab exercises the same length. However, most of the lab sets can be done in from one to one and a half hours.

A Comment section appears either when an extended explanation is necessary or when it is appropriate to summarize some concepts before proceeding with more activities.

At the end of each chapter, the important terms and concepts that are discussed in that chapter are listed. In addition, there are review exercises for each chapter.

Rather than describe in greater detail my approach to the course, I invite you to turn to the text itself and explore these mathematical and teaching ideas in the laboratory and classroom. If any of the equipment described in the text is unfamiliar to you, there is a description as well as a list of distributors on pages xvii–xviii. If you are using or are seriously considering using this book either as a primary text or as a supplement, you may want to look at the Instructor's Manual, which is available from the publisher. But that can be done later. The clearest initial impression of the book will come from browsing through it or, even better, trying out its ideas.

I would never have completed this book without the assistance of many people. I appreciate the patience of the many students who worked through the preliminary versions and helped me learn. I am also grateful to the many teaching assistants who helped both the students and me. In particular, Bob Grone and Clay Fickle gave many thoughtful suggestions. My good friend Margo Nanny went over a preliminary edition, and her insightful suggestions were extremely valuable. My wife Theresa's cheerful encouragement and excellent suggestions during the preparation of the final manuscript are much appreciated. The careful typing (and good humor) of Sonia Ospina, Ruth Hillard, Shirley Clarke, Marianne Braun, Margaret Ruhl, and Jill Weaver made my work more enjoyable. I am particularly indebted to Dee Brannon, who typed many preliminary versions and oversaw their production. Her competent assistance was invaluable. Finally, I thank Victor Klee, who encouraged me to submit the book to W. H. Freeman and Company, and the editorial staff for their assistance.

July 1979   Julian Weissglass

To the Student

A course or mathematics laboratory based on this text will be different from other mathematics courses you have taken. You will be working with manipulative materials and exchanging questions and answers with your classmates as you work. Learning this way is easy and natural, but it is not the usual approach in mathematics courses; therefore, I offer a brief explanation and some suggestions to help you.

Learning in small groups from your peers is common, natural, and enjoyable. We learn games this way. To a great extent this is the way we learn language. If children began the study of language by first studying all the rules of grammar, language would be a very distasteful subject. Mathematics is simpler, more orderly, and easier to comprehend than a language such as English. Much of mathematics is governed by a small number of abstract rules. This tempts us to use the shortcut of simply memorizing these rules rather than exploring and discovering mathematics in a concrete and human context.

This shortcut of memorization makes mathematics forbidding, deadens student interest, and can stand in the way of understanding. What do I mean by *understanding* here?

Understanding requires that new ideas and facts be related to those you already know and that all this information be remembered so that it can be used to solve old or new problems.

Notice that the understanding we will be seeking goes beyond mere memorization and includes the ability to generalize and to apply what has been learned in new contexts.

Human beings are inherently curious and enjoy learning and understanding. Have you ever watched a child explore a puddle of water or investigate a leaf? Have you seen children's joy in building towers of blocks or their fascination with words or numbers? I have designed this book so that you will find the information and experiences in this course enjoyable, so that in turn you will be able to provide pleasant mathematical learning experiences for your students in the future. It *is* possible to provide classroom learning experiences that are both significant and enjoyable. There are two conditions that are essential to achieving these desirable results.

First Condition: *The attitude in the classroom must be positive.*

If students are upset or disturbed or if they fear or dislike the subject, their full attention will not be available for learning. It is up to you, as a student and as a future teacher, to identify and deal with the fears, hesitations, and disturbances that get in the way of learning. While a class is not a therapy session, remember that from time to time we all experience anxiety about personal matters or classroom pressures. It is worth the effort to deal with these anxieties, both now with your classmates and in the future when you are teaching, because they destroy one of the preconditions for learning—a positive attitude toward the subject and the learning situation.

Second Condition: *Material must be presented in the proper context and at the proper rate for understanding.*

If ideas are presented out of context or too rapidly, they are not understood. In addition, such an experience is distressing to the learner and thus violates our first condition. Furthermore, the effect does not stop there. We have all experienced situations in which we did not understand something, both in and out of school. Some of these early experiences were very hurtful. Although we may not identify them as such, they were *also* defeats in attempting to learn. Many of the negative feelings and learning blocks that we have stem from these early experiences.

Fortunately, it is not necessary to have completely recovered from the effects of previous distressing experiences in order to learn. We can create situations in which adequate attention is available for learning and proceed in ways that increase the likelihood that new information will be presented in context and at the proper rate. Although many lecturers are quite successful at this, I have found that the small-group laboratory approach has many advantages in seeking to satisfy the two conditions above.

The Small-Group Laboratory Approach

In the small-group approach, people learn from one another. This has benefits that are related to the conditions for learning. First, learning from peers eliminates the fear of authority that often interferes with learning. Children experience much criticism from parents and teachers. Because of these past experiences, even a teacher who does not intend to criticize students will still raise the fear of criticism in many

Figure 1

students. For this reason it is easier to ask questions of or accept suggestions from a peer. Second, the communication of ideas that takes place in a small group is supportive of the learning process. The experience of helping someone learn will give confidence and overcome feelings of inadequacy that many people may have. Also, teaching others assists your own learning. One learns best what one is teaching. Finally, as a student, the small-group approach makes it easy for you to get to know your classmates. This will help you learn by reducing any feelings of isolation you may have. I encourage you to take advantage of this opportunity to get to know your classmates. One way of doing this is to take a few minutes to talk about something good that happened to you recently or about a good learning experience. Perhaps your instructor will suggest topics for you to talk about. Be sure that everyone gets a turn to speak and try to keep the discussion positive and cheerful. It will increase everyone's receptiveness to doing mathematics if attention is directed away from upsetting experiences and toward pleasant ones.

The small-group laboratory approach also has definite advantages with regard to the second condition—that new information must be presented in context. An obvious difficulty with lectures is that of presenting information in the proper context and at the proper rate for a diverse audience. It is clear that more individualization than lecturing can provide may be needed to meet the second condition. I have found that groups of four work best, as they increase the resources available within each group yet still allow individual interaction.

Asking Questions

Another advantage of a small group is that the student can assume a greater responsibility for the rate of presentation of new information. How fast and how

To the Student

Figure 2

well you learn depend largely on you. I think you will often be delighted and surprised and feel more confidence in yourself. However, sometimes you may feel frustrated or confused. This is also part of learning. The frustration and confusion can be reduced by the small groups themselves as they try to ensure that new ideas are presented in context and at the proper rate. One of the best ways to do this is by asking questions and encouraging others to do the same.

Here are some suggestions for encouraging question-asking in your group:

Do not criticize the questions of others nor hesitate to ask questions yourself for fear of criticism. Set an example for those in your group in this respect.

Encourage questions by being friendly and cheerful and by expressing thanks and appreciation to those who ask questions.

If someone is obviously confused, encourage that person to ask a question or ask a question yourself to try to clear things up.

Use a series of simple questions to bring out an explanation rather than simply telling someone the answer.

Remember that the *process* of learning is even more important than the *facts* learned.

Show respect for your group and keep up morale by being on time and prepared. Being prepared means knowing what questions you want to ask; it does not mean knowing all the answers.

Approach the group each day with aims of helping everyone else learn, asking interesting questions, and making the experience an enjoyable one.

If your group encounters difficulties in freely asking questions, it is helpful to pair off and examine the causes of this problem and then report to the full group. You might each try to remember when you were able to ask questions freely and why it was that you stopped.

Laboratory Equipment

There are several advantages to using the various kinds of manipulative equipment called for in this book. The material and equipment you will encounter are attractive and interesting in their own right. In addition, you will see their applicability to your own teaching. Even though we will not investigate in detail the procedures for teaching with this equipment, you will readily see how the mathematics you are learning and the equipment used to teach it relate to your future career.

Please remember as you read this book that I have not necessarily developed a topic in a way that would be best for beginning students. In fact, I have often chosen a more challenging approach. Frequently, the best approach for working with elementary-level students would be boring for you or be so familiar that you would not think about the processes involved.

Another advantage to using the equipment specified in this book is that it will supply mathematical experiences and ideas that you may have missed but that are necessary for understanding more abstract mathematical concepts. This is related to our second condition for learning. One source of our difficulty in understanding mathematics is that we are frequently presented with new information before we have an adequate foundation for it. Very basic experiences in the physical world that are a prerequisite for understanding mathematical concepts have often been missed or forgotten. As a result, we tend to rely on memorization without real understanding, and this knowledge may not be adaptable to new situations. The laboratory equipment enables us to have some basic experiences with the mathematical-physical world in a challenging way. These experiences will do more than simply facilitate our own learning. They will make it easier for us when we are teachers to stress the exploration of processes and the understanding of concepts rather than settling for rote memorization of techniques and the manipulation of symbols.

Using the Book

Many students have contributed their ideas and experiences to this book. This process should continue. If you have any suggestions for improvement, please send them to me in care of the publisher.

I hope you enjoy your learning. I have written this book to help you do so. If previous experiences with mathematics interfere, this obstacle is surmountable.

The Biographical Sketches

I believe everyone is born with the ability to learn and do mathematics. Obviously, not everyone preserves and uses this ability. For most people this is a result of bad experiences with mathematics, lack of information about opportunities, or lack of encouragement to continue their studies. For example, the oppression of women in society has resulted in women not achieving in mathematics proportional to their numbers, in the establishment of myths about women's ability for and interest in mathematics, and in ignorance about their achievements. Similar statements can be made about other groups. In the United States, Blacks, Chicanos, and Native Americans are notable examples.

In order to counteract the lack of information, I have provided throughout the text biographical information about some less well known mathematicians. I hope these sketches will encourage you to contribute toward the equalization of achievement in mathematics by minority groups and women. Information about better-known mathematicians is readily available in the books on the history of mathematics listed in the bibliography at the end of the book.

The Equipment

The equipment referred to in this book is described below. Any equipment not readily available is followed by one or more numbers. These refer to the list of commercial distributors at the end of this section and indicate at least one place where the equipment may be obtained. Much of the equipment can be constructed by students, and instructions for doing so are included below. A recommended equipment list for a class is included in the Instructor's Manual.

Abacus (1)
We will use the Chinese abacus. Inexpensive ones made of plastic are available.

Attribute Blocks (A-Blocks) (1, 4)
We use a set of 32 blocks (4 colors, 4 shapes, 2 sizes) that can be purchased or easily made from cardboard or construction paper. Another widely available set of A-blocks consists of 60 blocks (3 colors, 5 shapes, 2 sizes, and 2 thicknesses).

To make a set of A-blocks, trace 4 copies of the shapes in Figure 3 on the following page onto construction paper or cardboard. Color one set blue, one green, one red, and one yellow; or use colored construction paper.

Compass

Colored Rods (C-Rods) (1, 2, 4)
These are rods with each length (from 1 cm to 10 cm) a different color. They are available from many sources or can be simulated by colored strips constructed by the student. The text refers to the colors used by the Cuisenaire Company, which also makes an introductory set of 72 rods. When a container of colored rods is mentioned in the text, it should contain at least 20 white rods, 12 red, 9 light green, 4 purple, 5 yellow, 4 dark green, 4 black, 4 brown, 4 blue, and 4 orange.

To make a set of C-rods, cut strips from colored construction paper. The 20 white strips should be 1 cm $\times$ 1 cm; the 12 red strips 1 cm $\times$ 2 cm; and so on through the set of ten colors.

Geoblocks (optional) (1, 4)
These are a set of blocks to investigate three-dimensional configurations.

Geoboard (1, 2, 3, 4)
The geoboard we use is a square made of wood or plastic in which is implanted a 5 $\times$ 5 array of pins or nails. Rubber bands are used to outline geometric shapes.

Index Cards or Cardboard

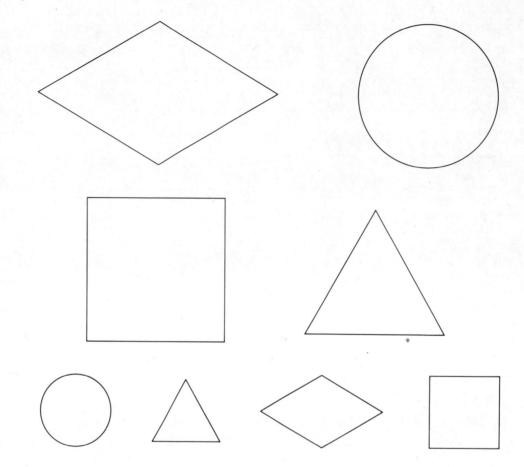

Figure 3

Meter Stick (1, 2, 3, 4)

**Metric Tape Measure (1, 3, 4)
or Adding-Machine Tape**

Mira (1, 2, 3)
The Mira is a semitransparent plastic device that helps students to discover and understand geometric relationships and properties.

Multibase Arithmetic Blocks (MBA-Blocks) (1, 3)
These blocks are available in Base 2, 3, 4, 5, and 10. Each base includes "units," "longs," "flats," and "blocks" that correspond to different powers of the particular base. For a photograph of Base 10 blocks, see Chapter 2.

Pattern Blocks (1, 4)
These are a collection of blocks in 6 different colors and shapes. We use them in Chapter 17 to investigate polygons in the plane.

Protractor

Red and Black Counters
These may be cut from construction paper.

Ruler

String

Tangrams (1)
Tangrams are easily constructed by the student. Instructions are included in Chapter 12. Plastic sets are available.

Trundle Wheel (optional) (1, 2, 3, 4)
This is used for measuring distances.

Commercial Distributors

1. Creative Publications, P.O. Box 10328, Palo Alto, CA 94303.
2. Cuisenaire Company of America, 12 Church St., New Rochelle, NY 10805.
3. Educational Teaching Aids, 159 W. Kinzie St., Chicago, Il 60610.
4. SEE, 43 Bridge St., Newton, MA 02195.

"There should be no element of slavery in learning. Enforced exercise does no harm to the body, but enforced learning will not stay in the mind. So avoid compulsion, and let your children's lessons take the form of play."—Plato, The Republic

"The principal agent is the object itself and not the instruction given by the teacher. It is the child who uses the objects; it is the child who is active, and not the teacher."—Maria Montessori

"It is in fact nothing short of a miracle that the modern methods of instruction have not entirely strangled the holy curiosity of inquiry; for this delicate little plant, aside from stimulation, stands mainly in need of freedom; without this it goes to wrack and ruin without fail."—Albert Einstein

1
Set Theory

"[T]he experience of beauty is not a simple but a complex experience. In mathematics it may be compounded of surprise, wonder, awe, or of realized expectation, resolved perplexity, a sense of unplumbed depths and mystery; or of economy of the means to an impressive result."—Henri Poincaré

Lab Exercises: Set 1

EQUIPMENT: One set of A-blocks for each group.

Since so much of mathematics is concerned with sets of objects, it is proper that we start with learning about the theory of sets. We will use A-blocks (*Attribute Blocks*) to begin our study, since they provide us with concrete experience (see Figure 1-1). In the Comment following this group of exercises, we will extend our consideration to more general sets.

1. Take out your A-blocks and play with them for a few minutes. What can you do with them? Can you make some interesting constructions and designs? Discuss the following with your group.

   (a) How would young people play with them? Would they play similarly to the way you play, or would they play differently?

   (b) If you were a teacher, how might you use these blocks in a classroom, based on what you know right now? Why might the loops be included? What might the cards be used for?

2. (a) Select one shape—for example, triangles. Take all the blocks of that shape and place them in one of the loops. While the rest of the group closes their eyes, one person remove a piece and hide it. The others open their eyes and try to name the piece that was removed. What was your thought process in

Figure 1-1

determining which block had been removed? If you were seven years old, how might you have solved the problem? Would it have helped to arrange the blocks in any particular way?

(b) Select a color, for example, blue, and repeat (a).

(c) Select a size and repeat (a).

3. (a) What are the shapes of the blocks? _____

_____

Make groups (subsets) of blocks so that each subset contains only pieces of the same shape. How many subsets do you have? _____ How many blocks are there in each subset? _____

(b) What are the colors of the blocks? _____

_____

Repeat (a) for colors. The number of subsets is _____. The number of blocks in each subset is _____.

(c) What are the sizes of the blocks? _____

_____

Repeat (a) for sizes. The number of subsets is _____. The number of blocks in each subset is _____.

4. Arrange the set of blocks into subsets so that each subset contains only those pieces with the same color *and* the same shape.

(a) How many subsets are there? _____

(b) How many blocks are in each subset? _____

(c) How do blocks within a subset differ from each other? _____

(d) Name some of the subsets. _____

_____

Definition: *A particular shape, color, or size is called a* value.

5. (a) Someone in the group silently thinks of two values (for example, *red* and *triangle*). Put all of the blocks having *either* of these values inside a loop, but leave one out. The group then tries to guess the missing block. (This game can be played by leaving the missing block on the table with all the others, or by hiding it from view while the group closes their eyes.)

(b) Do the same for three values.

6. Choose a value (a particular color, shape, or size). Enclose the subset of all pieces having this value (e.g., all yellow pieces) in a loop. Choose another value and enclose the subset of all pieces having this value (e.g., all small pieces) inside another loop. How many pieces belong inside both loops? _____

7. Each set of A-blocks contains several loops of string that can be used to enclose certain sets of A-blocks. There are also label cards (called *labels* for short) with values (red, small, circle, etc.) written on them. Arrange the loops and labels as shown in Figure 1-2.

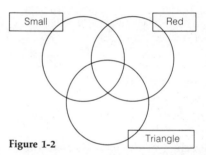

Figure 1-2

(a) Place blocks in the spaces where they belong.

(b) One person makes up a similar problem for the group using only cards with "not" on them. ("N" stands for "not.")

(c) Another person makes up a problem using labels with both positive and negative values.

You can play this game by yourself by choosing the cards at random.

8. (a) In Exercise 7, does any combination of labels leave no pieces outside the three loops? _____ If so, which one or ones? _____ Do any combinations leave one piece outside? _____ Write down a particular combination of labels that leaves the most pieces outside _____ (there may be more than one).

(b) Pick a number and see if you can label the loops so that there are exactly that number of pieces outside.

(c) Choose three labels at random. Try to predict the number of pieces that will be outside the loops before you place the pieces inside the loops.

9. *Game.* This is a game to be played after you have mastered Exercise 7. One person (called the *labeler*) chooses a positive label for each loop and places it face down by its loop. The aim of the game is for the other members of the group (called the *players*) to discover what the labels of the three loops are, by placing pieces in the spaces and finding out whether they belong there or not. The labeler may want to draw a diagram for his or her reference, showing the labels for each loop.

(*Note:* The loops divide the table into seven regions, and each block belongs in exactly one region. The only information given to the players is whether or not a block is correctly placed in its unique region.)

The players may want to figure out some way of recording the information they receive in each move.

A more difficult version of this game is created by using all negative labels. The most difficult is to have no restriction on the labels.

Comment: Operations on Sets

"Neglect of mathematics works injury to all knowledge, since one who is ignorant of it cannot know the other sciences or the things of the world."—Roger Bacon

It will be helpful for you to have a set of A-blocks available while reading this section. If you do not have a set, you can make one by following the instructions on page xvii.

Some terminology will be useful in discussing the theory of sets that you have been exploring with the A-blocks.

Definition: *A* set *is a collection of objects.*

For example, the set of students at your school, the set of large, square A-blocks, the set of books in the local library.

Figure 1-3

Definition: *The objects in a set are called the* elements, *or* members, *of the set.*

A set is specified once we can tell whether an element is in the set or not. For A-blocks we will often do this physically by putting all the red triangles inside a loop of string or in a box. Another way of identifying a set is by listing all its members within braces. For example, the set of the first three letters of the alphabet can be written as $\{a, b, c\}$. Usually, we will use capital letters to denote sets. Thus we write $A = \{a, b, c\}$, or $B = \{$wrench, screwdriver$\}$ or $A = \{$Bob, Mary, Sue$\}$.

Another way of specifying sets is to use a symbol and a description of the conditions the symbol must satisfy in order to qualify for membership. For example, let $A = \{n \mid n$ is a whole number greater than 2 and less than 7$\}$. Then $A = \{3, 4, 5, 6\}$. As another example, let $B = \{x \mid x$ was a United States President whose last name began with J$\}$. Then $B = \{$Jefferson, Jackson, Andrew Johnson, Lyndon Johnson$\}$. Sometimes we use dots . . . to indicate that the elements continue in the obvious pattern. For example, $A = \{0, 1, 2, 3, 4, \ldots\}$ indicates all whole numbers, $B = \{0, 2, 4, 6, \ldots\}$ indicates all even whole numbers.

You have before you a set of A-blocks. How many elements are there in this set? ____ Separate the blocks into groups according to shape. How many groups are there? ____ Each group is itself a set, for example, the set of triangles. We say that the set of triangles is a *subset* of the set of A-blocks. In general,

Definition: *A set B is a* subset *of a set C if every element of set B is an element of C. Two sets are* equal *if they have the same elements.*

The symbol $\subset$ is often used to denote "is a subset of." For example, $\{1, 2\} \subset \{1, 2, 3\}$.

10. For each of the following pairs of sets A and B, determine whether one is a subset of the other.
    (a) $A = \{a, b, c\}$ $B = \{b, c, d, e\}$
    (b) A = set of large A-blocks
        B = set of large red triangular A-blocks
    (c) $A = \{1, 2, 3, 4, 5, \ldots\}$
        $B = \{2, 4, 6, \ldots\}$
    (d) A = set of red A-blocks
        B = set of squares

11. (a) Describe some subsets of the set of A-blocks.
    _____
    (b) Place all the blocks that have four corners into one subset. How many elements are there in this subset? _____
    (c) Arrange the set of blocks into subsets so that each subset contains only those pieces that have the same color and the same size. How many subsets are there? _____ How many blocks are there in each subset? _____ Within each subset, in what ways do the blocks differ? _____ Name some of the subsets.
    _____
    (d) Do part (c) for pieces having the same shape and the same size.

■ Sometimes we will want to refer to a possible collection of elements without knowing whether there are any elements at all in the collection. For example, if you were not familiar with the A-blocks, you might refer to the subset of pentagons or the subset of pink triangles.

Definition: *We call a set with no elements an* **empty set.**

Since an empty set has no elements, any two empty sets could be considered to have the same elements and hence be equal. Thus, we assume that there is only one empty set and refer to it as *the* empty set. The empty set is a subset of every set. It is denoted by { }. (Some people use the symbol ∅ for the empty set.)

12. Which of the following describe the empty set? _____
    (a) The set of 23-foot tall, green-haired men.
    (b) The set of women Presidents of the United States.
    (c) The set of Presidents of the United States with last name beginning with the letter C.

    (d) The set of A-blocks that are triangular and circular.

■ Arrange two loops so that it is possible to put the red pieces inside one loop and the circles inside the other. The set of blocks that is inside both loops is called the *intersection* of the sets.

Definition: *If A and B are sets, then the* **intersection** *of A and B, written A ∩ B, is the set of elements in both A and B.*

For example, if you had chosen the set R of red blocks for the first set and the set C of circles for the second set, the intersection R ∩ C would be the set of red circles. The situation might look something like Figure 1-4.

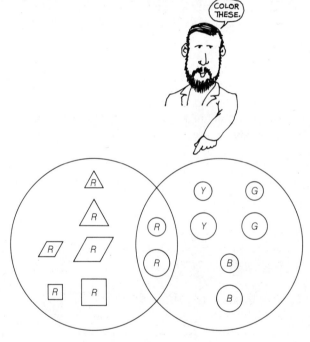

Figure 1-4

Observe that the intersection of two sets might be the empty set. Can you think of an example? Since we have adopted the convention that the empty set is a subset of every set, we can safely say the intersection of two subsets is always a subset.

13. Try to do the following problem without using the A-blocks. Suppose you were to put all the pieces that are either red or a circle into a box.

How many pieces would go into the box? _____ If you then took out of the box all the pieces that were not circles, in what way would the pieces that you took out be alike? _____ If you took out the pieces that were not red, how would they be alike? _____ Do the same exercise for two different values.

■ The set consisting of all pieces that are either red or a circle is called the *union* of the set of reds and the set of circles. This union is the set of all blocks shown in Figure 1-4.

Definition: *If A and B are sets, the* union *of A and B, written $A \cup B$, is the set of elements that are in A or in B.*

The "or" here does not exclude the possibility of being in both. For example, the red circles are in $R \cup C$.

14. Convince yourself that if A is the empty set, then $A \cup B = B$ and $A \cap B = A$ for all sets B.

■ In the rest of this chapter we will use the following letters to denote the indicated subsets of the set of A-blocks.

G = the set of green blocks
R = the set of red blocks
Y = the set of yellow blocks
B = the set of blue blocks
Q = the set of squares
L = the set of large blocks
S = the set of small blocks
C = the set of circles
T = the set of triangles
P = the set of parallelograms (or diamonds)
U = the set of all A-blocks
   (the universe, or universal set)

A diagram such as the one in Figure 1-4 is called a *Venn diagram* after John Venn (1834–1923), an English mathematician who used these diagrams in his book *Symbolic Logic*. Sometimes it is helpful to shade the specified region. For example, $R \cap C$ could be pictured as in Figure 1-5. Or, if we shaded R with horizontal lines and C with vertical lines, we would obtain

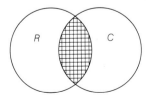

Figure 1-5

Figure 1-6, and $R \cap C$ would be the portion that is shaded both ways. What is $R \cup C$?

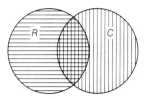

Figure 1-6

15. Sometimes it is necessary to consider three or more sets. Suppose we want to determine the result of forming the union of T and L and then intersecting that set with R. Use the A-blocks to find the result.

■ We write the set in Exercise 15 as $(T \cup L) \cap R$, using the parentheses to indicate that we form the union of T and L first and then intersect $T \cup L$ with R.

16. Find $T \cup (L \cap R)$ with the A-blocks and verify that $(T \cup L) \cap R \neq T \cup (L \cap R)$.

■ Now let us see how to draw the Venn diagram for the above situation. We draw three circles as shown in Figure 1-7. If we shade $T \cup L$ with ⁄⁄⁄ and R with ⦀⦀, then $(T \cup L) \cap R$ is ⊠⊠, as shown in Figure 1-8.

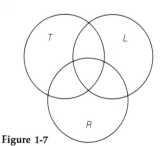

Figure 1-7

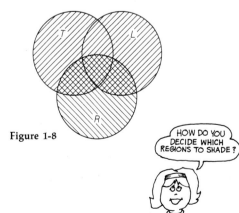

Figure 1-8

One way is to consider each of the eight regions in turn and ask the question: Does it get shaded with

▨? That is, is it in $T \cup L$? Then repeat the question for the other shading.

For example, if we number the regions for ease of reference as shown in Figure 1-9, we ask and answer the questions:

Is Region 1 in $T \cup L$? No. Don't shade it.
Is Region 2 in $T \cup L$? Yes. Shade it. ▨
Is Region 3 in $T \cup L$? Yes. Shade it. ▨
Is Region 4 in $T \cup L$? No. Don't shade it.
Is Region 5 in $T \cup L$? Yes. Shade it. ▨
Is Region 6 in $T \cup L$? Yes. Shade it. ▨
Is Region 7 in $T \cup L$? Yes. Shade it. ▨
Is Region 8 in $T \cup L$? Yes. Shade it. ▨

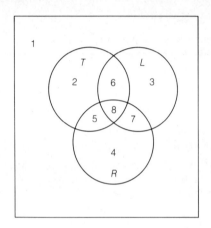

Figure 1-9

Now repeat the questions for R. When you are done, you will obtain the required diagram.

Now consider the diagram for $T \cup (L \cap R)$. If we shade T with ▦ and $L \cap R$ with ▭, then $T \cup (L \cap R)$ is the region that contains any shading at all (see Figure 1-10).

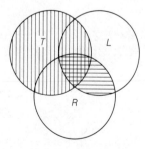

Figure 1-10

17. In each of the following, use the blocks and loops to form the indicated set, and draw a Venn diagram with the indicated set shaded. (If you have difficulty, number the eight regions in the Venn diagram and ask the questions indicated above.)

   (a) $P \cap Y$
   (b) $L \cup B$
   (c) $(G \cap S) \cap Q$
   (d) $(T \cap R) \cup (L \cap R)$
   (e) $(R \cup S) \cap P$
   (f) $(R \cap P) \cup (S \cap P)$

Lab Exercises: Set 2

EQUIPMENT: For each group, one set of A-blocks.

This set of exercises continues the study of set theory, using A-blocks. You will learn about complements and how to do "survey problems."

Definition: *Mathematicians often refer to the set under consideration as the* universe.

For example, if you are working with A-blocks, the set of 32 A-blocks is the universe; if you are working with whole numbers, then the universe is $\{0, 1, 2, 3, \ldots\}$.

18. Consider the universe to be the set of A-blocks. Put the A-blocks on the table and put all the triangles inside a loop. The set of pieces outside the loop is the subset of not-triangles. It consists of all the A-blocks that are not triangles, and is called the *complement* of the set of triangles. The universe is divided into two sets: the set of triangles and the complement of the set of triangles.

Definition: *The* complement *of a set A is defined to be those elements of the universe that are not in A. We shall use A^c to denote the complement of A.*

Some people use the notation $\sim A$ and some use A' to denote the complement of A.

19. From Exercise 18 you can see that T^c equals the union of the set of squares, the set of parallelo-

grams, and the set of circles. More briefly, $T^c = Q \cup P \cup C$. Use the words or symbols to indicate the complements below.

(a) What is the complement of L?

$L^c =$ _____

(b) What is the complement of P?

$P^c =$ _____

(c) What is the complement of Q?

$Q^c =$ _____

The Venn diagram for the complement of A^c of a set A is indicated in Figure 1-11. The shaded portion is A^c.

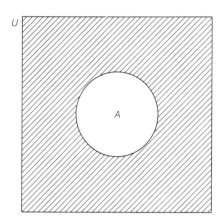

Figure 1-11

20. Make three labels $\boxed{R \cup T}$, $\boxed{R^c}$, and $\boxed{T^c}$. Arrange the loops and labels as shown in Figure 1-12. (Note: R^c is the same as not-R, and T^c is the same as not-T.)

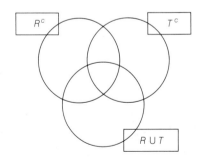

Figure 1-12

(a) Place the blocks in their appropriate spaces.

(b) Look at $(R \cup T)^c$. Look at $R^c \cap T^c$. Convince yourself that these sets are equal.

(c) In the Venn diagram in Figure 1-13, shade R^c with vertical lines and shade T^c with horizontal lines. Remember to use the questions if you have difficulty. What is $R^c \cap T^c$? In the diagram in Figure 1-14, shade $(R \cup T)^c$. Are you convinced that $(R \cup T)^c = R^c \cap T^c$?

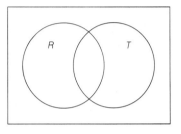

Figure 1-13

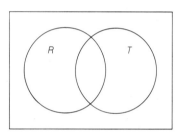

Figure 1-14

21. Find $(L \cup Y)^c$ with the A-blocks. Write $(L \cup Y)^c$ in terms of L^c and Y^c.

22. Put all the not-reds inside a loop, leaving the reds outside. Can you write the set R in terms of B, Y, G, using the operations of union and taking the complement? $R =$ _____

23. Make a label card with $T \cup R$ written on it. Label one loop $T \cup R$ and another L. Find $(T \cup R) \cap L$. Similarly find $(C \cap L) \cup T$ and $(P \cup L) \cap T^c$.

24. (a) Have one person secretly select three subsets from the list G, R, Y, B, L, S, C, T, Q, P, and write down an expression with parentheses that involves the subsets and the operations $\cup$, $\cap$. For example, $(C \cup T) \cap L$ or $(R \cap L) \cup (Q \cap L)$. Be sure to conceal the expression from the group. The person figures out the blocks in the set that have been written down and shows them to the group. They try to figure out an expression that gives the same set. (Note: There may be

more than one.) If the group needs a hint, tell them which three subsets are being used.

(b) Do the same exercise, including the use of operation ^{c}.

25. *Game—Optional.* Play the game below on the grid that appears on the following page.

The first player places a piece in any of the spaces. The second player may play in any space, but if the second player places a piece next to one that is on the board, it must be different from that piece *in two ways*. Can you fill up the board?

To make this game competitive, you may divide the pieces at random and require each player to draw from his or her own group of pieces. To further increase the challenge, play with only the large pieces or the small pieces. Make up some new rules for the game. For example, you may vary the number of differences required.

26. In a survey of 100 students, the number of students studying various subjects was found to be as follows.

English	56
History	38
Mathematics	30
English and history	14
English and mathematics	12
History and mathematics	9
All these subjects	5

Let E, H, and M represent the sets of students studying English, history, and mathematics, respectively. In the Venn diagram of Figure 1-15, we can enter the number 5 in the region corresponding to $E \cap H \cap M$. Also, since $E \cap H$ has 14 students, by subtracting 5 from 14 we can put a 9 in the region indicated. Continuing in this manner, we can assign a number to each of the eight regions in the diagram. (Each of the sets is a subset of the set of 100 students and one of the eight regions lies outside of the circles.) You can now answer the following questions.

(a) How many students were studying none of these subjects? _____

(b) How many had English as the only one of these subjects? _____

(c) How many studied English and history but not mathematics? _____

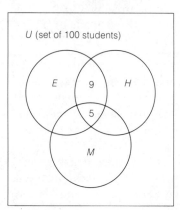

Figure 1-15

Important Terms and Concepts

Set	Intersection ∩
Subset	Union ∪
Element = member	Universe
Value of a block	Complement of $B = B^c$
Empty set { }	

Review Exercises

1. Arrange the set of A-blocks into subsets so that each subset contains only those pieces that have the same shape and the same size.

   (a) How many subsets are there? _____
   (b) How many blocks are there in each subset? _____
   (c) Within each subset, in what ways do the blocks differ? _____
   (d) Name some of the subsets. _____

2. Find two sets of A-blocks, each of which has a nonempty intersection with the set of red blocks but whose intersection is empty.

3. Match each set in the left-hand column with an equal set from the right-hand column. Write the letter in the blank.

   _____ L (a) P^c
   _____ $T \cup Q \cup C$ (b) L^c
   _____ { }   (c) $(R \cap P)^c$
   _____ $R^c \cup P^c$ (d) $T \cap C$
   _____ $(R \cap S) \cap Q$ (e) $(Y \cap B) \cup L$
   _____ S (f) $(Q \cap R) \cap S$

4. Decide whether each of the following statements is true or false. Write T or F in the blank.

   _____ (a) If A is a subset of B, then $A \cap B = A$.

   _____ (b) A is always a subset of $A \cup B$.

   _____ (c) $(A \cup B)^c = A^c \cup B^c$.

   _____ (d) If A is a subset of B, then $A \cup B = B$.

   _____ (e) If $A \cup B = B$, then A is a subset of B.

5. Let A be the set of red circles, B the set of red triangles, and C the set of circles. What is $A \cap B$?

   _____ $A \cap C$? _____

   $A \cup C$? _____

6. Use the A-blocks to illustrate that $(R \cap P)^c = R^c \cup P^c$.

7. Write $(S \cap B)^c$ in terms of S^c and B^c. Use the A-blocks to help you.

8. Let E be the set of large triangles and F the set of green triangles. What is $E \cup F$? _____

   What is $(E \cup F)^c$? _____ Use the blocks to show that $(E \cup F)^c = E^c \cap F^c$. If M and N are any subsets of blocks, what do you think $(M \cap N)^c$ is in terms of M^c and N^c?

9. What is $(B \cap Y)^c$?

10. Use the A-blocks to help you find an expression for $L \cup R$ in terms of S, B, G, Y, using the operations of union and taking the complement.

For Exercises 11–13, use the techniques of Exercise 26.

11. In a survey of dining habits, the following information was established.

    50 ordered salad
    65 ordered dessert
    50 ordered soup
    20 ordered salad and dessert
    15 ordered salad and soup
    25 ordered dessert and soup
    10 ordered salad, dessert, and soup
    20 ordered none of the three

    (a) How many people were in the survey? _____
    (b) How many people ordered salad and dessert, but not soup? _____
    (c) How many people ordered salad, but not dessert? _____
    (d) How many people ordered salad only? _____

12. A survey of 150 persons revealed the following information.

    55 liked classical music
    65 liked jazz
    75 liked rock
    15 liked classical music and jazz
    25 liked jazz and rock
    20 liked classical music and rock
    5 liked all three

    (a) How many people did not like any of the three? _____
    (b) How many liked classical music, but not jazz? _____
    (c) How many liked neither classical music nor jazz? _____
    (d) How many liked only classical music? _____

13. Let A, B, and C be as indicated in the Venn diagram of Figure 1-16.

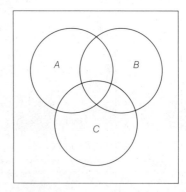

Figure 1-16

For each of the following, reproduce the Venn diagram and shade the specified region.

(a) $(A \cap B) \cap C$
(b) $A \cup C$
(c) $(A \cup B) \cap C$
(d) $(A \cap C) \cup (B \cap C)$
(e) $(A \cap C) \cap B$
(f) $A^c \cap B$
(g) $A \cap A^c$
(h) $A \cup A^c$
(i) $A^c \cap B^c$
(j) $(A \cup B)^c$
(k) $A^c \cup B^c$

14. *Puzzle Problem.* Can you arrange the 16 small blocks on the grid that appears on page 9 so that there is one piece of each color and one piece of each shape in every row and column? (You can make up a similar problem using the Aces, Kings, Queens, and Jacks from a deck of playing cards.)

© 1965 United Feature Syndicate, Inc.

Modern Women Mathematicians

Although women mathematicians are more numerous in the twentieth century, their numbers are still small compared to male mathematicians. It is always difficult to assess the contribution of contemporaries. However, the following women are noteworthy.

Hanna von Caemmerer Neumann received her doctorate from Oxford. She was a prolific researcher in algebra.

Olga Taussky Todd obtained her Ph.D. in Vienna and held positions at Bryn Mawr, Cambridge, and Gottingen. She is currently a professor of mathematics at the California Institute of Technology.

Sophie Picard, a Russian emigré to Switzerland, occupied the chair of higher geometry and probability theory at the University of Neuchatel.

Julia Robinson is a professor of mathematics at the University of California, Berkeley. She is the first woman mathematician elected to the National Academy of Sciences.

Maria Cibrario held the chair of mathematical analysis at the University of Medina and then at the University of Pavia.

Maria Pastorie worked in analysis and the theory of relativity. She was a professor of mathematics at the University of Milan.

Jacqueline Lelery-Ferraud was a professor of mathematics at the University of Paris.

Paulette Liberman, professor at the University of Rennes, did research in algebraic topology.

2
Counting and Numerals

"The number 2 is the first step in the boundless frontiers of counting."—Student modification of a quote from Karl Menninger

Lab Exercises: Set 1

EQUIPMENT: For each group, one set of multibase arithmetic blocks (MBA-blocks) (Base 2, 3, 4, or 5) and a die.

In this set of exercises we will discuss and explore the origin and meaning of words for numbers. We will use the multibase arithmetic blocks to investigate the underlying patterns of numeral systems.

Figure 2-1

1. Everyone in the group take a turn relating a memory having to do with counting or learning about numbers.

2. Each group should have one box of Base 2, 3, 4, or 5 (rectangular) MBA-blocks. Examine the blocks. Discuss the following questions in your group.

   (a) What do you think they might be good for?

   (b) What would a child do with these blocks?

   (c) Make a list of some of the things a teacher or child might do with these blocks.

3. (a) Set aside a handful of the smallest blocks and count them.

   (b) How would you explain what you did to an intelligent being from another planet who did not know any mathematics?

(c) What would a Spanish-speaking or French-speaking person say in part (a) above? (If anyone knows another language, ask that person to count in that language.)

(d) In what way is what they would do different from what you did?

(e) How is it similar?

(f) Discuss what human civilization would be like without words for numbers.

4. In Figure 2-2 are the number words an Australian tribe uses for the indicated sets. What do you think are the number words for the sets in (e) and (f)?

5. Each of the MBA-blocks has a certain volume.

(a) Select one block of each volume and arrange them in order of increasing volume.

(b) How would you describe the pattern of the arrangement to a blind person? That is, what is the rule for going from one block to the next block?

(c) Draw a picture of a set of Base 7 blocks.

6. Obtain a set of blocks of a different base and do Exercise 5.

■ The diagrams in Figure 2-3 illustrate Base 2 and Base 3 blocks. The names are suggested by the designer of the MBA-blocks. How would you extend this naming system to the next two blocks in the sequence? What do you think they would look like? Since it is impossible to use another dimension, we extend the pattern as exhibited in Figure 2-4 for Base 2.

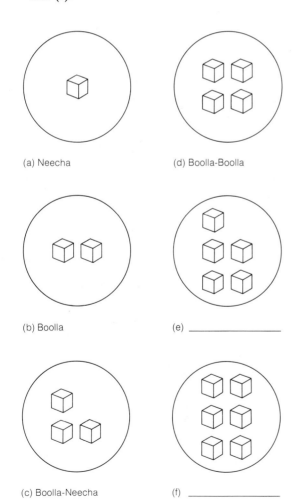

Figure 2-2

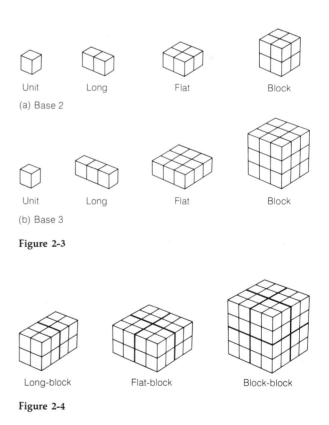

Figure 2-3

Figure 2-4

7. *Game: Win the Block.* For this game use a set of MBA-blocks of Base 3, 4, or 5, and a conventional die. Players alternate rolling the die and taking

as many units as indicated by the number on the top face of the die. They then may exchange units for longs, longs for flats, flats for blocks, always maintaining equal volume. The first player to obtain a block in this process wins the game.

It is *not* necessary to play this game in order to answer the following questions.

(a) What is the fewest number of rolls necessary to win a block? _____

(b) What is the greatest number of rolls necessary to win a block? _____

(c) Is it possible to get a long on the first roll? _____ A flat? _____

(d) Is it possible to get two longs on the first roll? _____

(e) Answer the same questions for the other bases.

8. *Game.* This is a game for two people. Use the unit MBA-blocks as counters. (Coins, buttons, or matchsticks would work as well.) Decide who is to go first. Form two groups of counters. In this game, a move consists of taking any number of counters from either one of the groups. At any time you can remove counters from only one group, but at each turn you may choose either group. The person who takes the last counter wins the game.

Play the game a few times, alternating the person who makes the first move. See if you can find a winning strategy. If you were going to move first and could arrange the piles any way you wanted, what would you do? It is likely that a young person could master the winning strategy before learning how to count. How would such a person describe the winning strategy? Try to describe it below without using the word "number" or "equal."

Comment: The Development of Counting

"Numbers have been one of the most important means by which humans have learned to master the environment."
—Karl Menninger

Once humans acquired the ability to speak, it was probably inevitable that sounds (words) would be assigned to certain sets of objects in order to describe the size of these sets. However, there was a time when humans did not do this, and even today there exist certain cultures that do not have words for numbers larger than three. Mathematical concepts develop just as do the other aspects of a civilization.

Cultures in early stages of mathematical development use different number sounds for different kinds of sets. One tribe in British Columbia, for example, has seven sets of number words. One set is for animals and flat objects, one for time and round objects, one for humans, one for trees and long objects, one for canoes, one for measures, and one for objects not in the other six categories. The separate category for canoes indicates the importance of this object to the tribe. A step toward complete abstraction is indicated by the existence of a set of words for objects not in the other six categories. There is evidence in the English language of the phenomenon of different number words for different categories of objects. We have several words to express the concept "two"—a brace of pheasants, a span of horses, a pair of shoes, a married couple, twins, and a duet, as well as all the words that are prefixed with "bi."

The use of one set of number words for all categories of objects represents a considerable achievement in the development of mathematics. To count canoes, apples, and pebbles all with the same words recognizes that a set of three canoes, three apples, or three pebbles has a common abstract property that we agree to express by a certain sound. We call this property *number*—the "number of elements in a set"—and we call the sounds *number names*.

The recognition that three apples and three canoes have a common property rests on a more fundamental concept. To see what the concept is, consider what a child might do with the blocks we used in Chapter 1. One thing might be to match up some of the subsets. Take out your set of A-blocks. Match the set of large triangles with the set of small triangles—the set of reds with the set of yellows; the set of squares with the set of circles. In each of the above write down how you matched the sets. Is there another way you could have done the matching?

One way of indicating the matching of the set of large triangles with the set of small triangles is shown in Figure 2-5. Or we could use words:

Large red triangle ↔ Small red triangle
Large yellow triangle ↔ Small yellow triangle
Large blue triangle ↔ Small blue triangle
Large green triangle ↔ Small green triangle

There are, of course, other ways to match the two sets of triangles. It is not necessary to match only by colors. For example, another match is shown in

2/Comment

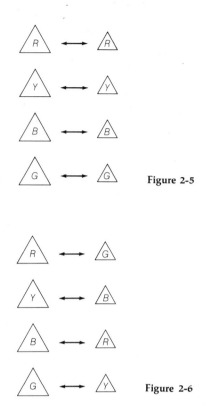

Figure 2-5

Figure 2-6

Figure 2-6. We say the sets match because we can pair the elements of one set with the elements of another set so that every element has one and only one element to which it is paired.

Definition: *We call such a pairing a* one-to-one correspondence *and say that two sets* match *if there is a one-to-one correspondence between their elements.*

9. Establish a one-to-one correspondence between the set of small A-blocks and the set of large A-blocks.

10. Draw a one-to-one correspondence between the elements of each pair of sets below.

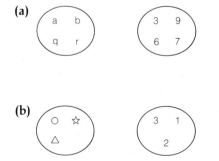

■ The concept of matching has some obvious but important properties.

Property (i): *It is always true that a set matches itself.*

For example, with the set of small triangles, we have the matching shown in Figure 2-7. Of course, there are other one-to-one correspondences in addition to the one in the example. How many different correspondences do you think there are? _____

Figure 2-7

Property (ii): *If A and B are sets and A matches B, then B matches A.*

This means that the concept of matching does not depend on the order in which we mention the sets. If we have a one-to-one correspondence between A and B, we also have a one-to-one correspondence between B and A.

Property (iii): *If A, B, and C are any three sets, and if A matches B and B matches C, then A matches C.*

11. Illustrate Property (iii) with A being the set of red squares, B the set of yellow squares, and C the set of blue squares.

■ For another example of Property (iii), suppose that every desk in your classroom is occupied by a student. They go out to recess. You want to give each student one cookie when they return. By Property (iii), you know that if you put one cookie on each desk, each student will be paired with a cookie.

12. Write down another example of how Property (iii) might be used.

■ The notions of a one-to-one correspondence and matching are of fundamental importance to the notion of counting. Review Exercise 8 above and convince yourself that the strategies could be mastered using these concepts by someone who does not know how to count.

The winning strategy for the game in Exercise 8 is to always have the groups such that the two sets match. If you line up the two groups in such a way as to establish a one-to-one correspondence, it is easy to achieve this.

The concept of one-to-one correspondence underlies the use of one set of sounds (number words) to count all types of objects. The same number word is assigned to all matching sets. As mentioned above, the number words represent an abstract property of the set that is the "number of elements in the set."

13. The Greenland Eskimos use the following number words for the indicated numbers.* Fill in what you think the missing sounds should be.

    1   Atauseq
    2   Marluk (the following)
    3   Penasut
    4   Sisamat
    5   Talimat (hand)
    6   Arfineq (other hand)
    7   Arfineq-machdlug (other hand two)
    8
    9
    10 Qulet (the tips of the fingers)
    11 Arqaneq (do it down there)
    12 Arqaneq-marluk
    13
    14
    15
    16 Arfersaneq (other foot)
    17
    18
    19
    20 Inuk navlugo (a person ended)

■ Although a number may have many names and symbols associated with it, the number concept itself does not depend on its name. The sounds *sisamat* (Eskimo), *quatre* (French), *cuatro* (Spanish), and *four* (English) all refer to the same number concept. This number concept may be defined as the property that is common to all sets that match the set

$$\{\bigcirc\ \triangle\ \square\ \square\}$$

Thus, the number concept assigns a number to each nonempty set of objects in such a way that matching sets are assigned the same number.

Definition: *The collection of these numbers is called the set of* counting numbers.

Since the set of counting numbers is infinite, we can use them to count very large sets (for example, the number of grains of sand on earth) without running out of numbers.

There is also a number concept for the empty set. This concept (i.e., the number of elements in the empty set) has the name *zero*.

Definition: *The counting numbers together with zero comprise the set of* whole numbers.

"I am zero, naught, one cipher,"
meditated the symbol preceding the numbers.
"Think of nothing. I am the sign of it.
I am bitter weather, zero.
In heavy fog the sky ceiling is zero.
Think of nowhere to go. I am it.
Those doomed to nothing for today
and the same nothing for tomorrow,
those without hits, runs, errors,
I am their sign and epitaph,
the goose egg : 0 :
even the least of these—that is me."
—Carl Sandburg, "The People, Yes"

Although number names describe an abstract property of sets and thus can be thought of as adjectives, these number names have a very special use. We use them to count the elements in a set. Thus we can count the set of A-blocks in Figure 2-8 by matching them with a sequence of sounds. If we use the sounds of the Australian tribe of Exercise 4, the counting would be as shown.

$$\underset{\text{Neecha}}{\bigcirc\!\!\!\!R}\qquad \underset{\text{Boolla}}{\triangle\!\!\!\!Y}\qquad \underset{\text{Boolla-Neecha}}{\diagup\!\!\!\!Y\!\!\!\!\diagup}$$

Figure 2-8

We would then use the last number word of the sequence to describe the set, saying that the set has "Boolla-Neecha elements." This point is significant, although somewhat subtle. Can you imagine a process for using color names to determine the color of a particular set of objects? Instead of searching through a collection of concepts to find the concept that applies to a given set, we use the number names themselves as a universal matching set to find the number that describes the set.

---
*L. L. Hammerich, *The Eskimo Language* (Oslo: Universitetsforlaget, 1970).

Using the sequence of number words as a matching set to count with enables us to efficiently communicate with each other. Of course, it is often possible to directly compare the size of sets. Two shepherds could pair off their sheep in order to determine whose herd was the smaller. The one who ran out of sheep first would have the smaller herd. However, this procedure is quite inefficient. By adopting a set of number words as a universal matching set, we greatly increase our efficiency in communicating.

Number words employ our ability to articulate sounds to communicate number concepts. As civilization became more complex, a more permanent method of communicating and storing numbers was needed. Various methods were used to make number concepts visible—knots in a string, scratches or notches on wood or stone, and indentations in clay. Before paper became available, the most commonly used devices were sticks of wood or bones with notches or other symbols carved upon them. These devices, called tally sticks, were easily transported and yet left a permanent record. An 8,000-year-old bone discovered in Zaire, Africa, shows evidence of a notational and counting system that one anthropologist says has its origins about 30,000 B.C.* Figure 2-9a shows a prehistoric notched bone tally stick, and Figure 2-9b an Exchequer tally stick used by the British Royal Treasury in the thirteenth century. These tally sticks, simple as they are, were a significant invention and contributed greatly to the development of civilization.

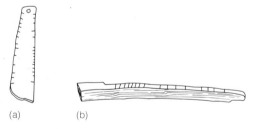

(a)   (b)

Figure 2-9

There is evidence for the widespread use of tallying and tally sticks in literature and folklore. From an old Norse song called "Hildebrand's Dirge" comes:

> At my head stands the broken shield
> On it are tallied ten times eight
> Proud men whose slayer I became.

In Shakespeare's *King Henry VI* one finds the statement, "Our forefathers had no other books than the score and the tally." The word "score" comes from the old Saxon "sceran," which means "to cut" or "to shear." Later it was used to denote 20 units. Recall Lincoln's famous address which starts, "Four score and seven years ago . . . ."†

By using a double tally stick, or a stick with a removable insert, two people could each have a record of a transaction (see Figure 2-10). Such tally sticks

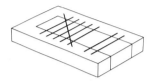

Figure 2-10   A double tally stick.

had legal standing as late as the eighteenth century. A 1719 statute in Basel reads

And if anyone is not good at settling accounts by writing and reading, they shall be satisfied with crudely made tallies or tickets: Then if one party brings to court a tally or ticket as evidence of the debt, and the other party produces the corresponding or matching ticket or wooden tally, and they are found to be the same, credance shall be given to them and the amounts stated on them shall be acknowledged.

As late as 1804 the *Code Civile* of Napoleon stated that tally sticks had the force of contracts between people who were accustomed to using them.

In spite of the official recognition of tally sticks, they are not the most convenient way to record number concepts and people very early developed alternative ways to symbolize numbers.

Definition: *Written symbols for numbers are called* numerals.

The earliest numeral systems show the close connection with the process of tallying. We shall examine a few of these systems in the next set of exercises.

Lab Exercises: Set 2

EQUIPMENT: None.

In the exercises that follow, you will study the numeral systems of the Egyptians, the Romans, the Mayans, and the Chinese.

---

*For information about African counting systems and other aspects of African mathematics, see Claudia Zaslavsky, *Africa Counts* (Boston: Prindle, Weber and Schmidt, 1973).

†For other reports from literature, and actual photographs of tally sticks, see Karl Menninger, *Number Words and Number Symbols* (Cambridge, MA: MIT Press, 1969), pp. 223–248.

Egyptian Numerals

As early as 3300 B.C. the Egyptians had developed a system of numerals whereby they could express numbers into the millions (see Figure 2-11).

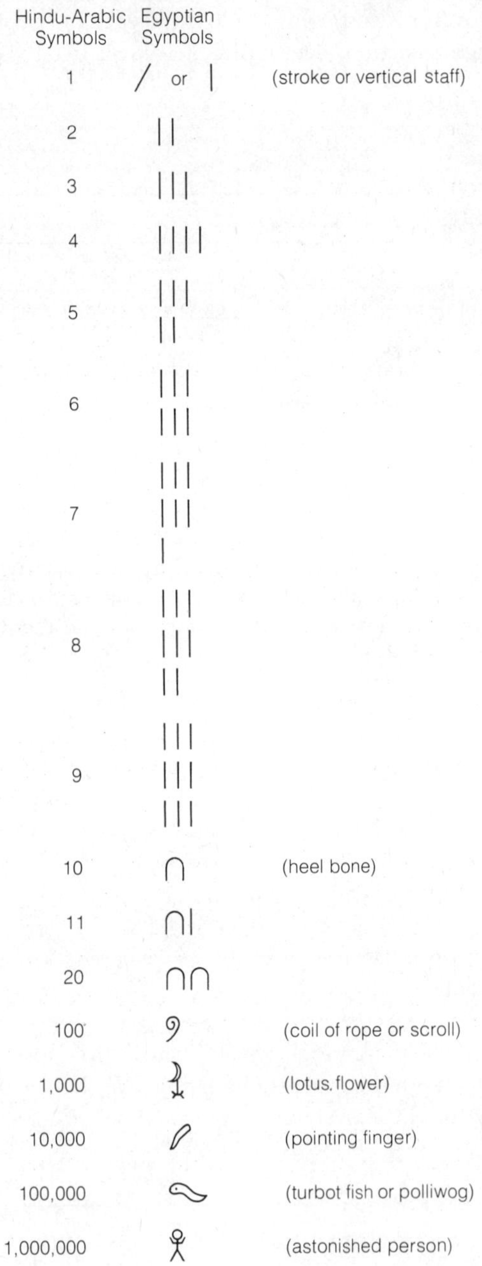

Hindu-Arabic Symbols	Egyptian Symbols	
1	/ or |	(stroke or vertical staff)
2	||	
3	|||	
4	||||	
5	||| / ||	
6	||| / |||	
7	||| / ||| / |	
8	||| / ||| / ||	
9	||| / ||| / |||	
10	∩	(heel bone)
11	∩|	
20	∩∩	
100	⌒	(coil of rope or scroll)
1,000	⚶	(lotus flower)
10,000	⌇	(pointing finger)
100,000	⌢	(turbot fish or polliwog)
1,000,000	⚚	(astonished person)

Figure 2-11

By using an additive principle, we can write any number.

Definition: *The additive principle states that the value of a set of symbols is the sum of the value of the symbols.*

Thus ⚶⚶ 99999 ∩∩ \|\| represents 1,000 + 1,000 + 100 + 100 + 100 + 100 + 100 + 10 + 10 + 10 + 1 + 1, or 2,532.

14. Listed below are some numerals in our Hindu-Arabic system. Express each as an Egyptian numeral.

   (a) 328
   (b) 5,343

15. Listed below are two Egyptian numerals. Express each one as a Hindu-Arabic numeral.

   (a) ⚚⚚ 99999 ∩∩∩∩∩ \|\|\| _____
   (b) ⌇⌇⌇ ⚶⚶⚶ ∩∩∩ _____

16. How would you express 123 in Egyptian numerals without using the symbol ∩?

■ Observe that the Egyptian system works on the principle of grouping ten symbols and replacing them with a new symbol. Thus ten \|'s are replaced by one ∩ and ten ∩'s by one 9 and so on. This is essential if we are to symbolize large numbers efficiently. The Egyptian system is said to be a Base 10 system, since each group of ten is replaced by a new symbol.

Roman Numerals

17. Everyone relate a memory of seeing or using Roman numerals.

■ The Roman numeral system was used in European bookkeeping as late as the eighteenth century and is still used in books and on clocks, and to denote dates on some buildings and movie copyrights. Roman numerals did not become standardized until after the invention of the printing press in 1438. The early and later forms of the Roman numerals are indicated below.

	Early Roman numerals	Late Roman numerals
1	I	I
5	V	V
10	X	X
50	⊥ or ↓	L
100	C	C
500	D	D
1,000	CIↃ	M
5,000	IↃↃ	IↃↃ
10,000	CCIↃↃ	CCIↃↃ

Any number can be expressed in terms of these symbols by using them additively. To write 364, for example, we write three C's, one L, one X, and four I's. Notice that the Roman system is not a pure grouping system. It is a Base 10 system with additional symbols for 5, 50, 500, etc. The use of these symbols markedly decreases the amount of writing one must do.

For example, 58 is written LVIII rather than XXXXXIIIIIIII. In the following exercises, use Late Roman numerals.

18. Write the following Hindu-Arabic numerals as Roman numerals.

    (a) 527 _____

    (b) 2,289 _____

    (c) 35 _____

    (d) 262 _____

19. Write the following Roman numerals as Hindu-Arabic numerals.

    (a) CXXXXIII

    (b) CCIƆƆ CCIƆƆ IƆƆ M M CCCC LXXIII

■ The Roman numeration system developed (but did not consistently use) a *subtractive principle*, in which a smaller number preceding a larger number would be subtracted. Thus VIIII was sometimes written as IX and IIII as IV.

20. Below are Roman numerals that have been written using the subtractive principle. Write each one as a Roman numeral without using the subtractive principle, and as a Hindu-Arabic numeral.

    (a) XL _____
    _____

    (b) XIX _____
    _____

21. Express each of the following Roman numerals in an equivalent form using the subtractive principle.

    (a) XXVIIII _____

    (b) CCCCLXXXIII _____

■ One drawback of the Roman system is the amount of writing one must do. Let us consider how it might be improved. Suppose we attempt to shorten the writing by using the Roman symbols I, X, C, M, . . . to indicate the basic groups (ones, tens, hundreds, etc.) and our symbols 1, 2, . . . , 8, 9 to indicate how many of these basic groups we have. For example, 5C3X4I represents five hundreds plus three tens plus four units or 534 in our system. And 8,243 in our system would be written as 8M2C4X3I. In this system a *multiplicative principle* has been adopted. Each pair of symbols represents the product of the two symbols. Thus 8M2C4X3I means $(8 \times 1{,}000) + (2 \times 100) + (4 \times 10) + (3 \times 1)$.

22. Express the following Hindu-Arabic numerals using the above system.

    (a) 39 _____

    (b) 248 _____

    (c) 4,962 _____

23. Express the following numerals as Hindu-Arabic numerals.

    (a) 3M2C9X4I _____

    (b) 6C8X3I _____

To my knowledge, this system has never been used. I introduce it only to illustrate the multiplicative principle that, for example, the Chinese used in developing their numeration system.

Chinese Numerals

The Chinese symbols are indicated in Figure 2-12. In this system, numerals are written vertically using the multiplicative principle, as in Figure 2-13.

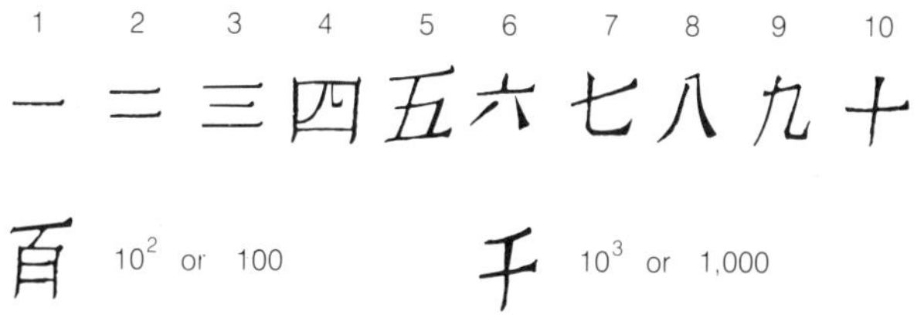

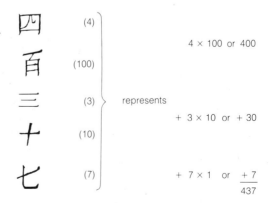

Figure 2-12

Figure 2-13

24. (a) Write the Chinese numeral as a Hindu-Arabic numeral.

Write the following Hindu-Arabic numerals as Chinese numerals.

(a) 279   (b) 2,097   (c) 2,907

■ The use of the multiplicative principle greatly simplifies the recording of numerals because it eliminates the repetition of symbols. A further simplification is to eliminate the second symbol of each pair and use the *position* of each symbol to indicate the size of each group. This type of system is called a *positional numeration system* and requires the use of a symbol for zero to keep track of a position when no digit is to be used. The Chinese system does not require a symbol for zero, although in the thirteenth century a circle was introduced for this purpose.

Mayan Numerals

The positional numeral system developed by the Mayas of southern Mexico and Central America is believed to be the earliest positional numeral system that used zero as a placeholder. Their numeral system was developed by the Mayan priests around 400 or 300 B.C. (The first recorded use of a zero in Hindu-Arabic numeration systems occurs in India in A.D. 870.)

Figure 2-14 shows some Mayan numerals and their Hindu-Arabic equivalents.

Figure 2-14

What would you guess the symbol for 20 to be? The logical guess would be ≡≡≡ since so far this has been a simple grouping system of Base 5. However, in writing numerals above 19, the Mayans used a positional system, where the value of the positions increases by multiples of 20, except for the third position which increases by a multiple of 18. (This will be explained shortly.) The Mayan symbol for zero is ⬮ (a shell) and thus the Mayan numeral for 20 is ⬮; for 21, it is ⬮ ; for 22 ⬮ ; for 25 ⬮ ; for 30 ⬮ ; and for 35 it is ⬮. For 40 (two twenties) it becomes ⬮, for 41 ⬮ , and so on. We see that the numbers are *written vertically, starting at the top*. The bottom position indicates the number of units and the second position the number of twenties.

EXAMPLE: Since 352 is seventeen "twenties" and 12 units, we can write the Mayan numeral for 352 as

Twenties	●● 17 × 20	340
Units	●● 12 × 1	+ 12
		352

EXAMPLE: The Mayan numeral

••••   14 × 20   280
          represents   + 14
••••   14 × 1   294

25. Write the Hindu-Arabic equivalent for each of the Mayan numerals below.

(a) ≡≡
    ⬮

(b) ≡
    •••

26. Write the Mayan numerals for each of the following.
   (a) 29 (b) 250 (c) 343

■ Since the Mayan calendar had 18 months of 20 days each, plus 5 "useless days," it was convenient to have the third position in their numeral system denote 18 × 20 or 360. Thus we have Figure 2-15.

Three hundred sixties	•	1 × 18 × 20	360
Twenties	⬭	represents + 0 × 20 or + 0 and
Units	⬭	+ 0 × 1	+ 0
			360

⸻ represents 5 × 360   1,800
•••   + 3 × 20 or   60
⸻   + 12 × 1   + 12
                                1,872

Figure 2-15

The advantage of having the third position represent 18 × 20 becomes clear when we attempt to represent the number of days in a certain number of Mayan years. Since 1 year has ⬭ (360) days, the number of days in 3 Mayan years is ⬭ and the number of days in 12 Mayan years is ⬭. Since the Mayans used their numeration system only for calculations concerning their calendar, this convention was very useful to them. The subsequent positions return to a Base 20 system, that is, the fourth position represents 20 × 18 × 20 = 7,200 and the fifth 20 × 20 × 18 × 20 = 144,000, and so on.

EXAMPLE:

7,200's	•
360's	⸻
20's	⬭
1's	•••

represents 7,200 + (5 × 360) + (0 × 20) + 3, or 9,003

27. Write the Hindu-Arabic equivalent for each of the Mayan numerals below.

   (a) •• (b) ⸻
       ⬭   •
       ⬭   ⬭
       •••

28. Write the Mayan numeral for the number of days in 20 Mayan years. (Don't count "useless days.")

29. Write the Mayan numerals for the following.
   (a) 8,000 (b) 9,999 (c) 463

30. One person writes an Egyptian numeral and passes it to another, who writes the Roman equivalent and passes it to a third, who writes the Mayan equivalent. The originator checks it with the original and if it disagrees, finds the error.

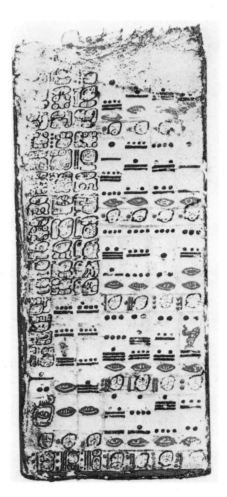

Figure 2-16 Mayan numerals, from the Dresden Codex. Photograph from S. G. Morley, *An Introduction to the Study of the Maya Hieroglyphics* (Bulletin 57, Bureau of American Ethnology, 1915).

Comment: Numerals

"The sequence of number words in a culture is built up long before any mature system of numerals becomes established."
—Karl Menninger

As we have seen, different civilizations developed and used different systems of numeration to symbolize and record their number concepts. Just as different number words can represent the same number concept, different symbols can also represent the same number concept. Thus IIII, IV, and 4 all represent the same number concept.

Definition: *We call the symbols for the number concepts* numerals *and reserve the word* number *for the concept itself (although we will not be rigorous in making this distinction).*

A numeration system that is efficient for both recording and calculating is a sophisticated development. An early advance in the development of a numeration system is the adoption of the *additive principle*, which says that if several symbols are written together, the final value is obtained by adding the value of each symbol. Without the additive principle each number concept would require a different symbol.

The device of *grouping* is also useful—when a certain number of symbols of one kind are written down, they are replaced by a new symbol. The Egyptian system uses a grouping system with Base 10—whenever 10 symbols of a given kind appear, they are replaced by a new symbol. The Roman system also employs grouping with Base 10, but adds the variation of using an additional symbol at the mid-points—i.e., 5, 50, 500, etc. Another variation that we saw in the Roman system is the *subtractive principle*. Both of these variations reduce the amount of writing that is necessary.

The Chinese system is quite different from the Egyptian and Roman systems. First, it uses a *code* or *cipher* to symbolize the first nine numbers. Each number from one to nine has its own symbol. Then, instead of a simple grouping system as employed by the Egyptians and Romans, a *multiplicative principle* is used. There are symbols for each of the powers of 10 (ten, one hundred, one thousand, etc.), and the number of times that symbol occurs is indicated by the numeral preceding the symbol. This system is considerably more efficient than the Egyptian or Roman system and is quite close to our system. Since the symbols in the Chinese system are unfamiliar—and difficult for Westerners to draw—let us examine the system we fabricated following Exercise 21 to see the similarity between this type of system and ours. In that system the number 327 would be written 3C2X7I. But do we need to use the labels C, X, I? Consider the number 307. We can easily interpret 3C7I, but if the labels C and I are omitted, we have "37," which is very ambiguous. What is needed is a *positional principle*, or *place-value notation*, in which the position of the symbol indicates the power of the base. To do this requires that we have a symbol to indicate nothing—namely a symbol for zero. That is a very significant advance in the system.

The Hindu-Arabic System

Our system is referred to as the "Hindu-Arabic system," since the symbols we used originated with the Hindus and the system was modified and introduced into Europe by the Arabs. However, many civilizations contributed in part to its development. The concept of place-value can be traced back to the Babylonians and Sumerians, although it was not extensively developed by them. The concept of *coding* (using different symbols for different digits) originates in another Egyptian system (and also appears in a number system developed by the Ionian Greeks, which we have not studied). Historians are not in agreement about the extent of the influence that these civilizations had on the development of the Hindu system.*

It is known that as early as the second century B.C. the Brahmans in India had developed a number script. These numerals appear in inscriptions on copper plates, temple walls, and in caves. They remained relatively unchanged for almost a thousand years. The symbols used by the Brahmans for the first nine numbers were

$$- = \equiv \mathcal{Y} \ \mathsf{h} \quad \mathsf{b} \ 7 \ \mathsf{y} \ \mathsf{?}$$

At first the Hindu system was not a positional numeration system. Historians estimate that around A.D. 600 a true positional numeration system appeared in India. That system used the first nine Brahman symbols and a little circle or a dot to denote zero. The earliest known written zero appears on the wall of a temple in central India. It was inscribed in

---

*For a complete discussion, see Menninger, *Number Words and Number Symbols*, pp. 391-399.

A.D. 870. The numerals in use at that time are shown below.

$$\neg 2\; 3\; 8\; 4 \quad 6\; 7\; 5\; 9\; 0$$

The Hindu system was introduced into the Arabian empire in the eighth century, and by the Arabs into Europe via Spain. By the fifteenth century the following numerals were standard throughout Europe.

$$1\; 2\; 3\; 2\; 4 \quad 6\; \wedge\; 8\; 9\; 0$$

This account does not do justice to the development of our number system.* An illustration of the complex process that led to its general adoption is an ordinance passed by the City Council in Florence, Italy, in 1299, which made it illegal to use these numerals for financial procedures. Since we have used these symbols all our lives, it is difficult for us to imagine how strange they must have appeared when they were first introduced.

Lab Exercises: Set 3

EQUIPMENT: For each group, one set of MBA-blocks (Base 2, 3, 4, or 5).

In these exercises you will use the MBA-blocks to study the principles involved in a place-value numeral system of any base. The use of MBA-blocks of bases different from 10 provides an unfamiliar situation that will challenge you and give you some idea of what your students will face when they learn arithmetic in our Base 10 system.

31. *Factory Game.*† Your group will keep production records for a certain factory. You may work with MBA-blocks of any base. Read the instructions *completely* before starting the exercise.

    Suppose that during the month of July the factory produced 1 unit block per day.

    (a) Have one person in your group count out the number of unit blocks produced during July. (Do you remember the rhyme, "Thirty days hath September, April, June, and November. . . ."?)

---

*For further information, see D. E. Smith and J. Ginsburg, *Numbers and Numerals* (Reston, VA: National Council of Teachers of Mathematics, 1969).
†Adapted with permission from W. Fitzgerald et al., *Laboratory Manual for Elementary Mathematics* (Boston: Prindle, Weber and Schmidt, 1973).

Figure 2-17

(b) Proceed clockwise around the group. Have the next person exchange the units for an appropriate number of longs. (There may be units left over.)

(c) The next person exchanges the longs for flats.

(d) The process continues, exchanging flats for blocks, blocks for long-blocks, long-blocks for flat-blocks, etc., until no more exchanges can be made.

(e) The *final result* is recorded in the first line of Figure 2-18 by entering the numbers of units, longs, flats, etc., present *after* all exchanges have been made. The entry for July has used the example of Base 2 blocks. For Base 3 blocks the entry would be 1011, for Base 4 blocks it would be 133, and for Base 5 it would be 111.

32. Fill in the second line of the production record chart for a factory producing 3 units per day for August.

33. Fill in the third line so as to show the combined production of July and August.

34. Write down the production record for a factory that produces 2 units a week for an entire year.

35. Write down the production record for a factory that produces 4 units a day for 31 days. Does the production record for this exercise agree with the record for Exercise 33? _____ Should it? _____

    Why or why not? _____

    _____

36. Repeat Exercises 32–35 with a different base. If you have not used Base 2 yet, you should do so now.

Figure 2-18 Production Record

Block-blocks	Flat-blocks	Long-blocks	Blocks	Flats	Longs	Units		Production information
		1	1	1	1	1	Base __2__	1 unit per day during July
							Base ____	
							Base ____	
							Base ____	
							Base ____	
							Base ____	
							Base ____	
							Base ____	
							Base ____	
							Base ____	

37. Suppose you were given the production record below. How many units are recorded? ____ Express the number in our system (Base 10).

Blocks	Flats	Longs	Units	
2	3	3	1	Base 5

38. We will use the shorthand notation 2331_{five} for the production record in Exercise 37. How many units of production are recorded by each of the following? (Express the number as a Base 10 numeral.)

    (a) 2107_{eight} _____

    (b) 3100_{four} _____

    (c) 1111_{three} _____

    (d) 625_{seven} _____

39. Suppose we had a set of Base 12 blocks.
    (a) How many units would make up a long? (Write the number word, not the numeral.)

    _____

    (b) How many units would make up a flat? (Write the number word, not the numeral.)

    _____

    (c) How many units would make up a block? (Write the number word, not the numeral.)

    _____

40. If we were to use Base 12 blocks to make a production record, we would need symbols for the number words "ten" and "eleven." Let T be the symbol for ten and E be the symbol for eleven. How many units are represented in each of the production records below?

	Blocks	Flats	Longs	Units		
(a)	1	0	3	T	Base 12	____
(b)	4	E	T	3	Base 12	____
(c)			E	E	Base 12	____

■ You have probably noticed by now that these production records can be looked upon as numerals in a given base. For example, 99 can be represented as a Base 4 numeral by 1203_{four}, which represents one group of 64, two groups of 16, zero groups of 4, and 3 units.

41. (a) Work in Base 3. Have one person write the Base 3 numeral for 1. Proceed clockwise around the group, each person writing the succeeding numeral, e.g., the first person

writing 1, the second person writing 2, the third writing 10, etc. Proceed until you have written the numerals for the numbers through 25.

(b) Repeat part (a), working in Base 2.

42. *Optional.* In Exercise 34, you found the production record for a factory that produces 2 units each week for a year. Suppose that this production is 2 longs each week.

(a) What would the production record be? _____

(b) What would the record be if the factory produced two flats each week for a year? _____

Can you formulate a rule that describes this phenomenon?

Comment: Positional (Place-Value) Numeration System

"Ten . . . this number was of old held in high honor, for such is the number of fingers by which we count."—Ovid

The system which is now used over most of the world is a Base 10 positional numeration system. The symbols 1, 2, 3, 4, 5, 6, 7, 8, and 9 are used to denote the first nine digits, and the position indicates which power of the base is indicated. Thus 327 symbolizes 3 hundreds, 2 tens, and 7 units.

The exercises with the MBA-blocks show how the basic idea of our positional numeration system can be applied to develop a system for any base. Thus, if we wanted to use a Base 7 system, we would use symbols for the first six digits and let the position of the numeral indicate the power of the base being referred to. For example, 526_{seven} means 5 groups of 49 (seven times seven), 2 groups of sevens, and 6 units. The Base 10 numeral for this number is 265.

43. Convert each of the following to a Base 10 numeral.

(a) 425_{six} _____

(b) 1101_{two} _____

(c) $11TE_{twelve}$ _____

(d) 987_{ten} _____

(e) 111_{four} _____

(f) 1270_{eight} _____

■ It is possible to convert a Base 10 numeral to a numeral in any base. For example, suppose that you wish to express 123 as a Base 4 numeral. One way to proceed would be to imagine using Base 4 MBA-blocks to represent 123 units. If you used 1 block ($64 = 4^3$ units), you need 59 more units. To represent 59 units, you would use 3 flats ($48 = 3 \times 4^2$), and still need 11 units. Then 11 units could be represented by 2 longs ($8 = 2 \times 4$) and 3 units. Thus 123 can be represented by 1 block, 3 flats, 2 longs, and 3 units, and the Base 4 numeral is 1323_{four}. We can express this procedure numerically in the form $123 = (1 \times 4^3) + (3 \times 4^2) + (2 \times 4) + (3 \times 1)$.

44. Convert the following Base 10 numerals to a numeral in the indicated base:

(a) $86 = $ _____ $_{two}$

(b) $113 = $ _____ $_{four}$

(c) $99 = $ _____ $_{three}$

(d) $963 = $ _____ $_{twelve}$

(e) $963 = $ _____ $_{eight}$

(f) $401 = $ _____ $_{five}$

45. *Optional.* Study the process in the factory game in order to develop a method for converting a Base 10 numeral to a Base 4 numeral, using successive division by 4. Extend the method to other bases and use it to do Exercise 44.

Figure 2-19

Important Terms and Concepts

Counting numbers (positive integers)
Matching—one-to-one correspondence
Number names
Numerals
Grouping
Additive principle
Base (of a numeration system)
Subtractive principle
Multiplicative principle
Positional principle
Place-value notation
Zero

Review Exercises

1. When we count on our fingers, we are using which one of the following. Write the letter in the blank. _____
   (a) Additive principle
   (b) Positional numeration system
   (c) One-to-one correspondence

2. When we count out loud, we are using which one of the following. _____
   (a) Numerals
   (b) Number names
   (c) Additive principle

3. What would the winning strategy be for the game in Exercise 8 if the rule was changed so that whoever took the last counter would lose the game? (Be sure to think this one through.)

4. Refer to the factory game in Exercise 31. If a group handed in 2043 as a production record in Base 4, would the supervisor know that the group did not complete its task? Why? _____

   Write the production record in a form the supervisor would accept. _____ four

5. If you were working with the blocks below, in which base would you be working? _____

      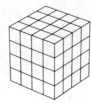

6. Change the Egyptian numerals shown below to Hindu-Arabic numerals.
   (a) 𓀀𓏲𓏲𓈖𓈖𓈖𓈖𓏤𓏤𓏤
   (b) 𓆱𓊖𓊖𓏲𓈖𓏤

7. Change the Hindu-Arabic numerals below to Egyptian numerals.
   (a) 403
   (b) 311,762

8. Express the following Roman numerals in an equivalent form that uses the subtractive principle.
   (a) CCCCXXIIII
   (b) MDCCCCVIIII

9. Write the following as Roman numerals.
   (a) 1,298
   (b) 14,321

10. Write the following numerals in the Chinese system. Explain why you do not need a symbol for zero to write 802 in the Chinese system.
    (a) 973   (b) 802

11. Write the following as Hindu-Arabic numerals.
    (a)   (b)

12. Write the Mayan numerals for the following.
    (a) 365   (c) 7,199
    (b) 7,200   (d) 42,222

13. Change each of the following to a Base 10 numeral.
    (a) 463_{seven} _____
    (b) 674_{eight} _____
    (c) 101101_{two} _____
    (d) 34102_{five} _____
    (e) $T4E_{twelve}$ _____

2/Review Exercises

14. Change each of the following Base 10 numerals to a numeral in the indicated base.

(a) 1243 = _____ five

(b) 481 = _____ five

(c) 999 = _____ three

(d) 904 = _____ twelve

(e) 100 = _____ two

(f) 882 = _____ two

15. For each of the following numeral systems, indicate which of the items below apply. (There can be more than one.) Write the letter(s) in the blank.

_____ Hindu-Arabic

_____ Chinese

_____ Roman numerals

_____ Egyptian

_____ Mayan

(a) Additive principle

(b) Subtractive principle

(c) Multiplicative principle

(d) Positional numeration system

(e) Base 10 system

16. What are the similarities between the Mayan numeration system and the system of number words developed by the Eskimos?

17. What are the similarities between the Mayan system of enumeration and ours?

18. Write the Base 4 numerals for the numbers between 1 and 25.

19. *Optional.*

(a) Describe a matching between the set of counting numbers and the set of even counting numbers—i.e., the set {2, 4, 6, . . . }.

(b) Describe a matching between the set of even counting numbers and the set of odd counting numbers.

By permission of Johnny Hart and Field Enterprises, Inc.

Benjamin Banneker (1731–1806) was a free black who lived in Maryland. For the most part, he educated himself in mathematics and astronomy. About 1773 he began to do astronomical calculations and correctly predicted a solar eclipse in 1789. Between 1792 and 1802 he published a widely acclaimed almanac. He was appointed by George Washington to the District of Columbia Commission, which did the surveying for the new capital of the United States.

3

Addition and Subtraction

"'Can you do addition?' The White Queen asked, "What's one and one and one and one and one and one and one and one and one and one?' 'I don't know,' said Alice, 'I lost count.' 'She can't do addition,' the Red Queen interrupted."—Lewis Carroll

Lab Exercises: Set 1

EQUIPMENT: For each group, one set of MBA-blocks and two containers of C-rods.

In these exercises we will use MBA-blocks and C-rods to investigate the operations of addition and subtraction. Although we will begin to develop methods for obtaining the correct answer, this chapter will concentrate on the properties of addition and subtraction. Methods will be studied in greater depth in Chapter 4.

1. (a) Two people each take a set of unit MBA-blocks. Express the number of elements in each set as a numeral in the base you are working with.

    ____ ____ Base____

    Now take the *union* (combination) of the two sets of blocks and represent the number of elements in the union as a numeral in the base you are working with.

    ____ Base____

    You have written the numeral for the *sum* of the two numbers.

    (b) Start with different sets of unit MBA-blocks and do (a) again. Record your results below.

    ____ + ____ = ____ Base____

2. Given two numerals, you can use the blocks to calculate the sum of the numbers they represent.

    (a) Write down two numerals in the base you are working with.

    ____ ____ Base____

    Represent each numeral using the MBA-blocks. Combine the piles for each numeral and make all the exchanges possible so as to minimize the number of pieces used to represent the numeral. Record the numeral of the resulting pile. ____You have written the numeral for the *sum* of the two numbers.

    (b) Try another problem and record your result below. (Be sure to use the blocks.)

    ____ + ____ = ____ Base____

3. Represent 62 with your MBA-blocks. Write the numeral in your base for this pile: 62_{ten} = ____ Save this pile. Next, represent 38 with your blocks. 38_{ten} = ____ Combine the piles (i.e., take the union of the two sets of blocks) and make all possible exchanges.

    Write down the numeral for the resulting

pile _____ Base _____ and convert it to a Base 10 numeral. _____ Is it the sum 62 + 38?

4. Can you develop a method (called an *algorithm*) for doing addition problems as in Exercise 3 with pencil and paper instead of with the blocks? Discuss the procedure you would follow. (You should be able to explain why your algorithm works.)

5. Make up a difficult problem. Have two people do it with MBA-blocks and two other people do it with pencil and paper, using the algorithm developed in Exercise 4. Compare the results.

■ If you are using the MBA-blocks, your computation will be easier if you use the "counting boards" on pages 30 and 31, for Base 2 and Base 3, respectively. You can devise your own counting boards for other bases.

6. Using a set of MBA-blocks of a different base, repeat Exercises 2 through 5. Record your results below.

7. Make up a subtraction problem in the base you are working with (e.g., for Base 5, $432_{five} - 213_{five}$):

   _____ − _____. Represent the *minuend* (the first number) by a pile of blocks, and then remove a pile of blocks equal to the *subtrahend* (the second number). Write down the numeral of the resulting pile: _____. This numeral represents the *difference* of the numbers.

8. (a) Do two more problems:

   (i) _____ − _____ = _____

   (ii) _____ − _____ = _____

   (b) See if you can develop an algorithm for doing subtraction problems in your base with only a pencil and paper. You should be able to explain why your algorithm works. Make up a subtraction problem and have two people use MBA-blocks to solve it, while two other people use pencil and paper. Compare the results.

In Exercises 9 through 16 we continue our investigation of addition and subtraction using C-rods. For a description of C-rods, see page xvii.

9. Put the C-rods on the table. Play with them for a few minutes. What can you do with them? Discuss the following questions:

   (a) How might young people play with the C-rods?

   (b) Would they play differently than you do? Explain.

   (c) If you were a teacher, how might you use the rods in a classroom?

Figure 3-1

■ A straight line of one or more C-rods, as in Figure 3-2, will be called a *train*. Two trains will be equal if they have the same length.

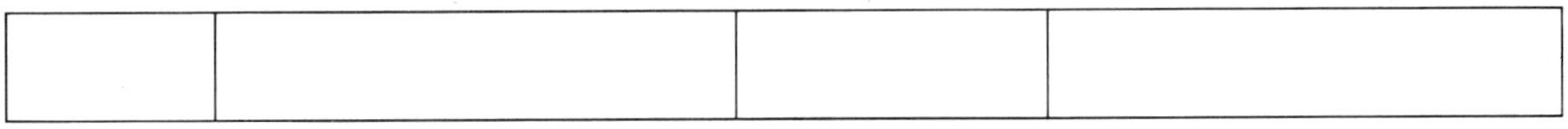

Figure 3-2

Counting Board for Base 2 MBA-Blocks						
Block-blocks	Flat-blocks	Long-blocks	Blocks	Flats	Longs	Units

3/Lab Exercises: Set 1

Counting Board for Base 3 MBA-Blocks				
Long-blocks	Blocks	Flats	Longs	Units

10. (a) Find two different ways of making a train equal to the orange rod.
    (b) In how many different ways can you make a train having the same length as the purple rod? _____ The yellow rod? _____ (The ways are considered to be "different" if either the rods composing the train are different or the rods are the same but their order is different.)

11. (a) Place some white rods and red rods in front of you. Make a train that does not contain more than one rod of the same color. We will call such trains *no-duplicate trains*. How many different-length no-duplicate trains can be made from the white and red rods? _____
    (b) Now place some white rods, red rods, and light-green rods in front of you. How many different-length no-duplicate trains can you make from these rods? _____
    (c) *Optional.* If you were to continue this investigation by successively increasing the number of rods, you would need a systematic way to record the information. Can you think of a way you might do this? Try to find a pattern in your results. Can you predict how many different-length no-duplicate trains can be formed from the set of all ten C-rods?

■ In order to write efficiently about the rods, we will use the following symbols:

W	for white	DG	for dark green
R	for red	Bk	for black
LG	for light green	Br	for brown
P	for purple	Bl	for blue
Y	for yellow	O	for orange

If t_1 and t_2 are two trains, let $t_1 + t_2$ be the train obtained by placing t_2 to the right of t_1. For example, R + P is formed as shown in Figure 3-3.

Figure 3-3

12. (a) Verify that R + P = DG.
    (b) Find each of the following sums, using the rods.
       (i) W + DG = _____
       (ii) Br + W = _____
       (iii) R + LG = _____
       (iv) Y + P = _____
       (v) LG + R = _____
       (vi) R + R = _____
    (c) Let t_1 = Br + Y, t_2 = O + P, t_3 = Bl + DG, t_4 = Bk + Bk.
       Verify that $t_1 + t_2 = t_3 + t_4$.

13. Form a train t from a yellow rod, a purple rod, and a black rod. Form another train s from a red rod and a blue rod. Place s adjacent to t, as shown in Figure 3-4. Find a train that, when joined to s, will make a train equal in length to t. This train is called the *difference* of the two trains and is denoted $t - s$.

    What is $t - s$? _____

14. Find the following differences.
    (a) O − R = _____
    (b) Bl − Bk = _____
    (c) (O + DG) − Bk = _____

15. Illustrate each of the following statements, using trains.
    (a) If t_1 is longer than t_2, then $(t_1 - t_2) + t_2 = t_1$.
    (b) $(t_1 + t_2) - t_2 = t_1$.
    (c) $(t_1 + t_2) - t_1 = t_2$.
    (d) If t_1 is longer than $t_2 + t_3$, then $t_1 - (t_2 + t_3) = t_1 - t_2 - t_3$.

16. Pick four rods, a, b, c, and d, so that a is longer than c and b is longer than d. Use the rods to show that $(a + b) - (c + d) = (a - c) + (b - d)$.

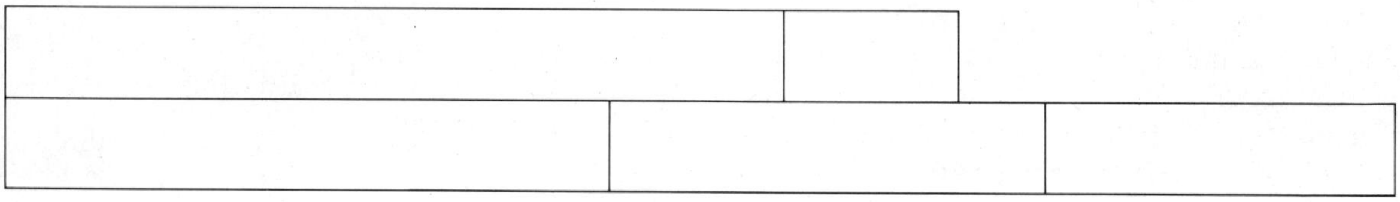

Figure 3-4

Comment: Properties of Addition and Subtraction

"And the boy in Winnetka, Illinois, who wanted to know: 'Is there a train so long you can't count the cars? Is there a blackboard so long it will hold all the numbers?'"
—Carl Sandburg

In Chapter 2, we studied various numeral systems. These systems were developed in order to record numbers, but the effectiveness of a numeral system is much greater than simply its ability to record and communicate number concepts. It can also be of great use in calculating with numbers. We shall see that different numeral systems vary in their usefulness for calculations, but first, it is helpful to realize that it is possible to calculate without numerals. In fact, the word "calculate" is derived from the Latin word *calculus* which means "pebbles." How can one use pebbles to calculate?

Suppose that the white C-rods are pebbles and that you are a shepherd counting your flock in the meadow depicted in Figure 3-5. Suppose you find five sheep behind some trees in Region A and three sheep behind some rocks in Region B. When you see the sheep at A, you pick up one pebble for each sheep and put the pebbles into your left pocket.

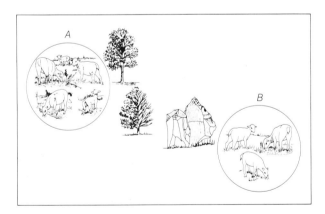

Figure 3-5

(*Note:* You are establishing a one-to-one correspondence between the set of sheep at A and a set of pebbles.) When you get to B, you again pick up one pebble for each sheep there and put the pebbles into your right pocket. What must you do to calculate the total number of sheep in the meadow? Clearly, you need to put the pebbles in each pocket together and count them. In order to add the number of elements in two sets that have no element in common, put the two sets together and count the set. This is what addition is about.

17. Decide whether the following statement about A-blocks is true or false and justify your answer. Write T or F in the blank. If necessary, use a set of A-blocks to assist you.

    ____ Since C (the set of circles) has 8 elements, and Y the set of yellow blocks) has 8 elements, the set $Y \cup C$ has $8 + 8 = 16$ elements.

Addition

We now wish to give a formal definition of addition for the set of whole numbers, namely the set $\{0, 1, 2, 3, \ldots\}$.

Definition: *To add two numbers, m and n, find a set A with m elements and a set B with n elements such that A and B have no elements in common. Then $m + n$ is the number of elements in the union of A and B.*

For example, to add $3 + 4$, we can take a jar with three marbles (Figure 3-6a) and a jar with four marbles (Figure 3-6b), take the union of the two jars, and see that it has seven marbles (Figure 3-7). Hence $3 + 4 = 7$.

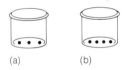

(a)   (b)

Figure 3-6

Figure 3-7

18. (a) Why is it important that A and B have no elements in common?

    (b) State the condition, "A and B have no elements in common," using the operation of intersection.

The Commutative and Associative Properties of Addition

The definition of addition given above enables us to verify all of the familiar facts about addition. For example, since the number of elements in the empty set { } is zero and { } ∪ B = B for any set B (see Exercise 14 in Chapter 1), we have

$$0 + b = b \quad \text{for all whole numbers } b$$

Illustrate this fact using a picture of jars with marbles.

In Exercises 10 and 11 you discovered that sometimes two trains had the same length. In particular, the order the rods are positioned does not affect the length of the train. For example, the trains in Figure 3-8 have the same length. Hence we can write W + R = R + W.

Figure 3-8

In Exercise 11 you saw that R + LG = LG + R. Illustrate with some other examples that if t_1 and t_2 are trains, then $t_1 + t_2 = t_2 + t_1$. This illustrates a fundamental principle of addition.

Commutative property of addition: *If m and n are any whole numbers, then m + n = n + m.*

Suppose we wish to add together three trains, say R, P, and Y. Since we only have defined the addition of two trains, we must add together two of the trains and then add the remaining train to the result. There are essentially two ways to do this. We can form R + P and then add Y to this train, obtaining (R + P) + Y = __?__ + Y. Or we can form P + Y and add it to R, obtaining R + (P + Y) = R + __?__. Use C-rods to show that the result is the same. Then show that for some other trains t_1, t_2, and t_3 we have $(t_1 + t_2) + t_3 = t_1 + (t_2 + t_3)$. This illustrates another fundamental principle of addition.

Associative property of addition: *If m, n, k are any whole numbers, then (m + n) + k = m + (n + k).*

Subtraction

We start our study of subtraction by again considering some concepts involving sets.

Consider the set of all A-blocks. Call it U. Remove T (i.e., the set of all triangles) from U. The set that remains is called the *difference* of U and T. It is the set of blocks that are not triangles.

Definition: *If B is a subset of A, then* the difference of A and B *is the set of elements of A that are not in B. We write A\B for the difference of A and B.*

19. Use your A-blocks to find the indicated differences.

    (a) U\R

    (b) U\L

    (c) (L ∪ T)\T

    (d) (L ∪ T)\R

20. Let A = {a, b, c, d, e, f, g, h}, B = {a, c, e}, C = {a, f, g, h}, D = {a}, and E = { }. Find the indicated differences.

    (a) A\B _____

    (b) A\C _____

    (c) A\D _____

    (d) C\D _____

    (e) A\E _____

    (f) D\E _____

Definition: *To subtract a number n from a number m, find a set A with m elements and a subset B of A with m elements. Then m − n is the number of elements in A\B. (Observe that you may not be able to perform the subtraction for any pair of numbers.)*

For example, in order to compute 5 − 2, find a set A that has five elements,

$$A = \{\boxed{R}, \ \textcircled{R}, \ \triangle\!R, \ \diamondsuit\!R, \ \textcircled{R}\}$$

and choose a subset having two elements,

$$B = \{\textcircled{R}, \ \textcircled{R}\}$$

and find the difference of A and B:

$$A\backslash B = \{\boxed{R}, \ \triangle\!R, \ \diamondsuit\!R\}$$

which illustrates that 5 − 2 = 3. (I hope you are not surprised.)

Two familiar properties of subtraction follow from

our definition. The first is that if n is any whole number, then

$$n - 0 = n$$

For example, choose a set A with n elements. A subset with 0 elements (in fact, the only subset with 0 elements) is the empty set { }. Now $A \setminus \{\ \} = A$. Hence $n - 0 = n$.

The second property is that if n is any whole number, then

$$n - n = 0$$

This follows since if A is any set $A \setminus A = \{\ \}$.

Now let us look at another approach to subtraction. Consider again the way we defined the difference of two trains of C-rods in Exercise 13. The train $t - s$ is the train that when added to s equals t. We could restate this as, "$t - s$ is the train that if put in the □ would make the equation $s + □ = t$ correct."

If we now wanted to define the difference of two numbers, we could proceed similarly and say that $a - b$ is that number n which solves $b + □ = a$, i.e., the number that makes the equation correct when put into the □.

Another way of saying this is that the two statements, $a - b = n$ and $n + b = a$, mean exactly the same thing. In fact, if we were using rods, we would illustrate the two rod statements $s + t = r$ and $r - t = s$ by the one situation shown in Figure 3-9.

Figure 3-9

An example involving numbers is to observe that since $3 + 6 = 9$, we know that $9 - 3 = 6$. Of course, since $3 + 6 = 6 + 3$, we also know that $6 + 3 = 9$, and so $9 - 6 = 3$. Thus, because of the commutativity of addition, each statement about addition gives us *two* subtraction statements.

21. Write down the two subtraction statements that are equivalent to the given addition statement.

    (a) $9 + 6 = 15$ _____ _____

    (b) $132 + 11 = 143$ _____ _____

    (c) $R + Y = Bl$ _____ _____

■ The properties for trains illustrated in Exercise 15 also hold for numbers.

We summarize below some properties of subtraction.

$$\begin{array}{ll} n - 0 = n & \text{for all whole numbers } n \\ n - n = 0 & \text{for all whole numbers } n \\ (m - n) + n = m & m \text{ greater than } n \\ (m + n) - m = n & \\ (m + n) - n = m & \end{array}$$

The last property indicates that subtraction of a number "undoes" addition of that number. Mathematicians call such operations *inverse operations*. Can you think of any similar examples from everyday life? How about tying-untying your shoelaces, opening-closing doors, walking forward-walking backward?

One of the things you might have done while playing with the C-rods in Exercise 9 was to build a staircase, as in Figure 3-10.

Figure 3-10

The staircase makes it immediately clear that numbers can be compared in size. The best way of stating this precisely is to say that train s is longer than train t when there is a train k such that $s = t + k$. For numbers this becomes:

Definition: *The whole number m is greater than the whole number n when there is a nonzero whole number k such that $m = n + k$.*

If we introduce the symbol $>$ for "is greater than" (we read $m > n$ as m *is greater than* n), we can write this rule succinctly as

$$m > n \Leftrightarrow m = n + k \quad \text{for some nonzero whole number } k$$

where the symbol $\Leftrightarrow$ means that each statement implies the other.

22. Verify each of the following.

    EXAMPLE: $10 > 7$ since $10 = 7 + 3$

    (a) $8 > 2$ _____

    (b) $11 > 9$ _____

    (c) $103 > 89$ _____

    (d) $17 > 0$ _____

23. Form a train t with the C-rods. Form another train s such that $t > s$. Convince yourself that if u is any other train, $t + u > s + u$.

■ Once we know about subtraction, we can say that $m > n \Leftrightarrow m - n$ is a nonzero whole number. The statement $n < m$ is read: n *is less than* m, and holds precisely when $m > n$.

24. Verify each of the statements in Exercise 22 by means of a subtraction statement.

    EXAMPLE: $10 > 7$ since $10 - 7 = 3$ is a nonzero whole number.

■ Again look at the C-rod staircase and imagine that it goes on forever. Given any two whole numbers m, n, both appear somewhere in the staircase. There are three possibilities: Either they are the same step ($m = n$), m appears to the left of n ($m < n$), or n appears to the left of m ($n < m$). This is called the *trichotomy law*.

Trichotomy law: *If m and n are whole numbers, then one of the following holds:*

   (1) $m = n$
   (2) $m < n$
   (3) $n < m$

Important Terms and Concepts

Addition
Subtraction
Algorithm
Commutative property of addition
Associative property of addition
Difference of two sets
Inverse operations
$m > n$
Trichotomy law

Review Exercises

1. Explain how you might use the model of the marbles in the jars to explain the associative property of addition to a classmate.

2. Let $A = \{1, 2, 3, 4, a, b, c\}$, $B = \{3, 4, a, b\}$, $C = \{4\}$, $D = \{\ \}$. Determine the following.

   (a) $A \backslash B = $ _____
   (b) $A \backslash C = $ _____
   (c) $A \backslash D = $ _____
   (d) $B \backslash C = $ _____
   (e) $B \backslash D = $ _____
   (f) $C \backslash D = $ _____
   (g) $B \backslash B = $ _____

3. Match the statements about numbers with the statements about sets. (There are extra statements about sets.) Write the letter in the blank.

   _____ $a + b = b + a$
   _____ $a - 0 = a$
   _____ $(a + b) + c = a + (b + c)$
   _____ $(a + b) - b = a$
   _____ $a + 0 = a$

   (a) $(A \cup B) \cup C = A \cup (B \cup C)$
   (b) $(A \backslash B) \cup B = A$
   (c) $A \cup B = B \cup A$
   (d) $(A \cup B) \backslash B = A$
   (e) $A \backslash \{\ \} = A$
   (f) $A \cup \{\ \} = A$
   (g) $A \cap B = B \cap A$
   (h) $(A \cap B) \cap C = A \cap (B \cap C)$

In Exercises 4–6, decide whether the statement is true or false. Write T or F in the blank.

4. _____ If $R \cup W = R$, then W is a subset of R.

5. _____ If $E \cap F = \{\ \}$, then it must be true that the number of elements in $E \cup F$ equals the sum of the number of elements in E and the number of elements in F. More briefly, Number $(E \cup F) = $ Number $(E) + $ Number (F).

6. _____ Is the operation of subtraction associative? That is, is the statement $(a - b) - c = a - (b - c)$ valid for all counting numbers?

7. *Optional*. Use the definition of $<$ ($m < n$ if there exists a counting number k such that $n = m + k$) to show that if $p < q$ and $q < r$, then $p < r$.

8. *Optional*.

   (a) Let us use Number (A) to denote the number of elements in A. Show that if A and B are any finite sets, then Number $(A \cup B) = $

Number (A) + Number (B) − Number (A ∩ B).

(b) At Lincoln School 156 students take the school bus. Of these 49 participate in after-school activities. A total of 84 students participate in after-school activities. If all of the students that either take the school bus or participate in after-school activities come to an assembly, how many students will attend?

© 1965 United Feature Syndicate, Inc.

Modern Black Mathematicians

Elbert Francis Cox (1896–1969) was the first black to obtain a Ph.D. in mathematics. He received his degree from Cornell University in 1925. He was head of the mathematics department at Howard University for 32 years.

J. Ernest Wilkins (b. 1923) received his Ph.D. in 1942 from the University of Chicago at 19 years of age. He worked on the Manhattan Project, and is now a professor of applied mathematics and physics at Howard University.

David Blackwell (b. 1919) received his Ph.D. from the University of Illinois in 1941. He was the first black elected to the National Academy of Sciences in any field. He is now a professor of statistics at the University of California, Berkeley.

4
Algorithms for Addition and Subtraction

"Children should be taught so that they see the relationship of mathematics to the physical world and understand that mathematics is a powerful instrument for describing and interacting with the environment."—Julian Weissglass

Lab Exercises: Set 1

EQUIPMENT: One Chinese abacus for each pair of students.

In these exercises you will learn to teach each other addition and subtraction with Egyptian and Mayan numerals. This will increase your understanding of how *our* numeral system works. You will also learn how to use the abacus to do addition and subtraction. Some of the exercises will require you to teach another person in the group. To prepare for this, each person in the group should tell about a good teacher they knew and what qualities he or she possessed.

In Exercises 1–4, you will work in pairs. One person is to play the role of teacher in the numerical system indicated, and the other person is to play the role of student. If you have difficulties, consult another pair in your group.

Egyptian

1. The teacher is to teach the student how to do the following addition problems. Do not use Hindu-Arabic numerals in your explanation; you can use our names for numbers but not the symbols.

   (a) ||| + || =

   (b) ||||||| + ||||| =

   (c) ∩∩∩||| + ∩∩|| =

   (d) ∩∩∩∩∩∩∩∩|||||||| + ∩∩||||| =

   (e) 99∩∩∩∩∩∩∩|||||| + 9∩|||| =

2. As a group, try to formulate rules for addition using Egyptian numerals. Since the point of this exercise is to discover how to operate with an unfamiliar numeral system without converting to our system, try to make your explanation as independent of our system as you can. For example, how does "carrying" work? Is Egyptian addition more or less complicated than addition in our system? Why?

3. (a) The teacher is to teach the student how to do the following subtraction problems.

Do not use Hindu-Arabic numerals in your explanation.

(i) |||||| – ||| =

(ii) ∩ – ||| =

(iii) ∩∩||| – ∩|||||| =

(iv) 99∩∩∩∩ – 9∩||| =

(v) ℓ999 – 99999||| =

(b) Check your answers to (iv) and (v) by converting to our system. As a group, try to formulate subtraction rules for Egyptian numerals. How does "borrowing" work? Is Egyptian subtraction more or less complicated than subtraction in our system? Why?

Mayan

4. (a) The teacher is to teach the student the following addition problems:*

(i) •••• + •••• =

(ii) —••• + —•••• =

(iii) ═• + —•• =
(Remember place value.)

(iv) •••• + ═ =
       ═   ••••
(These are two-digit Mayan numerals.)

(v) ═   ••
     —•• + ⟨𝟎⟩ =
     —••   ═
(These are three-digit Mayan numerals.)

(b) Did you discover that you can compute with Mayan numerals without knowing what numbers they represent? If not, review the

---
*One group of students solved these problems using the C-rods. They let a white rod represent a dot and a yellow rod represent a line and calculated with the rods. You might try this if you have difficulty with the numerals.

above exercises with this idea in mind. As a group, formulate rules for Mayan addition, making your explanation as independent of our system as possible.

5. (a) The teacher is to teach the student the following subtraction problems.

(i) •• – ••• =

(ii) —••• – ••• =

(A one-digit numeral is subtracted from a two-digit numeral.)

(iii) •
       ⟨𝟎⟩ – •• =

(iv) —• – —• =
      •••

(v) —••
     ••• – ═ =
     —••   ⟨𝟎⟩
              ═
(The difference of three-digit Mayan numerals.)

(b) Formulate a rule for Mayan subtraction. How does "borrowing" work?

6. Discuss in your group the similarities and differences between the way you usually add and subtract and the way you add and subtract with Mayan numerals. Write down at least one similarity and one difference.

The Abacus

In general, ancient civilizations did not calculate with numerals. Instead, they used various counting devices. One of these is the *abacus*, which we shall now investigate.

The abacus has several different forms. The Chinese abacus, which is the one we will be working with, is shown in Figure 4-1. Arrange the beads as shown

Figure 4-1

in the photograph. This is called *clearing* the abacus. To use the abacus, we assign each of the columns a place value. Which column should we assign to be the unit column? _____ Since we will be working with whole numbers, we assign the right-hand column to the *units* column. We assign the next column to the left to be the *tens* column, and so on, as indicated in Figure 4-2.

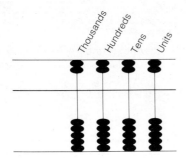

Figure 4-2

In each column, the five beads below the horizontal bar each have a value of 1, and the two beads above the horizontal bar each have a value of 5. In order to represent numbers on the abacus, we move beads on the columns toward the horizontal bar; for example, to represent:

1—Raise 1 lower bead in the units column to the horizontal.
2—Raise 2 lower beads in the units column to the horizontal.
3—Raise 3 lower beads in the units column to the horizontal.
4—Raise 4 lower beads in the units column to the horizontal.
5—Raise 5 lower beads in the units column to the horizontal.

An alternate way of representing 5 is to lower 1 bead to the horizontal. Two representations of 5 on the abacus are shown in Figure 4-3.

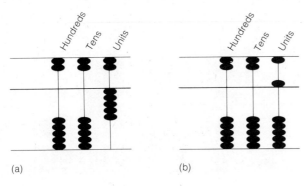

Figure 4-3

Continuing, to represent:

6—Lower 1 upper bead, raise 1 lower bead.
7—Lower 1 upper bead, raise 2 lower beads.
8—Lower 1 upper bead, raise 3 lower beads.
9—Lower 1 upper bead, raise 4 lower beads.
10—Lower 1 upper bead, raise 5 lower beads or lower 2 upper beads.
11—Lower 2 upper beads, raise 1 lower bead.
12—Lower 2 upper beads, raise 2 lower beads.

The above scheme can be extended to all columns, the beads indicating the number of times the value of each column is to be taken. Thus, Figure 4-4 represents 368, since the beads indicate 3 hundreds, 6 tens, and 8 units.

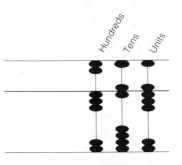

Figure 4-4

7. Determine the number represented by each of the following. Observe that it is possible to have two different representations of the same number, since there is a total of 15 units in each column.

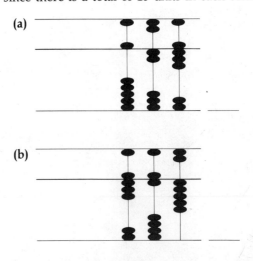

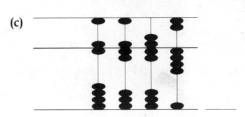

(d)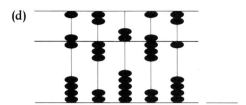

(e)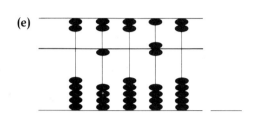

(f)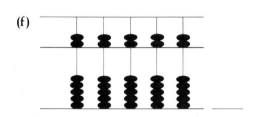

8. (a) Can you think of a reason for having 15 units in each column?

   (b) What is the minimum number of beads the abacus would need in each column in order to be able to represent any number using a Base 10 system? _____

9. Represent each of the following numerals on your abacus.

   (a) 29   (c) 999
   (b) 168   (d) 18,625

10. Addition on the abacus is quite simple. Try to find a method of using the abacus to calculate the sum of two numbers; then use the abacus to calculate the following sums. (If you have trouble finding such a method, see the rules that follow, but first try to find one yourselves.)

    (a) 23 + 68   (c) 7,238 + 1,164
    (b) 872 + 129

11. Now find a way to calculate subtraction problems on the abacus, and attempt to solve the following problems. (The rules follow, but first try to find a method yourselves.)

    (a) 65 − 24   (c) 98,423 − 2,965
    (b) 728 − 174

■ It is not necessary or even advisable to read the following rules if you have successfully developed your own method for doing addition and subtraction on the abacus.

Comment: Rules for the Abacus

1. Represent a number, i.e., 143, on the abacus.

2. Working from the right, add the digits of the other number, i.e., 374, to each column.

3. To add the following numbers to any column when the sum is less than 10, if the number is:

   1—Raise 1 lower bead *or* lower 1 upper bead and remove 4 lower beads.
   2—Raise 2 lower beads *or* lower 1 upper bead and remove 3 lower beads.
   3—Raise 3 lower beads *or* lower 1 upper bead and remove 2 lower beads, etc.

4. To add the following numbers to any column when the sum is greater than or equal to 10, if the number is:

   1—Cancel 9, forward 1 to the next column on the left.
   2—Cancel 8, forward 1 to the next column on the left.
   3—Cancel 7, forward 1 to the next column on the left, etc.

5. Whenever there is a total of 5 lower beads in a column, they can be lowered and replaced by 1 upper bead, as shown in Figure 4-5.

Figure 4-5

EXAMPLE: Add 143 + 374.

Step I. Enter 143 on the abacus (Figure 4-6).

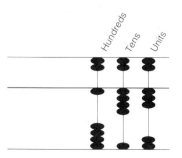

Figure 4-6

Step II. Add 4 to the unit column by lowering 1 upper bead and removing 1 lower bead (Figure 4-7).

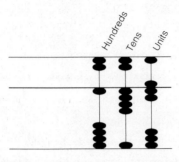

Figure 4-7

Step III. Add 7 to the tens column by raising 1 lower bead in the hundreds column and removing 3 lower beads in the tens column (Figure 4-8).

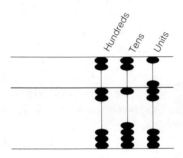

Figure 4-8

Step IV. Add 3 to the hundreds column by raising 3 lower beads *or* by lowering 1 upper bead and removing 2 lower beads. The two possible results are shown in Figure 4-9. Both represent 517.

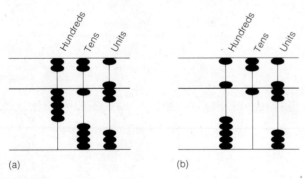

(a)   (b)

Figure 4-9

Rules for Subtraction

1. Represent the minuend (the a in $a - b$) on the abacus.

2. Starting with the left digit of the subtrahend (the b in $a - b$), deduct the number of indicated beads from the required columns.

3. If in subtracting there is in any column an insufficient number of beads, use the following procedure. To subtract:

   1—Borrow 1 from the column to the left, return 9 to the indicated column.
   2—Borrow 1 from the column to the left, return 8 to the indicated column.
   3—Borrow 1 from the column to the left, return 7 to the indicated column, etc.

EXAMPLE: Compute $463 - 335$.

Step I. Enter 463 on the abacus (Figure 4-10).

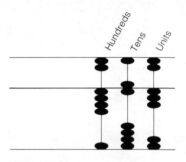

Figure 4-10

Step II. Remove 3 from the hundreds column (Figure 4-11).

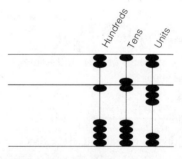

Figure 4-11

Step III. Remove 3 from the tens column (Figure 4-12).

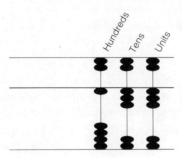

Figure 4-12

Step IV. Remove 5 from the units column by borrowing 1 from the tens column and returning 5 to the units column (Figure 4-13).

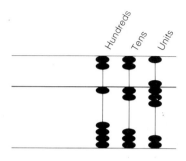

Figure 4-13

Lab Exercises: Set 2

"When nothing remains (in subtraction), put down a small circle so that the place be not empty, but the circle must occupy it, so that the number of places will not be diminished when the place is empty and the second be mistaken for the first."
—Al-Khwarizmi (A.D. 830)

EQUIPMENT: One set of MBA-blocks for each group.

In this set of exercises you will explore various ways of doing addition and subtraction. Of course, you already know how to add and subtract. Most people have learned (or memorized) one method for each operation. Nevertheless, it is important that teachers realize there are many different ways to solve a problem, and that they understand how different methods work. Such teachers will be able to encourage students to discover their own methods. A method for doing a computation is often called an *algorithm*, a word we shall often use.

12. **(a)** Someone in the group add 43 + 52 out loud by first adding the tens and then the units. Do you ever add this way? Some people use this method when computing in their heads.

   (b) Someone in the group add 486 + 231 out loud, starting with the hundreds and progressing from left to right.

   (c) Make up a method for using such a left-to-right algorithm on paper.

13. **(a)** A student was asked to explain how she added 57 + 38. She said, "Thirty-eight is almost forty. I added 57 plus 40 to get 97, and took away 2 to get 95." Is this a valid method of addition? _____

   (b) Have different people in the group do the following problems using this method.

       (i) 44 + 29 _____

       (ii) 126 + 99 _____

       (iii) 81 + 97 _____

14. **(a)** Someone in the group do the subtraction problem 85 − 23 out loud by first counting backwards two tens from 85 and then counting backwards three.

   (b) Someone in the group do the subtraction problem 71 − 33 out loud by first counting backward three tens from 71 and then counting backwards three.

■ Notice that the "counting-backwards" method described above is a left-to-right algorithm which eliminates the need for "borrowing" or "converting." Some children will naturally use this method when dealing with a subtraction problem.

   (c) Find a way to use the counting-backwards method with a pencil and paper and do the following problems.

       (i)   65
            − 24
       ——

       (ii)   92
            − 43
       ——

       (iii) 426
            − 134

   (d) Suppose a student handed in the following computation:

          615
       − 234
       ——
          415
          385
          381

   How would you explain the student's reasoning to a skeptical colleague who wanted to know why you weren't "dulling"* your students by using the "standard" computational technique?

   (e) Left-to-right algorithms are also possible for addition. Suppose a student with a preference for left-to-right algorithms did the computation on the following page. How would you explain the algorithm to the same skeptical

---

*The typist misread my handwritten "drilling" as "dulling." How fortuitous!

colleague? (There is usually one in every school!)

$$243$$
$$+\ 435$$
$$\overline{643}$$
$$673$$
$$678$$

15. Use the MBA-blocks to do (a) and (b) below. (It might be helpful to keep track of your work by placing the blocks on the appropriate counting boards we discussed in Chapter 2.)

    (a) (Base 2 blocks) Represent 11101_{two} by a pile A of blocks and 110_{two} with a pile B of blocks. By inspecting the piles, form a pile X of blocks so that $B + X = A$. Write the Base 2 numeral for the pile X. _____
    Use the blocks to verify that this numeral solves $11101_{two} - 110_{two} = X$. Someone explain to the rest of the group why this works.
    Now try to use the same method to compute

    $$11101_{two}$$
    $$-\ \ \ 110_{two}$$

    without blocks; that is, find by inspection, the Base 2 numeral X_{two} such that $110_{two} + X_{two} = 11101_{two}$.

    The answer is $X_{two} = $ _____.
    Make up some additional Base 2 subtraction problems for this algorithm.

    (b) (Base 3 blocks) Represent 12011_{three} by a pile A of blocks and 1221_{three} by a pile B of blocks. By inspecting the piles, form a pile X of blocks so that $B + X = A$. Write the Base 3 numeral for pile X. _____ Use the blocks to verify that this numeral solves $12011_{three} - 1221_{three} = X_{three}$. Someone in the group explain why.
    Now try to use the same method to compute

    $$12011_{three}$$
    $$-\ \ 1221_{three}$$

    without blocks. That is, find by inspection the Base 3 numeral X_{three} such that $1221_{three} + X_{three} = 12011_{three}$. The answer is $X_{three} = $

_____. Make up some additional problems in Base 3.

■ This algorithm is called the *additive* (or *Austrian*) *method*.

16. Try to develop a way to use this algorithm with Base 10 numerals. Do the following problems using the additive method.

    (a) 723   (b) 256   (c) 732
        $-\ 265$ $-\ 121$ $-\ 168$

17. Discuss in the group other subtraction algorithms that you know. Develop a different one.

18. *Optional*. Do (a) and (b) using MBA-blocks.

    (a) Represent 11101_{two} by a pile A of blocks and 110_{two} by a pile B of blocks. By inspection, find a pile C such that $B + C$ represents 100000_{two} (the number of zeros equals the number of digits in the minuend). Combine $A + C$ and remove the largest block. Represent the remaining pile as a Base 2 numeral. Verify that this numeral represents $11101_{two} - 110_{two}$ by converting to Base 10. Magic? No! Try it again for the problem $1011_{two} - 101_{two}$. Now see if you can do it without the blocks for the problem $111011_{two} - 1111_{two}$. Check your answer by computing with Base 10 numerals. Try to explain why it works.

    (b) Do (a) for $12011_{three} - 1221_{three}$.

■ The above algorithm has been called the *complementary method*. It is the method that computers use for subtraction. For the method to work easily, we choose a C such that $B + C$ is a "nice" numeral (1 followed by an appropriate string of zeros). We then compute $A + B$ and can easily subtract $B + C$, since it is of the form 10 ... 0.

19. *Optional*. Do the following Base 10 problems using the complementary method and explain the process to your partner.

    (a)   657   (b) 56,397
        $-\ 283$ $-\ 8,628$

Comment: Algorithms for Addition and Subtraction

"The duke himself often came into the treasury, and no computations were ever concluded without his presence or his official seal. He himself would sit at one end of the table and would move counters and calculate like the others; and there was no difference between their reckoning and his, except that the duke worked with golden counters and the others with silver ones."—From the memoirs of Olivia de la Marcke (1474)

After doing the lab exercises above, you are probably convinced that Egyptian numerals are not very conducive to the performance of extensive written calculations. Roman numerals are also quite cumbersome in this regard, as a little experimentation will quickly verify.

Although little is known about the actual methods that ancient civilizations such as the Babylonian, Egyptian, Hindu, Greek, and Roman used for their calculations, it is known that until the sixteenth century almost all calculations were done with the assistance of some sort of mechanical device. Anthropologists hypothesize that counting boards were used, and one large counting board from ancient Greece has been preserved. A counting board is a board engraved with lines and number symbols. To perform the calculations, loose counters were moved. It seems likely that these counting boards were used in a manner similar to the way you used the abacus in the last section, and that there were also methods for multiplication and division.

The Chinese form of the abacus is called the *suan pan* (literally "reckoning board"). It appeared in China around the twelfth century. Before that time counting boards were used. Students in China still learn how to use the abacus in elementary schools, and it is used extensively by shopkeepers and merchants (Figures 4-14 and 4-15).

The Japanese abacus, called the *soroban*, was intro-

Figure 4-14 Students in an elementary school classroom in Kuangchow (Canton).

Figure 4-15 Bookkeeper using a suan pan in a Peking arts and crafts factory.

duced from China around the sixteenth century (Figure 4-16). It differs from the suan pan in that it has only one bead above the horizontal, and four beads below. (Why is this enough?)

Figure 4-16 A soroban.

After working with the abacus, one can appreciate its popularity. It provides a quick and easy way to do calculations without pencil and paper and with little need to retain many facts in one's memory. The speed that can be developed is amazing. After the Americans occupied Japan in 1945, they organized a calculating contest to show the superiority of American methods. The following is a quote from the *Reader's Digest*, March 1947:

The contest matched 22-year-old Kiyoshi Matsuzaki, a Japanese Communications Ministry clerk with seven years' special abacus lessons, against 22-year-old Pvt. Thomas Ian Wood of Deering, Mo., an Army finance clerk with four years' experience on modern machines. Matsuzaki, who flipped the wooden beads with such lightning dexterity that he was immediately nicknamed "The Hands," used an ordinary Japanese *soroban*, selling for about 25 cents before the war. Wood's electric machine cost $700.

The abacus won the addition event—columns of four- to six-digit figures—taking all six heats and finishing one of them more than a full minute ahead of Wood. The abacus also won in subtraction. Wood staged a rally in multiplication, since abacus multiplication requires many hand motions; but Matsuzaki was out in front again in division and in the final composite problem. "The Hands" also made fewer mistakes.

One reason why Matsuzaki won is that like all abacus veterans, he does the simplest arithmetic in his head, pegging the results on the abacus and going on from there.*

In a more recent contest sponsored by an Australian department store an experienced operator using a Chinese abacus was the winner over a U.S. calculator executive using an electronic calculator. Modern technology has not been sufficient to overcome the speed of the abacus.

However, technology has managed to lower the price of electronic calculators, and that is some consolation to those of us who have not developed much skill with the abacus, especially since speed with the abacus requires years of practice.

The success of the abacus operators, even against

---

*Adapted from TIME, The Weekly Newsmagazine. Reprinted by permission.

electronic machines, helps to explain why written computation with the Hindu-Arabic numerals was slow to replace the abacus. As late as the eighteenth centruy, bookkeepers in Europe were using Roman numerals and abacuses for their work. In my opinion, it seems pedagogically sound to have children learn how to use the abacus for the operations of addition and subtraction, before asking them to use pencil and paper. For one thing the abacus is much quicker than hand calculations. Second, using the abacus involves the concepts of computing with a place-value system (i.e., borrowing and carrying) in a direct physical way.

As we mentioned in Chapter 3, our numerals originated around A.D. 600 in central India. The Hindus possessed great skill in mathematical calculation, and pursued complicated problems with great fervor. Their methods were studied by the mathematicians from the Arabian empire and in A.D. 830 the earliest, and one of the most influential, books explaining how to do arithmetic appeared. It was written by Al-Khwarizmi. His name, translated into Latin as Algorithmus or Algorismus, became a household word all over Europe and is the origin of our word "algorithm." Al-Khwarizmi was a Persian mathematician who became a scholar in the court of Baghdad, which was a great center of learning in his time. His book *Concerning the Art of Hindu Reckoning* was originally written in Arabic, although the only surviving copy is a twelfth-century Latin translation. European readers of the translations of Al-Khwarizmi's book attributed the whole numeration system to him and thus led to the misconception that the numerals were Arabic in origin. A quotation from his book is at the beginning of Lab Exercises: Set 2.

The numerals and methods contained in Al-Khwarizmi's book were not readily adopted in Western Europe. A sixteenth-century woodcut (Figure 4-17) shows a contest between someone using a counting board and someone using Hindu-Arabic numerals. There were many reasons for the reluctance to adopt Hindu-Arabic numerals—the effectiveness of the existing methods that used the counting board, the familiarity with Roman numerals, the scarcity of paper, the cumbersomeness of the early algorithms, and the conceptual difficulties of accepting 0 as a numeral. In addition, the numbers in use at the time were small and Roman numerals were adequate to deal with them. The complexity of the place-value concept itself must also have served as a hindrance to the adoption of the new system. Your work with the Mayan system has given you some idea of the difficulty of learning a system that is unfamiliar to you—and you are already familiar with the concept of place value! Eventually, however, the new system

Figure 4-17 Woodcut *Margarita philosophica nova*, by Gregorius Reisch (1512). Copyright: Museum of the History of Science, Oxford University.

became accepted by the merchants and bookkeepers of Western Europe because of its increased efficiency and capacity to do complex computations.

We will only briefly explain how to do arithmetic with Hindu-Arabic numerals. We defined addition in Chapter 3. If we now wish to actually calculate the sum or difference of two numbers, we need a method that will enable us to arrive at the answer rapidly. We do not want to return to the definitions each time, find two appropriate sets, and take their union in order to be able to compute a sum. In the lab exercises you studied various techniques for performing addition and subtraction. We call these techniques *algorithms*. A good algorithm enables us to do something efficiently, without thinking about the meaning of each step. As Alfred North Whitehead points out, in *An Introduction to Mathematics* (London: Williams and Northgate, 1911),

It is a profoundly erroneous truism, repeated by all copybooks and by eminent people when they are making speeches, that we should cultivate the habit of thinking of what we are doing. The precise opposite is the case. Civilization advances by extending the number of important operations which we can perform without thinking about them. Operations of thought are like cavalry charges in a battle—they are strictly limited in number, they require fresh horses, and must only be made at decisive moments.

It is important, however, when students first learn how to use an algorithm that they understand how it works. Before they develop the ability to perform operations without thinking about them, it is essential to do the thinking necessary to understand how the algorithm works. Otherwise they will be manipulating symbols according to meaningless rules they have memorized. This has some bad effects. It gives the students the wrong idea of what mathematics is. It demeans them by underestimating their intelligence. And it does not provide them with the concepts necessary to build upon as they proceed to new situations.

As you have seen in the lab exercises, there are often different algorithms that work for the same situation. It is useful to keep this in mind when teaching arithmetic. Students should be encouraged to develop their own algorithms.

Let us see what is required to justify the addition algorithm used in Exercise 14(e). Recall that the following was written

$$\begin{array}{r} 243 \\ + 435 \\ \hline 643 \\ 673 \\ 678 \end{array}$$

Breaking down the sequence of reasoning further, we could write

$$\begin{aligned} 243 + 435 &= 243 + (400 + 30 + 5) \\ &= (243 + 400) + (30 + 5) \\ &= 643 + (30 + 5) \\ &= (643 + 30) + 5 \\ &= 673 + 5 \\ &= 678 \end{aligned}$$

Perhaps a further justification of $243 + 400 = 643$ is needed. Consider

$$\begin{aligned} 243 + 400 &= (200 + 43) + 400 \\ &= 200 + 400 + 43 \\ &= 600 + 43 \\ &= 643 \end{aligned}$$

To justify $643 + 30 = 673$, we write

$$\begin{aligned} 643 + 30 &= (600 + 43) + 30 \\ &= 600 + (43 + 30) \\ &= 600 + (40 + 30 + 3) \\ &= 600 + (70 + 3) \\ &= 670 + 3 \\ &= 673 \end{aligned}$$

Notice that in justifying each of these steps, we are using the commutative and associative laws of addition. Also notice that we are starting with a relatively complex addition problem (with two 3-digit numerals) and must first build skill in handling simpler problems. Thus we assumed in the above that you know that $40 + 30 = 70$. If that were not known, it would be necessary first to develop an algorithm for adding multiples of ten (4 tens + 3 tens = 7 tens) and so on. This discussion does not completely justify the algorithm, but should give you an idea of what is involved.

20. Do the following addition problem with the method you usually use, and then write a justification of it similar to the one above.

$$\begin{array}{r} 296 \\ + 148 \\ \hline \end{array}$$

■ Now consider the problem of how to do subtraction. In order to see how the algorithms are justified, first look at how you would justify simple subtraction to a child. One way to justify that

$$8 - 3 = 5$$

is to form a set with eight elements (Figure 4-18a), remove three elements (Figure 4-18b), and then count what is left.

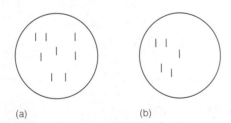

Figure 4-18

Of course, we could count backwards as we removed the elements. The step would be

|||||||| ⟶ ||||||| ⟶ |||||| ⟶ |||||
  8   take   7   take   6   take   5
        away   away   away
        1 is   1 is   1 is

Thus, to do simple subtraction problems like this, all that is necessary is to know how to count backwards by ones.

If we wrote the above symbolically, we would have

$$\begin{aligned} 8 - 3 &= 8 - (1 + 1 + 1) \\ &= (8 - 1) - (1 + 1) \\ &= 7 - (1 + 1) \\ &= (7 - 1) - 1 \\ &= 6 - 1 \\ &= 5 \end{aligned}$$

Note that we are using here the property $a - (b + c) = (a - b) - c$.

21. Illustrate this property using C-rods.

■ This is not the only algorithm to do subtraction, but it does work. If we were to extend our facility with counting backwards to multiples of tens, hundreds, etc., we could do any subtraction problem without any need for "borrowing" or "converting."* For example, we can reason that

$$\begin{aligned} 87 - 40 &= 87 - (10 + 10 + 10 + 10) \\ &= 77 - (10 + 10 + 10) \\ &= 67 - (10 + 10) \\ &= 57 - 10 \\ &= 47 \end{aligned}$$

Now, returning to the problem of Exercise 14(d), we reason

$$\begin{aligned} 615 - 234 &= 615 - (200 + 30 + 4) \\ &= 415 - (30 + 4)) \\ &= 385 - 4 \\ &= 381 \end{aligned}$$

which is vertically written

$$\begin{array}{r} 615 \\ - 234 \\ \hline 415 \\ 385 \\ 381 \end{array}$$

Observe that we are relying on the property $a - (b + c) = a - b - c$.

Of course, one has to develop some mental skill in counting backwards in order to effectively use this algorithm. Notice that this algorithm starts at the left and works toward the right. The most common algorithm used for subtraction starts at the right and goes to the left and employs converting. For example, to calculate $82 - 46$, one thinks of 82 as 8 tens and

---

*The term "borrowing"—once in common use—is being replaced by "converting," which we shall use from now on in this text.

2 ones, and 46 as 4 tens and 6 ones. Then

$$\begin{array}{r} 82 \\ -46 \end{array}$$ becomes $$\begin{array}{r} 8\text{ tens} + 2\text{ ones} \\ -(4\text{ tens} + 6\text{ ones}) \end{array}$$

which becomes $$\begin{array}{r} 7\text{ tens} + 12\text{ ones} \\ -(4\text{ tens} + 6\text{ ones}) \\ \hline 3\text{ tens} + 6\text{ ones} \end{array}$$

Eventually, you end up writing something like

$$\begin{array}{r} {}^7 8^1 2 \\ -4\ 6 \\ \hline 3\ 6 \end{array}$$

The process of thinking of 8 tens + 2 ones as 7 tens and 12 ones is called converting.

You have now worked with three or four algorithms for subtraction (the additive method, counting backwards, the common converting method, and if you did the optional Exercise 18, the complementary method). Perhaps the most important lesson here is that there is no one "right" way to do a computation. An algorithm is a technique that is effective and understandable. If we understand why an algorithm works, we have some insight about the underlying principle of the mathematical situation. Then we can use the algorithm with confidence and vary our approach if a new and different problem confronts us. If we simply memorize techniques, we may not be able to respond flexibly to new situations.

22. In the following, each letter represents a numeral (and each occurrence of the same letter represents the same numeral). What numeral does each letter represent?

$$\begin{array}{r} RR \\ +\ KK \\ \hline EKE \end{array}$$

Important Terms and Concepts

Abacus
Algorithm
Counting backwards
Left-to-right algorithms
Additive method

Review Exercises

1. Do the following problems in the system in which they are given (do not convert to Hindu-Arabic numerals).

Addition

(a) 999∩∩∩∩|||||| + 9∩∩∩∩||||| =

(b) 9999∩∩∩||| + 9999∩∩∩∩|||||| =

(c) [dot diagram] + [dot diagram] =

(d) [dot diagram] + [dot diagram] =

(e) [dot diagram] + [dot diagram] =

Subtraction

(f) 9∩∩∩||||| − ∩|||||| =

(g) 9 − | =

(h) [dot diagram] − [shell symbol] =

(i) [shell symbols] − [dot diagram] =

(j) [dot diagram] − [dot diagram] =

2. What would the following abacus representations be equivalent to in our Hindu-Arabic system?

(a)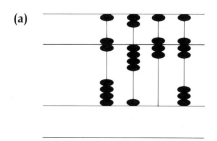

(b)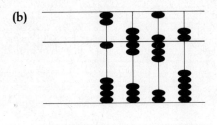

(c)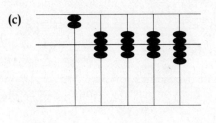

3. The following pictures depict addition and subtraction problems using the MBA-blocks in a given base to represent numbers. State the base that each problem is in and answer the problem.

Addition

(a)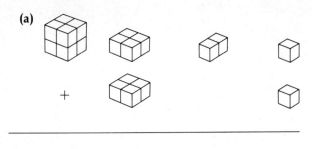

(b)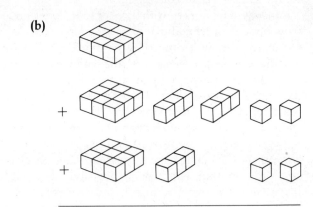

Subtraction

(c)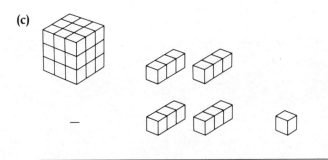

(d)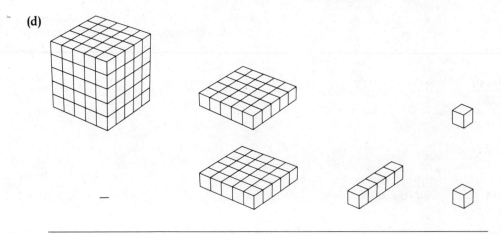

(e)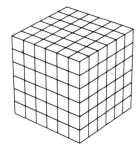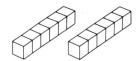

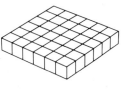

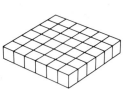 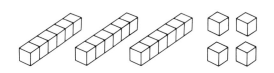

4. Do the following subtraction problem in three different ways.

$$\begin{array}{r} 513 \\ -241 \\ \hline \end{array}$$

5. Explain the ecological advantage of the abacus over paper and pencil.

6. Do the following problems in the base in which they are given, using any algorithm.

Addition

(a) $\begin{array}{r} 11001_{two} \\ + 11101_{two} \\ \hline \end{array}$

(b) $\begin{array}{r} 11111_{two} \\ + 1011_{two} \\ \hline \end{array}$

(c) $\begin{array}{r} 3212_{four} \\ + 2321_{four} \\ \hline \end{array}$

(d) $\begin{array}{r} 5426_{seven} \\ + 1265_{seven} \\ \hline \end{array}$

Subtraction

(a) $\begin{array}{r} 2341_{five} \\ - 1423_{five} \\ \hline \end{array}$

(b) $\begin{array}{r} 5432_{six} \\ - 3434_{six} \\ \hline \end{array}$

(c) $\begin{array}{r} 10000_{two} \\ - 1111_{two} \\ \hline \end{array}$

(d) $\begin{array}{r} 3121_{four} \\ - 333_{four} \\ \hline \end{array}$

7. In what number base or bases between 2 and 10 are each of the following calculations valid?

(a) $\begin{array}{r} 46 \\ + 25 \\ \hline 104 \end{array}$

(b) $\begin{array}{r} 42 \\ + 14 \\ \hline 24 \end{array}$

(c) $\begin{array}{r} 512 \\ + 523 \\ \hline 1235 \end{array}$

(d) $\begin{array}{r} 21 \\ - 10 \\ \hline 11 \end{array}$

(e) $\begin{array}{r} 21 \\ + 13 \\ \hline 34 \end{array}$

(f) $\begin{array}{r} 132 \\ - 42 \\ \hline 40 \end{array}$

8. Jim finds the following computation on the floor:

$$\begin{array}{r} 123 \\ 244 \\ \hline 312 \end{array}$$

He tries to figure out which base it is in and

decides that the computation is incorrect. What is his reasoning?

9. *Puzzle Problem.* Follow the instructions in Exercise 22 to solve the following.

$$\begin{array}{r} SEND \\ + \ MORE \\ \hline MONEY \end{array}$$

Chicano Mathematicians

Access to higher education by Chicanos in the United States has been so limited that only recently have any Chicanos received a Ph.D. in mathematics, and there are very few of these. Two Chicanos, Richard Griego and David Sánchez are professors at the University of New Mexico; the former is chairperson of the department. Of course, there are eminent mathematicians from Latin America, but these have not experienced the same restriction of opportunity as Chicanos in the United States.

5

Multiplication

"Here begynnes the chapter of multiplication, in the quych thou must know four thynges. Furst, quat is multiplication? The secunde, how mony cases may hap in multiplication. The thyrde, how mony rewes of figures there most be. The 4 what is the profet of this craft."—From The Crafte of Nombrynge, *a fifteenth-century English arithmetic treatise*

Lab Exercises: Set 1

EQUIPMENT: Two containers of C-rods for each group.

In this lab set we will define multiplication by the use of trains of C-rods, and will use the C-rods to illustrate the *commutative, associative,* and *distributive* properties of multiplication.

Please remember that I am not suggesting that you teach multiplication in this manner to elementary students. What follows is a challenging way for *you* to investigate the properties of multiplication using an unfamiliar concrete model.

1. (a) Place a purple rod on top of a DG rod as shown below. Then underneath the P rod make a rectangle of DG rods equal in width to the P rod. Put aside the P rod and form a train from DG rods. This train is called P "cross" DG and written P × DG.

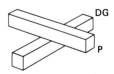

(b) Find LG × Y by placing an LG rod on top of a Y rod forming a rectangle of yellow rods equal in width to the LG rod and then forming the associated train.

(c) Find Bl × R (this is a train of *red* rods).

■ In finding the product of two rods, we use the *second* rod named to form the rectangle. The other side of the rectangle has length equal to the length of the *first* rod. For example, Y × R is the train formed from a rectangle composed of five red rods, while R × Y is a train formed from a rectangle composed of two yellow rods.

2. Form the trains Bl × LG and LG × Bl. Bl × LG is a train of _____ (what color?) rods. LG × Bl is a train of _____ (what color?) rods. Compare the lengths of the two trains. What do you notice?
   _____

3. Form the trains Y × Bk and Bk × Y. Y × Bk is a train of _____ (what color?) rods. Bk × Y is a train of _____ (what color?) rods. What do you notice about the lengths? _____

4. Verify that the trains Y × P and P × Y have the same length.

■ Let us agree to say that two trains are equal if they have the same lengths. Then we can write, for example, **Bl × LG = LG × Bl, Y × Bk = Bk × Y, P × Y = Y × P**. This illustrates the *commutative property of multiplication*.

5. Another way to illustrate the commutative property of multiplication is to form the rectangles associated with **LG × DG** and **DG × LG** and verify directly that the same number of white blocks would be needed to construct each rectangle. Do this by forming the rectangles with your C-rods and then placing one rectangle on top of the other.

6. We can translate **LG × DG = DG × LG** into the following statement about numbers: $3 \times 6 = 6 \times 3$. Translate each of the following into statements about numbers.

   (a) **Bl × LG = LG × Bl** _____

   (b) **Y × Bk = Bk × Y** _____

   (c) **P × Y = Y × P** _____

7. (a) Form the train **O + R**. Using rods of the *same color*, make a train equal in length to **O + R**. See if you can do this in five different ways.

   (b) Line up the trains you obtained next to **O + R**. Translate this situation into multiplication statements about numbers.

8. Make a train of red rods equal in length to the orange rod. Then make a rectangle of these red rods.

   (a) Which rod measures across this rectangle? You have solved

   $$\square \times R = O$$

   Use the same method to solve the following.

   (b) $\square \times Y = O$

   (c) $\square \times LG = DG$

   (d) $\square \times R = O + DG$

   (e) $\square \times Y = O + O$

   (f) $\square \times R = O + Br$

9. (a) Form the train **P × R** with your rods. Multiply **LG** by this train; that is, find **(P × R) × LG**. (It will be a train of LG rods.) Save this train.

   (b) Now form the train **R × LG**. Multiply this train by **P**; that is, find **P × (R × LG)**. (It will be a train of dark green rods.)

   (c) Compare the trains **(P × R) × LG** and **P × (R × LG)**. What do you notice? _____

   You have verified that **(P × R) × LG = P × (R × LG)**.

   (d) Use rods to verify that **(DG × R) × Y = DG × (R × Y)**.

   (e) Translate each of the above into statements about numbers and verify arithmetically.

■ Exercise 9 illustrates the *associative property of multiplication*.

10. (a) Form the train **LG + Y** and then find **R × (LG + Y)**. Save this train.

    (b) Form the two trains **R × LG** and **R × Y**. Then form their sum **(R × LG) + (R × Y)**.

    (c) Verify that **R × (LG + Y) = (R × LG) + (R × Y)**.

    (d) Verify in a similar manner that **LG × (P + W) = (LG × P) + (LG × W)**.

    (e) Translate (c) and (d) above into statements about numbers and verify arithmetically.

■ Exercise 10 illustrates the *distributive property of multiplication over addition*.

11. Find the following.

    (a) **W × DG =** _____

    (b) **W × O =** _____

    (c) **W × Bk =** _____

    (d) **W × W =** _____

    If t is any train, argue that $W \times t = t$. What does this mean for numbers? _____

■ This illustrates that 1 is an *identity element* for multiplication.

Comment: Properties of Multiplication

"The rules for multiplication and division, to be sure, require more diligence for their mastery, but their meaning will still be understood very quickly by those who give to them their full attention. These skills, of course, like all others, must be sharpened by practice and experience."—From a lecture by Philip Melanchton to students at the University of Wittenberg (1517)

There are two ways of thinking about multiplication. You saw, for example, that **P** × **LG** was another way of writing **LG** + **LG** + **LG** + **LG**. One approach, then, is to consider multiplication as an operation that results from repeated addition. Thus, $4 \times 3 = 3 + 3 + 3 + 3$. Each 3 appearing in the sum is called an *addend*, and there are four addends in the sum. In general, if a and b are whole numbers, then $a \times b$ is the result of the successive addition of a addends of b. $a \times b$ is called the *product* of a and b.

Using this approach, it is easy to compute the product of any two numbers. You have only to perform a series of addition problems. It is possible to work in any numeration system. For example, to compute $|| \times |||$ in Egyptian numerals, we write

$$|| \times ||| = ||| + ||| = |||||$$

12. Perform the following computations using Egyptian numerals. (Remember to regroup.)
    (a) $|| \times ||||||| =$
    (b) $||| \times \cap \cap | =$
    (c) $|||| \times 9 \cap \cap \cap ||| =$

■ In our work with the C-rods, we saw that the multiplication of trains had some very nice properties. For example, the commutative property states that if s and t are trains, then $s \times t = t \times s$. But it is by no means obvious that this should follow for products of numbers when multiplication is looked at in terms of repeated addition. Why in the world should

$$13 + 13 + 13 = 3 + 3 + 3 + 3 + 3 + 3 + 3 + 3 + 3 + 3 + 3 + 3 + 3?$$

A better interpretation of multiplication is needed. Therefore, let us consider again multiplication of C-rods. **P** × **LG** can also be interpreted as the number of units in the rectangular array of width **P** and length **LG**. In terms of numbers, this would be the number of units in an array having four rows and three columns. (The horizontal lines are called *rows*. The vertical lines are called *columns*.) Such an array is illustrated in Figure 5-1. As we shall see, this is a

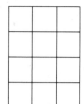

Figure 5-1

much more useful approach to multiplication. So we adopt the following definition.

Definition: *Given whole numbers a and b, construct a rectangular array of objects having a rows and b columns. Then the* product *of a and b, written $a \times b$, equals the number of objects in the entire array.*

This definition has the advantage that it connects multiplication directly to sets of objects rather than by way of another operation. As the following exercise shows, this interpretation encompasses the previous approach.

13. Outline a 3×4 array of squares on the dots below. Divide the rectangle you have outlined so as to illustrate that $3 \times 4 = 4 + 4 + 4$.

■ Since a similar procedure works for any product, it is clear that the number of objects in an array having a rows and b columns equals the number obtained by successively summing a addends of b.

14. Consider the 3×4 rectangle of Exercise 13. Rotate the page 90°. What product is now illustrated? _____

■ We see that the commutative property of multiplication, when thought of geometrically, is quite obvious. To see that the number of objects in an a by b array equals the number of objects in a b by a array, it is only necessary to rotate the first array 90°. Thus we can state the *commutative property of multiplication*.

Commutative property of multiplication: *If a and b are whole numbers, then* $a \times b = b \times a$.

After doing Exercise 9, you should find it easy to believe that multiplication will be associative. Let us see how this property follows from our definition. For example, how do we verify that $(3 \times 5) \times 4 = 3 \times (5 \times 4)$? We can easily verify $(3 \times 5) \times 4 = 15 \times 4 = 60$ and $3 \times (5 \times 4) = 3 \times 20 = 60$, but instead let us proceed by the definition. To calculate 3×5, we would construct a 3 by 5 array of objects (let us use white C-rods). Then to calculate $(3 \times 5) \times 4$, we would construct a (3×5) by 4 array. We can visualize this second array as a three-dimensional box, as shown in Figure 5-2. Now it is

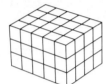

Figure 5-2

clear that we could have constructed this box by first building a 5 by 4 array and then constructing a (5×4) by 3 box. So the box also represents the product $(5 \times 4) \times 3$, which by the commutative property is the same as $3 \times (5 \times 4)$. The same argument would work for any three whole numbers. Thus we can state the associative property of multiplication.

Associative property of multiplication: *If a, b, and c are whole numbers, then* $a \times (b \times c) = (a \times b) \times c$.

The associative property permits us to write the product of three numbers, a, b, c as $a \times b \times c$, without parentheses.

By using algebraic arguments, the associative principle can be extended to more than three numbers. Since the commutative property can then be applied, we can conclude that the product of any finite string of whole numbers can be rearranged in any way. For example, $(a \times c) \times (b \times d) = ((a \times d) \times c) \times b = (a \times b) \times (c \times d)$, etc.

15. Compute each of the following, rearranging to simplify your calculations.

(a) $2 \times 15 \times 3 \times 5$ _____

(b) $5 \times 7 \times 6 \times 8 \times 2$ _____

■ We saw in Exercise 11 that the white rod has a special role in rod multiplication. The corresponding property for numbers is that 1 is an identity element for multiplication.

Identity element for multiplication: *For all whole numbers a,* $1 \times a = a \times 1 = a$.

This follows from the definition of multiplication, since an array of 1 row and a columns has a elements.

The number 0 also acts in a special way with respect to multiplication. If a is any whole number, then $0 \times a = 0$. This follows since an array with zero rows and a columns has zero elements, since it does not have any rows at all.

The last property of multiplication we will study is the *distributive property*. This provides the connection between multiplication and addition. It is not surprising that there is such a connection, since we have seen that multiplication can be viewed as repeated addition.

In Exercise 10 you illustrated the distributive property with C-rods. In terms of numbers it is stated as:

Distributive property of multiplication over addition: *If a, b, and c are whole numbers, then* $a \times (b + c) = (a \times b) + (a \times c)$.

Observe that since multiplication is commutative, a product written as $(b + c) \times a$ can be computed as

Courtesy of Al Capp; © 1967 News Syndicate Co., Inc.

$(b + c) \times a = a \times (b + c) = a \times b + a \times c = b \times a + c \times a$. Hence $(b + c) \times a = b \times a + c \times a$, and the distributive property works from both sides.

16. Illustrate that $4 \times (3 + 5) = 4 \times 3 + 4 \times 5$ by indicating below the appropriate rectangular arrays.

. . . .

. . . .

. . . .

. . . .

. . . .

. . . .

. . . .

17. Use rods to illustrate that $LG \times (Br - Y) = LG \times Br - LG \times Y$.

■ This illustrates that multiplication also distributes over subtraction.

Distributive property of multiplication over subtraction: If a, b, and c are whole numbers, then $a \times (b - c) = (a \times b) - (a \times c)$.

Important Terms and Concepts

Product

Commutative property of multiplication

Associative property of multiplication

Identity element for multiplication

Array

Distributive property of multiplication over addition

Distributive property of multiplication over subtraction

Review Exercises

1. Find $P \times Y$ and $Y \times P$ using C-rods and show that $P \times Y = Y \times P$. What property of arithmetic does this illustrate? _____

2. Use the distributive property to compute the following. (*Hint:* $28 + 72 = 100$.)

   $(68 \times 28) + (68 \times 72)$ _____

3. (a) Does addition distribute over multiplication? That is, if a, b, and c are whole numbers, does $a + (b \times c) = (a + b) \times (a + c)$? Justify your answer. (Can you see that this is different from the question: Does multiplication distribute over addition?)

   (b) Determine whether addition distributes over subtraction.

4. Why is it correct to say $5 \times 3 = 3 \times 5$, but it is incorrect to say $5 - 3 = 3 - 5$?

5. Suppose we define a new operation $*$ by $a * b = a^b$, where a and b are counting numbers. For example, $2 * 5 = 2^5 = 2 \times 2 \times 2 \times 2 \times 2 = 32$ and $3 * 3 = 3^3 = 3 \times 3 \times 3 = 27$.

   (a) Is $*$ commutative? _____

   (b) Is $*$ associative? _____

   (c) Does $*$ distribute over addition? _____

6. John has three shirts (red, green, and blue) and two pairs of pants (black and brown). How many different ways can he get dressed? _____ Draw a rectangular array that exhibits the possibilities.

7. Suppose you were making a set of A-blocks and you were going to use three sizes, three colors, and four shapes. How many blocks would be in your set? _____ How could you use the notion of arrays in describing your answer?

8. Use repeated addition to compute the answers to the following problems. Do them in the system in which they are given.

   (a) ||| × 99∩∩∩∩||||| =

   (b) ⎯•⎯ × ⎯••⎯ =

   (c)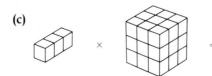

9. Suppose you were trying to teach multiplication to a fourth-grader. Describe two different ways that you could explain the process of multiplication.

10. Decide whether the following statements are true or false. Write T or F in the blank.

　　____ (a) According to the method used in this chapter, the picture below could be an illustration of the problem 5 × 4.

```
*  *  *  *
*  *  *  *
*  *  *  *
*  *  *  *
*  *  *  *
```

　　____ (b) The identity element for multiplication is zero because 0 × a = 0.

11. Illustrate that 5 × (2 + 4) = 5 × 2 + 5 × 4 by indicating below the appropriate rectangular array.

© 1967 United Feature Syndicate, Inc.

Theano (?–?), the wife of Pythagoras, was a teacher in the school founded by the Pythagorean order. One source indicates that at least 28 women were classified as Pythagoreans. Since none of the Pythagorean accomplishments was attributed to individual members, it is impossible to ascertain the individual contributions of the women mathematicians.

6
Algorithms for Multiplication

"The fatal pedagogical error... to throw answers like stones at the heads of those who have not yet asked the questions."—Paul Tillich

Lab Exercises: Set 1

EQUIPMENT: One set of MBA-blocks (Base 3, 4, and 5) for each group.

There are many algorithms for multiplication. In these exercises we will explore duplation (developed by the Egyptians) and the lattice method. We will also attempt to develop an algorithm for multiplying numerals by use of the MBA-blocks.

In Chapter 5 we explored how with Egyptian numerals we might use repeated addition to multiply. The Egyptians, however, did not use this method. They developed a method called *duplation*, which is a process of multiplying by doubling and then summing. In this method, the product of two numbers is obtained by multiplying one number by the powers of two necessary to express the other number, then adding the partial products. For example, to compute 19 × 22 using duplation with Hindu-Arabic numerals, we would form two columns of numbers starting with 1 and 22 by repeatedly doubling each number as shown.

1	22
2	44
4	88
8	176
16	352

We then determine which numbers in the left-hand column add up to 19 and mark them with a check. We also mark their corresponding numbers in the right-hand column:

√ 1	22 √
√ 2	44 √
4	88
8	176
√ 16	352 √
1 + 2 + 16 = 19	22 + 44 + 352 = 418

We then add the checked numbers in each column. The left-hand column gives 19 as a sum of powers of two. The right-hand column gives us the product 19 × 22.

With Egyptian numerals, the computation of ∩||||||||| × ∩∩|| is as shown below

```
\|         ∩∩||/
\||        ∩∩∩∩||||/
||||       ∩∩∩∩∩∩∩||||||||
||||||||   9∩∩∩∩∩∩∩||||||
\∩|||||||  999∩∩∩∩∩∩||/
∩|||||||   999∩∩∩∩∩∩∩∩∩|||||||||
```

or by making the exchanges 9999∩|||||||| .

1. Complete the computation of 21 × 34 begun below by duplation.

1	34
2	68
4	136
8	
16	

2. Use duplation to compute ∩||| × ∩∩||, using Egyptian numerals.

3. (a) Given a multiplication problem, we wish to develop a procedure for solving it using MBA-blocks. To begin, represent some number less than 50 with the blocks. How would you multiply this number by 2? Be sure to explain the procedure to your group.

   (b) How would you multiply a number by 3 with the blocks? Again explain your procedure to the group.

4. Now consider 2-digit multipliers. Consider the Base 4 problem below. (This is a 2-digit multiplier in which the second digit is zero.) One procedure would be to convert the long to four units and proceed as you would before by repeating the multiplicand once for each unit block.

   ×

   Do this or a similar problem in this way. Clearly, if we are to multiply large numbers, we need a more efficient method. Your group should try to find a procedure for doing this. You might find it helpful to consider the following questions.

   (a) What is the product of a long and a unit?

   (b) What is the product of a long and a long?

   Why? (It is not enough to state that 4 × 4 = 16. Explain your answer in terms of rectangular arrays of MBA-blocks.)

   (c) What is the product of a long and a flat?

   Why?

   (d) What is the product of a flat and a flat?

   Why?

   (e) What is the product of two longs and one long?
   Why?

   (f) What is the product of two longs and a flat?

   Why?

   (g) Does your procedure tell you how to multiply a 3-digit number by a 2- or 3-digit number using the blocks? Make up such a problem and do it with the blocks.

   (h) In the space below write a description of the algorithm your group developed.

5. (a) What is the product (2F + 2L + 1U) × L?*
   _____

   (b) What is the product (2F + 2L + 1U) × F?
   _____

   (c) How does this relate to multiplying in Base 10 by 10 and 100?

6. Use the algorithm you developed for multiplication with MBA-blocks to develop a paper and pencil algorithm for multiplying two numbers in a given base. Using your algorithm, do each of the following problems in the space provided. Do not convert to Base 10; instead, check your answers using blocks.

   (a) 220_{three} (b) 341_{five} (c) 123_{four}
       × 21_{three} × 43_{five} × 201_{four}

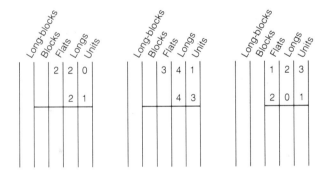

7. One person in the group explain the algorithm for multiplication indicated below.

   ```
         526
         387
      150,000
        6,000
        1,800
       40,000
        1,600
          480
        3,500
          140
           42
      203,562
   ```

   How does it relate to the following?

   526 × 385 = (500 + 20 + 6) × (300 + 80 + 7)

*F = flat, L = long, U = unit.

■ A very old method of multiplying is the *grating* or *lattice method*. Probably developed in India, it is illustrated in Figure 6-1, using the same problem as in Exercise 7. The multiplicand is written across the top and the multiplier down the right side. The products of the digits are written in their appropriate cells as shown.

Figure 6-1

We add along each diagonal starting with the farthest on the right-hand side and carrying when necessary. For example, the first sum is 2. The second sum is 8 + 4 + 4 = 16. The third sum is 8 + 4 + 6 + 1 + 5 + 1 (carried) = 25. The last digit of each sum is written below the diagonal, as shown in Figure 6-2. Find the three remaining digits. The answer 203,562 is read down the left side and across the bottom.

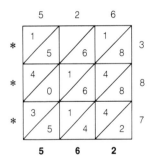

Figure 6-2

8. Compute each of the following, using the lattice method. Compare your results with the rest of your group.

   (a)   729   (b)   681
       × 268   × 754

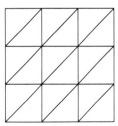
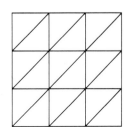

9. Do the following problems in the base indicated, using the lattice method.

   (a) 132_{four}
       $\underline{322_{four}}$

   (b) 1101_{two}
       $\underline{1110_{two}}$

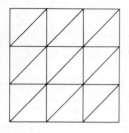
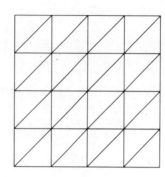

10. Discuss in your group the advantages and disadvantages of the lattice method. Write your conclusions below.

11. *Optional.* Develop a method for multiplication on the abacus.

Comment: Algorithms for Multiplication

"I will sette as I doe often in worke use, a pair of paralleles, or Gemove [twin] lines of one length, thus: ====, because no 2 thynges can be moore equalle."—Robert Recorde, 1557 (the first appearance of the equality sign)

You have seen that the MBA-blocks can be used effectively and accurately to perform any multiplication. By the use of such a manipulative device, the process of computing the product of two numbers can be understood in terms of a physical reality. Of course, this procedure requires considerable thought (as well as time), and as the quotation from Whitehead in Chapter 4 reminded us, our aim is to acquire the ability to perform operations without thinking.

Although it is possible to coerce students into developing their ability by memorizing rules (and many of us did it this way), such coercion is unnecessary. In order for us to be flexible in our teaching, we must understand the principle underlying the processes of computing a product. To achieve this, we shall study a few of the many algorithms for multiplication. It is interesting to note that as early as 1494, Luca Pacioli, an Italian mathematician, described eight different methods of multiplication in his book *Summa de Arithmetica*.

Duplation

One of the earliest algorithms was the method of duplation used by the Egyptians. To see why this method works, consider again the product 19×22 that we studied at the beginning of this chapter. If we were to write 19 as a Base 2 numeral, we would have $19 = 10011_{two}$, which can be interpreted as

$$19 = 1 \times 2^4 + 0 \times 2^3 + 0 \times 2^2 + 1 \times 2 + 1$$

Thus, to compute 19×22, we can write

$$19 \times 22 = (2^4 + 2 + 1) \times 22$$
$$= 2^4 \times 22 + 2 \times 22 + 1 \times 22$$

The numbers $2^4 \times 22 = 352$, $2 \times 22 = 44$, and $1 \times 22 = 22$ are the same numbers that were checked on the list obtained by repeated doubling.

12. Multiply the following using the duplation method.

    (a) 14×217 (b) 98×21 (c) 146×263

Multiplication with MBA-Blocks

To simplify our discussion, we will use

   U to denote unit
   L to denote long
   F to denote flat
   B to denote block
   LB to denote long-block
   FB to denote flat-block
   BB to denote block-block

In order to develop an algorithm for multiplying with MBA-blocks, we first realize that a representation of any number by blocks can be considered as a sum of units, longs, flats, etc. For example, $221_{four} = F + F + L + L + U$. We can then use the distributive law to reduce any product to a sum of products of the form L × L, L × F, F × B, etc. For example, to multiply $23_{four} \times 21_{four}$ using blocks, we would have

(2L + 3U) × (2L + 1U)
$$= (2L + 3U) \times (L + L + U)$$
$$= (2L + 3U) \times L + (2L + 3U) \times L$$
$$+ (2L + 3U) \times U$$
$$= (2L \times L) + (3U \times L) + (2L \times L) + (3U \times L)$$
$$+ (2L \times U) + (3U \times U)$$

This product (and, similarly, any product) can therefore be computed if we know U × U, L × U, L × L, etc. But these products are the same in any base:

U × X = X   for all X
L × L = F
L × F = B
L × B = LB
F × F = F × (L × L) = (F × L) × L = B × L = LB

Hence

(2L + 3U) × (2L + 1U)
$$= 2F + 3L + 2F + 3L + 2L + 3U$$
$$= 4F + 8L + 3U$$
$$= 1B + 2F + 3U \quad \text{(in Base 4)}$$

Therefore, $23_{four} \times 21_{four} = 1203_{four}$.

Now this is a rather lengthy algorithm, and you would certainly want to shorten it if you were going to do block multiplication very often. One way of doing this would be to notice, for example, that multiplying by a long can be accomplished by moving the digits in the multiplicand one place to the left (see Exercise 5). So, multiplying by 2L can be accomplished by first moving the digit and then multiplying by 2. Using these shortcuts, our computation of (2L + 3U) × (2L + 1U) might look like Figure 6-3.

B	F	L	U	
		2	3	
		2	1	
		2	3	(2L + 3U) × U
	4	6		(2L + 3U) × 2L = (2F + 3L) × 2
	4	8	3	Adding
	6	0	3	Changing longs to flats
1	2	0	3	Changing flats to blocks

Figure 6-3

If we were doing the problem with pencil and paper, we would probably want to shorten our work even more by working completely with Base 4 numerals. This means that the 483 and 603 would not appear in the calculations above. Instead it would look like

$$\begin{array}{r} 23_{four} \\ 21_{four} \\ \hline 23 \\ 112 \\ \hline 1203_{four} \end{array}$$

where the 112 is obtained by mentally calculating 23 × 2 in Base 4, and moving the digits one place to the left. This involves some facility in mentally computing with Base 4 numerals.

13. Do the following in the indicated base.

(a) 324_{five} × 23_{five} (b) 1011_{two} × 110_{two} (c) 3122_{four} × 213_{four}

The Partial-Products Algorithm

In Exercise 7 the product 526 × 387 was computed by summing the partial products. This algorithm is easily justified by writing 526 = 500 + 20 + 6 and 385 = 300 + 80 + 7. Then, applying the distributive law repeatedly, we obtain

526 × 387 = (500 + 20 + 6) × (300 + 80 + 7)
$$= (500 + 20 + 6) \times 300 + (500 + 20 + 6)$$
$$\times 80 + (500 + 20 + 6) \times 7$$
$$= (500 \times 300) + (20 \times 300) + (6 \times 300)$$
$$+ (500 \times 80) + (20 \times 80) + (6 \times 80)$$
$$+ (500 \times 7) + (20 \times 7) + (6 \times 7)$$
$$= 150{,}000 + 6{,}000 + 1{,}800 + 40{,}000$$
$$+ 1{,}600 + 480 + 3{,}500 + 140 + 42$$

which explains the calculation

$$\begin{array}{r} 526 \\ 385 \\ \hline 150{,}000 \\ 6{,}000 \\ 1{,}800 \\ 40{,}000 \\ 1{,}600 \\ 480 \\ 3{,}500 \\ 140 \\ 42 \\ \hline 203{,}562 \end{array}$$

Another way to form partial products is to start from the right and work toward the left. The problem would look like

$$\begin{array}{r} 526 \\ \underline{387} \\ 42 \\ 140 \\ 3{,}500 \\ 480 \\ 1{,}600 \\ 40{,}000 \\ 1{,}800 \\ 6{,}000 \\ \underline{150{,}000} \\ 203{,}562 \end{array}$$

This is the basis of the algorithm commonly used today. Usually the writing is somewhat condensed. For example, when computing the first three partial products, do the necessary carrying mentally so as to have only one line representing $(500 + 20 + 6) \times 7$. Our calculation would become

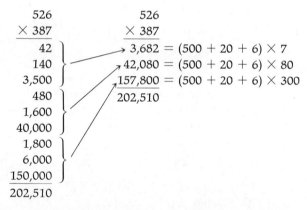

14. Do the following problem twice, once writing out out all the partial products, and again using the common algorithm. Draw a diagram similar to the one above, showing the correspondence between the two methods.

(a)   627   (b)   627
      $\times$ 149 $\times$ 149

The Gelosia or Lattice Method

We became acquainted with this method in Exercise 8 above. It is one of the oldest of the multiplication algorithms. When or where it originated is not known, but it was in use in India by the twelfth century, so that is the likely origin. The name *gelosia* was attached to it by the Italians. Pacioli (writing in 1494) states,

The sixth method of multiplying is called gelosia or gratticola . . . because the arrangement of the work resembles a lattice or gelosia. By gelosia we understand the grating which it is the custom to place at the windows of houses where ladies or nuns reside, so they cannot easily be seen. Many such abound in the noble city of Venice.

Consider again the lattice for 526×387 (Figure 6-4).

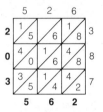

Figure 6-4

One factor is written along the top and the other down the right side. The numbers in the squares are the partial products, with the diagonals keeping track of the place value. Once the grating has been drawn, this method is easier and quicker than ours. The only carrying needed is when adding along the diagonals, and there is little likelihood of error. However, the necessity of drawing the lattice is a major drawback.

15. Calculate the following using the lattice method.

(a)   298   (b)   216
      $\times$ 76 $\times$ 1,743

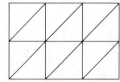
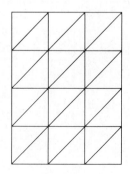

(c) $\begin{array}{r}122_{three}\\ \times\ 211_{three}\end{array}$

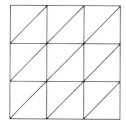

Important Terms and Concepts

Duplation   Partial products
Lattice (grating) method

Review Exercises

1. Use duplation to calculate the following multiplication problems in the system indicated.

   (a) 18×14

   (b) ∩||||| × ∩∩|

   (c)

2. Do each of the following with the MBA-blocks.
   (a) $(3F + 2L + 3U) \times (2F + 2F + 2U)$ Base 4
   (b) $(1B + 1L) \times (1LB + 1F + 1U)$ Base 2
   (c) $(2B + 2L + 1U) \times (2F + 2U)$ Base 3

3. Do the following multiplication problems in the indicated base.

   (a) $\begin{array}{r}11011_{two}\\ \times\ \ \ 101_{two}\\ \hline \text{two}\end{array}$

   (b) $\begin{array}{r}2312_{four}\\ \times\ \ \ 32_{four}\\ \hline \text{four}\end{array}$

   (c) $\begin{array}{r}234_{five}\\ \times\ 213_{five}\\ \hline \text{five}\end{array}$

   (d) $\begin{array}{r}562_{seven}\\ \times\ \ 43_{seven}\\ \hline \text{seven}\end{array}$

4. Do the following problems using the partial-products method and then using the lattice method.

   (a) $\begin{array}{r}476\\ \times\ 385\end{array}$

   (b) $\begin{array}{r}6{,}123\\ \times\ \ \ 468\end{array}$

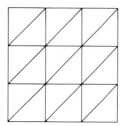 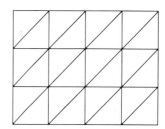

5. Explain why the lattice method works. In Review Exercise 4 what is the relationship between the numbers in the squares and the partial products?

6. Use the lattice method to calculate the following problems.

   (a) $12312_{four} \times 12312_{four}$

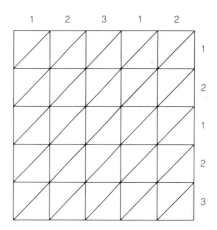

   (b) $1221_{three} \times 1202_{three}$

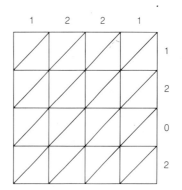

7. In each of the following lattice-method problems, determine the base and complete the problem.

(a)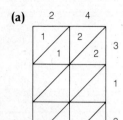

(b)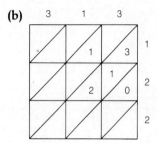

(c)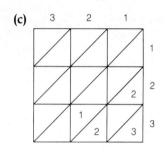

8. (a) If you were using Base 10 MBA-blocks, the product L × L would be equivalent to the numerical problem _____.

(b) Determine the following products of Base 10 numerals.

   (i) 10 × 12 = _____
   (ii) 10 × 35 = _____
   (iii) 10 × 463 = _____

(c) Find each of the following for Base 10 MBA-blocks.

   (i) L × (3L + 4U) = _____
   (ii) L × (2B + 7L + 6U) = _____

(d) What is the connection between (b) and (c) above?

9. If you see the computation 3L × 2L = 1B + 1F, you know that Base _____ blocks are being used.

10. Decide whether the following statements are true or false. Write T or F in the blank.

   _____ (a) The partial-products method of doing multiplication relies on the distributive law.

   _____ (b) For MBA-blocks of any base, F × F = a long block.

11. *Optional.* If "triplation" were defined analogously to duplation, would it work? Justify your answer.

Hypatia (370–415) lived and worked in Alexandria. She was the author of several treatises on mathematics, but unfortunately only fragments remain. Students from Europe, Asia, and Africa came to hear her lecture on number theory.

7
Division

"Before the introduction of the Arabic notation, multiplication was difficult, and the division even of integers called into play the highest mathematical faculties. Probably nothing in the modern world could have more astonished a Greek mathematician than to hear that, under the influence of compulsory education, the whole population of Western Europe . . . could perform the operation of division for the largest numbers. This fact would have seemed . . . a sheer impossibility. . . . Our modern process of easy reckoning with decimal fractions is the most miraculous result of a perfect notation."—Alfred North Whitehead

Lab Exercises: Set 1

EQUIPMENT: Two containers of C-rods and one set of MBA-blocks (Base 3, 4, or 5) for each group.

These exercises explore the division of whole numbers, using the C-rods and the MBA-blocks. As in Chapter 5, we will be trying to set up a learning situation that you will find challenging and instructive, and not advocating any particular approach for elementary students.

1. How many red rods are there in a blue rod? _____ Make a train of red rods that comes closest to being equal in length to the blue rod but remains shorter than the blue rod. What rod do you need to make your train equal to the blue? _____ We can summarize the result by writing

$$B = 4 \times R + \boxed{W}$$

We say there are four red rods in a blue rod, with a remainder of a white rod. In each of the following, use the rods to find the number that belongs on the _____ and the rod that belongs in the $\Box$. Be sure to make the train with the rods, even if you can calculate the answer in your head.

(a) $O = \underline{} \times LG + \Box$

(b) $O + W = \underline{} \times P + \Box$

(c) $O + Bl = \underline{} \times R + \Box$

(d) $Br = \underline{} \times R + \Box$

(e) $O + Y = \underline{} \times P + \Box$

(f) $O = \underline{} \times R + \Box$

■ In each of the above you have answered a question of the form "How many times can you subtract one type of rod from another type of rod, and what kind of rod remains?" In some of the examples no rod remained. We will investigate this situation in Exercises 2–7.

2. (a) Consider the brown rod. Make a train of red rods equal in length to the brown rod. Then form a rectangle of the red rods, as shown on the next page. Find the rod that measures the length of this rectangle; that is, find the rod that would make the following true: $\Box \times R = Br$. By experiment, we see that it is **P**. We say that the brown rod *divided by* the red rod is the purple rod, and write $B \div R = P$.

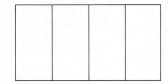

(b) Find (O + R) ÷ LG by making a train of LG rods equal in length to O + R, forming a rectangle, and then determining the rectangle's width.

$$(O + R) \div LG = \square$$

Observe that you are solving $\square \times LG = O + R$. Save the trains; you will need to use them again.

(c) Using the rods, solve the following.

$$(Br + Bk) \div Y = \square$$

Write down the associated equation involving rod multiplication.

We see that if t_1 and t_2 are trains, then $t_1 \div t_2 = t_3$ means that $t_1 = t_3 \times t_2$. We adopt this as our definition of division for trains of C-rods. t_2 is called the *divisor* and t_3 the *quotient*.

3. (a) Form the train P × LG; that is, form the rectangle of LG rods having width P, and then make a train of the LG rods. Call this train t. What is $t \div LG$? We would need to form a train of LG rods equal in length to t, make this train into a rectangle, and then find the rod that measures this rectangle. This will be P. Do you see it? We have shown that (P × LG) ÷ LG = P. Compare with Exercise 2(b).

(b) Show similarly that (LG × Y) ÷ Y = LG. Compare with Exercise 2(c).

4. Form the train LG × P (this is a train of purple rods). Call this train s. Use the rods to find $s \div LG$. (Make a train of LG rods equal in length to s. Make this train into a rectangle and find the rod that measures its length.) What do you notice? (LG × P) ÷ LG = _____.

5. Suppose we let Z stand for a rod of length 0 (zero). (What color would you choose for it?)

What would you say is the answer to each of the following?

(a) Z ÷ Y = _____
(b) Y ÷ Z = _____
(c) Z ÷ Z = _____
(d) R ÷ (R + Z) = _____
(e) (LG × Z) ÷ LG = _____

6. Let t = DG + Bl. Find $t \div$ LG, DG ÷ LG, and Bl ÷ LG. Verify with the rods that $t \div$ LG = (DG ÷ LG) + (Bl ÷ LG).

7. Represent 54_{ten} with your MBA-blocks. Write its numeral in the base you are using: 54_{ten} = _____ Base _____

(a) Discuss in your group how to divide 54_{ten} by 2 using the blocks. The result in your base is _____ ÷ 2 = _____ Base _____

(b) Discuss how to divide 54_{ten} by 18_{ten} using the blocks. Write down your procedure.

(c) Was it the same or different from the process used in (a)? _____

■ There are two possible physical interpretations of $a \div b$. One can think of a objects divided into b equal piles (and possibly a remainder), or how many b's are contained in a?

Which interpretation is easier to use when dividing by 2? _____; when dividing by 18? _____

In the following exercise we will attempt to develop an algorithm for division with blocks using the second interpretation. However, it may happen that there is not an exact number of b's contained in a. In that case we agree to find the largest number of b's contained in a. This number is called the *quotient*. The number left over is called the *remainder*. For example, the quotient of 54 ÷ 16 is 3, with a remainder of 6.

We shall illustrate how to use the MBA-blocks to solve division problems. Go over Example (a) or (b) below, depending on the base of the MBA-blocks you have available.

EXAMPLE (a): (Base 5) Find $142_{five} \div 4_{five}$.

Represent 142_{five} with the blocks. We could divide it into four equal piles. Instead, we determine how many 4's are contained in the pile.

7/Lab Exercises: Set 1

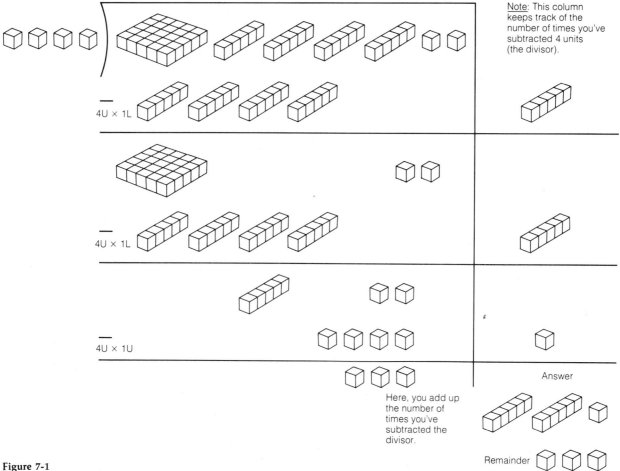

Figure 7-1

LONG METHOD: Exchange the pile representing 142_{five} into units and repeatedly remove groups of four, keeping track of how many groups you remove.

SHORT METHOD (subtraction of partial products): At each step (see Figure 7-1), subtract the product of "something" times the divisor (4 in this example). The "somethings" are then added to obtain a quotient. Of course, there may be a remainder. Observe that you do not need to subtract the largest possible multiple of the divisor at any stage.

Remember that your goal here is to understand how the algorithm works. The MBA-blocks are just unfamiliar enough to require you to think about what you have been doing routinely all these years. Your students will be as unfamiliar with the process of division as you are with using the blocks.

EXAMPLE (b): (Base 4) Find $221_{\text{four}} \div 3_{\text{four}}$.

Represent 221_{four} with the blocks. We could divide it into three equal piles, making exchanges as necessary. Instead, we determine how many 3's are contained in the pile.

LONG METHOD: Exchange the pile representing 221_{four} into units and repeatedly remove groups of four.

SHORT METHOD (subtraction of partial products): At each step (see Figure 7-2), we subtract the product of "something" times the divisor. The "somethings" are then added to obtain the quotient. Of course, there may be a remainder. Observe that you do not need to subtract the largest possible multiple of the divisor at any stage.

8. Use this method to do the following division problems with the blocks. Take turns explaining how to do a problem to a partner.

   (a) Base 5

   (i) $323_{\text{five}} \div 4_{\text{five}}$ _____

   (ii) $244_{\text{five}} \div 13_{\text{five}}$ _____

   (iii) $3143_{\text{five}} \div 243_{\text{five}}$ _____

   (b) Base 4

   (i) $312_{\text{four}} \div 3_{\text{four}}$ _____

   (ii) $331_{\text{four}} \div 13_{\text{four}}$ _____

   (iii) *Optional.* $3231_{\text{four}} \div 213_{\text{four}}$ _____

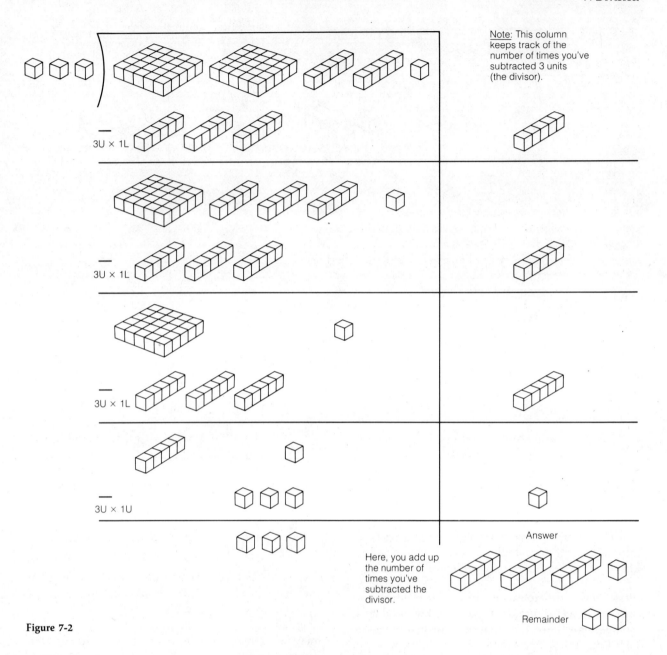

Figure 7-2

9. Based on what you have done in Exercise 8, develop a paper and pencil method using numerals to do division problems in a given base. Use your method to do three of the problems you did in Exercise 8. For each problem compare the numerical method with the MBA-block method.

Comment: Understanding and Doing Division

"I hear and I forget, I see and I remember, I do and I understand."—Chinese proverb

Just as there is an operation, subtraction, that undoes the addition of a number, there is an operation, division, that undoes the multiplication of a number. It is a very useful operation. One is often faced with problems of the type "I have 18 apples to give to 6 children. How many apples does each child get?"

What is desired is the solution to the equation $\square \times 6 = 18$.

Definition: *We will say that $a \div b$ is that number which solves the equation* $\square \times b = a$.

The first thing to notice is that it is not always possible to find a whole number that solves the equation. If we want to compute $7 \div 3$ according to the above definition, we must find a solution to $7 = \square \times 3$. But it is clear (by inspecting arrays if necessary) that there is no whole number that solves this equation. Because of this we say that $7 \div 3$ is not defined in the set of whole numbers. We could of course express $7 \div 3$ as a quotient and remainder, but that would be different from giving a whole number as a solution to the equation $7 = \square \times 3$. Eventually we will introduce *rational numbers* (fractions) so that there will always be a number that solves $\square \times b = a$ (whenever $b \neq 0$).

10. For which of the following pairs of whole numbers is division defined?

    (a) $18 \div 5$ (d) $3 \div 9$

    (b) $0 \div 6$ (e) $18 \div 1$

    (c) $27 \div 3$ (f) $17 \div 1$

■ Since division is defined in terms of multiplication, it is understandable that we use our discussion of multiplication for assistance in understanding division.

In Chapter 5 we saw that multiplication could be interpreted in two ways: either the number of objects in an array, or the result of repeated addition. Similarly, there are two corresponding ways of interpreting division. One way is by *partitioning a set into subsets*. We think of $a \div b$ as the result of partitioning a set with a elements into b subsets that have the same number of elements and asking how many objects there are in each subset. (For example, with regard to $24 \div 4$ you could take 24 objects and form 4 equal columns. How many objects would be in each column?) A second way is to interpret division as *repeated subtraction*. We think of $a \div b$ as answering the question, "How many b's can we subtract from a?" (For example, how many times can 4 objects be subtracted from 24?) Both these interpretations are useful, as illustrated in the following exercise.

11. State which interpretation of division applies in each of the following (it is not necessary to calculate an answer to the question).

    (a) A community is allotted 12,000,000 cubic feet of natural gas for the year. There are 3,000 homes in the community. If the gas were allotted equally, how much would each home receive?

    (b) A university wants to build a new residence hall. The Water Board is restricting the university to 9,600 gallons per year for the building. The engineers calculate that each student uses 300 gallons of water per year. What is the maximum number of students the hall can hold?

12. Make up a problem illustrating each of the interpretations of division.

■ Let us now consider the numbers 0 and 1. These played a special role in multiplication. We know that $0 \times a = 0$ and $1 \times a = a$ for all whole numbers. What about $0 \div a$? For example, consider $0 \div 5$. From the definition, $0 \div 5$ is that number that solves $\square \times 5 = 0$. There is only one number that solves this equation, namely 0. Hence $0 \div 5 = 0$. Using the same reasoning we see that for any $a \neq 0$, $0 \div a = 0$. Now consider the case $a \div 0$. For example, what is $5 \div 0$? By the definition of division, such a number would solve the equation $\square \times 0 = 5$. But there is no number that solves this equation. Hence we say that $5 \div 0$ is *not defined*. What about $0 \div 0$? There are many solutions to $\square \times 0 = 0$. For example, $5 \times 0 = 0$ and $29 \times 0 = 0$. In fact, any whole number will work. Hence $0 \div 0$ is ambiguous and we say it is not defined. Thus we say that for any whole number a, $a \div 0$ is not defined.

A similar analysis can be applied to study the role of 1 in division. We see that $a \div 1 = a$ since $a \times 1 = a$. Now consider $1 \div a$. It is the solution to $\square \times a = 1$. Since if $a \neq 1$ there is no whole number that solves $\square \times a = 1$, we say that, if $a \neq 1$, $1 \div a$ is not defined for whole numbers. If $a = 1$, we have $1 \div 1 = 1$, since $1 \times 1 = 1$.

13. Which of the following are defined in the set of whole numbers?

    (a) $6 \div 3$ (b) $0 \div 5$

(c) $5 \div 0$
(d) $1 \div 0$
(e) $0 \div 1$
(f) $1 \div 1$
(g) $2 \div 1$
(h) $1 \div 2$

Properties of Division

The first property of division is illustrated by the C-rods in Exercise 3. This property shows that division by a number "undoes" multiplication by that number; that is,

$$(a \times b) \div b = a$$

This follows immediately from the definition, since $(a \times b) \div b$ is the solution to $\square \times b = a \times b$. We say that division by a number is the *inverse operation* to multiplication by that number. Since multiplication is commutative, we also have

$$(b \times a) \div b = (a \times b) \div b = a$$

But what about the commutativity and associativity of division?

14. (a) Find whole numbers a and b such that $a \div b$ is a whole number and $b \div a$ is not.

Certainly, then, division is not commutative.

(b) Find whole numbers a, b, and c, such that $(a \div b) \div c \neq a \div (b \div c)$, but with all the quotients defined.

We can conclude that division is not associative.

15. For which of the following is $(a \div b) \div c = a \div (b \div c)$?

(a) $16 \div 4 \div 2$ _____
(b) $15 \div 5 \div 1$ _____
(c) $7 \div 1 \div 1$ _____
(d) $30 \div 10 \div 3$ _____
(e) $24 \div 6 \div 2$ _____

Make a guess as to when $(a \div b) \div c = a \div (b \div c)$.

■ Now let us consider whether division is distributive over addition. In Exercise 6 you verified that $(DG + Bl) \div LG = DG \div LG + Bl \div LG$. This is an example of division *distributing* over addition. If we translate into numbers, with $W = 1$, we get $(6 + 9) \div 3 = (6 \div 3) + (9 \div 3)$. But the property is limited by the fact that often some of the divisions won't be defined. For example, although $(7 + 8) \div 3 = 5$, $7 \div 3$ and $8 \div 3$ are not whole numbers. We must be content with the following statement:

Property: *If $(a + b) \div c$ is a whole number and $a \div c$ and $b \div c$ are whole numbers, then $(a + b) \div c$ is a whole number and $(a + b) \div c = (a \div c) + (b \div c)$.*

Thus division satisfies a "right-hand" distributive property when all the quotients are defined.

16. *Optional.* Use the definition of division to verify the right-hand distributive property above.

17. What about the "left-hand" property? Does $a \div (b + c) = (a \div b) + (a \div c)$? Find an example where $a \div (b + c)$, $a \div b$ and $a \div c$ are all whole numbers but where $a \div (b + c) \neq (a \div b) + (a \div c)$.

Hence the "left-hand" distributive property does not hold.

18. *Optional.* Use the definition of division to show that if $m \div k$ is a whole number and $n \div k$ is a whole number and $m > n$, then $(m - n) \div k$ is a whole number and $(m - n) \div k = (m \div k) - (n \div k)$.

Quotient and Remainder

We have seen that $a \div b$ is not always defined in the set of whole numbers since there is not always a whole-number solution to $\square \times b = a$. However, if we consider the division problem as the answer to how many b's make up a, we can approximate the answer and also determine the remainder.

For example, given 33 and 5, we can say that there are six 5's in 33 with a remainder of 3. We write $33 \div 5 = 6$ remainder 3, or,

$$33 = (6 \times 5) + 3 \quad \text{or} \quad 5\overline{)33}^{\,6 \text{ remainder } 3}$$

Given any two whole numbers a and b, we can find whole numbers q (the quotient) and r (the remainder) such that $a = q \times b + r$, and r is less than b. We write

$$a \div b = q \text{ remainder } r \quad \text{or} \quad b\overline{)a}^{\,q \text{ remainder } r}$$

19. For each of the following pairs a, b, find the quotient q and the remainder r such that $a = q \times b + r$.

(a) $a = 18$, $b = 5$ _____

(b) $a = 67$, $b = 8$ _____

(c) $a = 72$, $b = 9$ _____

Algorithms for Division

Although there are several ancient methods of division, including some that use the abacus, we will limit ourselves to studying our familiar long division. You might want to try to develop a method for long division on the abacus. If you are interested in pursuing this subject further, many books explore different algorithms for division.

Our method of long division is an adaptation and condensation of the method you used with MBA-blocks in Exercise 8. Given two numbers a and b, we want to find q and r such that $a = q \times b + r$ with $r < b$. For example, $33 = 6 \times 5 + 3$ and $3 < 5$. To find the quotient and remainder in more complicated situations, we subtract multiples of b from a until we have a remainder smaller than b. For example, to compute $187 \div 12$, we write $12\overline{)187}$. Suppose we subtract $10 \times 12 = 120$. We write

$$12\overline{)187}$$
$$-120 \quad 10$$
$$\overline{67} \leftarrow$$
$$-60 \quad 5$$
$$\overline{\text{remainder } 7 \quad 15}$$

At this stage we have $187 = 10 \times 12 + 67$. But 67 is greater than 12, so we proceed.

Here we have
$$187 = 10 \times 12 + 5 \times 12 + 7$$
$$= (10 + 5) \times 12 + 7$$
$$= 15 \times 12 + 7$$

Therefore
$$12\overline{)187} \quad \text{15 remainder 7}$$

20. Do each of the following using the method illustrated above.

(a) $14\overline{)203}$ (b) $62\overline{)895}$ (c) $23\overline{)1,081}$

■ This procedure can be condensed into

Step I.
$$\begin{array}{r} 10 \\ 12\overline{)187} \\ -120 \leftarrow 10 \times 12 \end{array}$$

Step II.
$$\begin{array}{r} 10 \\ 12\overline{)187} \\ -120 \\ \hline 67 \end{array}$$

Step III.
$$\begin{array}{r} 5 \\ 10 \\ 12\overline{)187} \\ -120 \\ \hline 67 \\ -60 \leftarrow 5 \times 12 \end{array}$$

Step IV.
$$\begin{array}{r} 15 \text{ quotient} \\ 5 \\ 10 \\ 12\overline{)187} \\ -120 \\ \hline 67 \\ -60 \\ \hline 7 \text{ remainder} \end{array}$$

Eventually it could be condensed further into

Step I.
$$\begin{array}{r} 1 \\ 12\overline{)187} \\ -120 \\ \hline 67 \end{array}$$

Step II.
$$\begin{array}{r} 15 \\ 12\overline{)187} \\ -120 \\ \hline 67 \\ -60 \\ \hline 7 \text{ remainder} \end{array}$$

Observe that in Step I above, the 1 represents 10 and the 120 is a result of computing 12×10.

21. Suppose the following example appears in a textbook.

$$\begin{array}{r} 129 \\ 13\overline{)1{,}685} \\ 13 \\ \hline 38 \\ 26 \\ \hline 125 \\ 117 \\ \hline 8 \end{array}$$

A student asks you why the "13" is under the "16" rather than under the "85." What would you say? _____

_____

22. Explain to an elementary student how to check if the result in the problem 1,685 ÷ 13 = 129 remainder 8 is correct.

Important Terms and Concepts

Divisor
Quotient
Remainder
Physical interpretation of division
Subtraction of partial-products algorithm
"Undefined" expressions

Review Exercises

1. Decide if the following statements are true or false, in general. Write T or F in the blank.

   _____ (a) $b \div a = a \div b$

   _____ (b) $(a \div b) \div c = a \div (b \div c)$

   _____ (c) $(b \times a) \div b = b$

2. Find q and r such that $a = q \times b + r$ with r less than b.

   (a) $a = 24$, $b = 3$ _____

   (b) $a = 14$, $b = 6$ _____

   (c) $a = 8$, $b = 10$ _____

3. Solve $212_{three} \div 2_{three}$ with Base 3 numerals, using the procedure developed with the MBA-blocks. _____

4. Choose a number between 1 and 10. Add 9. Subtract 2. Multiply by 4. Divide by 2. Divide again by 2. Subtract the original number. Your answer is 7. Why?

5. Divide ∩∩∩∩ ||| by ∩|. What interpretation of division did you use? _____

6. $15 \overline{)231}$ 14 remainder 21

   (a) In what sense is this solution correct? _____

   (b) In what sense is it not correct? _____

7. Under what conditions is $(a + b + c) \div d = (a \div d) + (b \div d) + (c \div d)$ true? _____

8. Describe what is meant by, "Division by a number is the inverse operation to multiplication by that number."

9. If you want to divide a large number a by a small number b (in any base), which interpretation of division would be easier to use, and why?

10. Recall that if b and c are whole numbers, $b \times c$ is the total number of items in b groups, where each group contains c items. Thus

    $$b \times c = \underbrace{c + \cdots + c}_{b \text{ times}}$$

    Use this approach to multiplication and pay close attention to the *meaning* of the order in which you multiply as you consider the following situations.

    (a) If you wish to distribute 12 cookies among 4 children, then you will give each child $\boxed{3}$ cookies, because

    _____ (i) $4 \times \boxed{3} = 12$

    _____ (ii) $\boxed{3} \times 4 = 12$

    Choose whether (i) or (ii) is the better completion of the sentence.

    (b) Suppose that you can save $200 a month. If you want to buy a $2,400 automobile, then you will have to save for $\boxed{12}$ months, because

    _____ (i) $200 \times \boxed{12} = 2,400$

    _____ (ii) $\boxed{12} \times 200 = 2,400$

    Choose whether (i) or (ii) is the better completion of the sentence.

    (c) In dealing 52 cards into 4 equal hands, you would deal $\boxed{13}$ rounds, because

    (i) $4 \times \boxed{13} = 52$

    (ii) $\boxed{13} \times 4 = 52$

    Try to think of this situation in two ways,

finding interpretations that correspond to both (i) and (ii).

11. Fill in the blanks.

   (a) $4\overline{)93}$ with quotient 23 remainder 1

   means $93 = (\underline{\quad} \times \underline{\quad}) + \underline{\quad}$

   (b) $\underline{\quad} \div \underline{\quad} = \underline{\quad}$ remainder $\underline{\quad}$

   means $100_{three} = 2_{three} \times 11_{three} + 1_{three}$

   Check the results of the following division problems by using the meaning of division with remainder in terms of multiplication and addition.

   (c) $17\overline{)400}$ with quotient 23 remainder 9

   (d) $12_{four}\overline{)331_{four}}$ with quotient 22_{four} remainder 1_{four}

12. Do the following division problems. If the numbers are not given in Base 10, do not convert to Base 10. You may wish to use MBA-blocks (or imagine that you are using them) at first. Do some without the blocks, using the division algorithm you developed in Exercise 9.

   (a) $231_{four} \div 2_{four}$ (d) $1011_{two} \div 11_{two}$
   (b) $937_{ten} \div 48_{ten}$ (e) $1342_{seven} \div 26_{seven}$
   (c) $343_{five} \div 123_{five}$ (f) $2102_{three} \div 12_{three}$

13. Tell which of the following problems have a definite answer and which represent an undefined operation. If the problem has an answer, find it. If the operation is undefined, tell why.

   (a) $6 \div 3 = \square$ (d) $0 \div 0 = \square$
   (b) $4 \div 0 = \square$ (e) $7 \div 3 = \square$
   (c) $0 \div 7 = \square$

14. How would you explain to an elementary school student why division by zero is not allowed?

15. Suppose a, b, and c are counting numbers and $a \div b = c$. What is $a \div c$? _____ Justify your answer.

16. You might see the following in an elementary mathematics textbook:

$$168 \div 7 = \square$$

$$\begin{array}{r} 168 \\ -\ 70 \quad \text{\textcircled{10}} \\ \hline 98 \\ -\ 70 \quad \text{\textcircled{10}} \\ \hline 28 \\ -\ 28 \quad \text{\textcircled{4}} \\ \hline 0 \end{array}$$

$$168 \div 7 = \boxed{24}$$

How would you explain this to your students?

17. *Optional.* Suppose m and n are counting numbers. Suppose that 5 divided into m leaves a remainder of 3, and 5 divided into n leaves a remainder of 4. What is the remainder after dividing 5 into $m + n$? _____

18. *Optional.* Use the fact that $a \div b$ is the number that solves $\square \times b = a$ to show that if m, n, and k are whole numbers and $m \div n$ is defined, then $(m \times k) \div (n \div k) = m \div n$.

8

Number Theory

"All things which can be known have number; for it is not possible that without number anything can be either conceived or known."—Philolaus, disciple of Pythagoras

Lab Exercises: Set 1

EQUIPMENT: Two containers of colored rods (C-rods) for each group.

The properties of whole numbers have fascinated people for a long time. In these exercises we explore *prime numbers* and *factors*.

1. Each person in the group make a train of C-rods. Can the rods of this train or a train of equal length be used to make a rectangle of width greater than 1? All the rods in the train must be used. (*Hint:* Find a one-color train having length equal to the length of your train.)

    For example, one rectangle associated with the train O + R would be as shown in Figure 8-1, which we could record as P × LG. What other rectangles are associated with O + R?

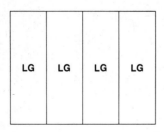

Figure 8-1

Have the group examine each person's train in turn and see how many different-sized rectangles can be formed. Record your results below, using the color abbreviations for the rods given in Chapter 3.

Train	Rectangle formed from trains of equal length
EXAMPLE: O + R	W × (O + R), P × LG, LG × P, R × DG, DG × R, (O + R) × W
(a)	
(b)	
(c)	
(d)	

Observe that the rectangles occur in pairs, and that for completeness we include the rectangles of width 1.

2. Assign the white rod length 1. For each of the trains in Exercise 1, record the lengths of the trains and the dimensions of the rectangles you made.

	Train length	Dimensions of rectangles
EXAMPLE:	12	1 × 12, 4 × 3, 3 × 4, 2 × 6, 6 × 2, 12 × 1
(a)		
(b)		
(c)		
(d)		

Observe that the product of the dimensions equals the length of the original train. If there is a whole number n such that $n \times a = b$, we say that a is a *factor* of b or a is a *divisor* of b (n is also a factor of b). For example, the factors of 12 are 1, 2, 3, 4, 6, and 12 since $12 = 12 \times 1 = 2 \times 6 = 3 \times 4$.

3. Make five trains that cannot be made into rectangles that have a width greater than 1. Record their lengths. _____
These numbers (except for 1) are examples of *prime numbers*.

Definition: *A whole number other than 1 is* prime *if no number divides it except itself and 1.*

Another way of saying this is that a prime number is a whole number that has exactly two different factors, namely itself and 1. For example, 3, 17, and 2,311 are prime.

Definition: *A whole number other than 0 or 1 that is not prime is called a* composite *number.*

The numbers 6, 20, and 2,310 are composite numbers. If a number is composite, it can be written as a product of factors. For example, $6 = 3 \times 2$, $20 = 4 \times 5$, and $2,310 = 10 \times 231$. We call the process of doing this *factoring*. Let us use the C-rods to factor some composite numbers.

Suppose we want to factor the number 60 using the C-rods. We can start by making a train equal in length to 60 whites using rods of the same color. One possibility is to make a train of 6 orange rods. We then form a rectangle from this train (Figure 8-2) and find the rod such that $\square \times O = O + O + O + O + O + O$.

The rod is DG and, as in Chapter 5, we can represent the product DG × O with the rods, as shown in Figure 8-3. Now factor the DG rod. DG = $\square \times \square$. Then factor the O rod. O = $\square \times \square$.

We can represent the product of all the factors of our original train of length 60 as a tower of 4 rods, as shown in Figure 8-4. Since none of these rods can be factored, we have obtained a factorization of the original train into the product of prime rods R ×

Figure 8-2

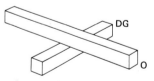

Figure 8-3

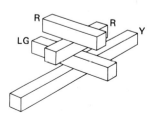

Figure 8-4

R × LG × Y. This corresponds to the prime factorization 60 = 2 × 2 × 3 × 5.

4. Use rods to represent each of the following numbers as a product of primes. Record both the rod and numerical factorizations. *Example:* R × R × LG × Y or R^2 × LG × Y, 60 = 2 × 2 × 3 × 5 = 2^2 × 3 × 5.

	Rod factorization	Numerical factorization
(a) 72		
(b) 126		
(c) 168		
(d) 360		
(e) 336		

Check your answers with the other members of your group.

5. (a) Represent 42 as a tower of prime rods.
   (b) Factor 2,520 into a product of prime numbers and represent the product as a tower. Show that this tower equals the tower you would obtain by combining the tower for 42 and the tower for 60. Why should this be so?

■ Prime numbers have always interested mathematicians. One thing they would like to find is a way to list prime numbers. In the next Comment you will explore a method developed by Eratosthenes. Since the method is rather cumbersome, mathematicians have tried to develop an algebraic expression that always produces prime numbers.

What is an *algebraic expression*?

Suppose we have a machine that operates on numbers. If we put a number *n* in this machine, it puts out some number (Figure 8-5). For example, one machine might add 5 to the number we put in, so that if 3 goes in, 8 comes out, as in Figure 8-6. Another machine might multiply the number by itself (i.e., square the number), so if 5 goes in, 25 comes out (Figure 8-7).

What happens if 8 goes in? _____
The particular machine we are concerned with

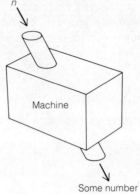

Figure 8-5

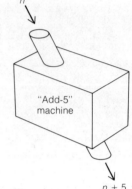

Figure 8-6

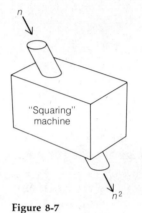

Figure 8-7

today operates according to the rule: If we put in a number *n*, the number that comes out is $n^2 - n + 41$. In other words, the machine squares the number, subtracts the number from its square, and adds 41. For example, if we put in 3, out comes $3^2 - 3 + 41 = 9 - 3 + 41 = 47$. If we put in 1, out comes $1^2 - 1 + 41 = 1 - 1 + 41 = 41$.

The expression $n^2 - n + 41$ is called an *algebraic expression*. Recall that we are looking for an algebraic expression that always produces prime numbers.

6. (a) Put the numbers 2, 4, 5, 6, and 7 into the machine that corresponds to the algebraic expression $n^2 - n + 41$. Record the output in the chart below.

n	$n^2 - n + 41$
1	41
2	
3	47
4	
5	
6	
7	

(b) Verify that all the numbers in (a) that came out of the machine are prime.

If we put 8, 9, ... 40 into the machine, the output would still be prime. (Verify a few more.) That is, $n^2 - n + 41$ is prime for $n = 1, 2, \ldots 40$.

(c) Verify that $n^2 - n + 41$ is not prime when $n = 41$.

Thus this expression does not work. But maybe there is one that does. An interesting expression in this regard is $2^p - 1$.

7. (a) Find $2^p - 1$ for $p = 2, 3, 5$, and 7, and record the results below.

p	$2^p - 1$
2	
3	
5	
7	

(b) If p is prime, is $2^p - 1$ always prime? Test the results above.

Have we found the algebraic expression we are searching for? If $p = 11$, $2^p - 1 = 2^{11} - 1 = 2,047$. Find a factorization of 2,047.

_____

8. (a) Compute the algebraic expression $2^{(2^n)} + 1$ for $n = 1, 2, 3$.

n	$2^{(2^n)} + 1$
1	
2	
3	

Pierre Fermat (1601-1665), a brilliant French mathematician, believed that this expression always gave a prime number, but was unable to prove it.

(b) Verify that $2^{2^n} + 1$ is prime for $n = 1, 2, 3$.

Fermat knew that $2^{2^4} + 1 = 16 + 1 = 65,537$ is prime, but he did not know that $2^{2^5} + 1 = 2^{32} + 1 = 4,294,967,297$ is not prime. So Fermat was wrong, and in fact no additional primes of this type have been found, even with the aid of computers.

We have failed in our search for an algebraic expression that always yields primes. But don't feel too bad. No one has yet found an expression that works. If you should do so, you will become famous. In the meantime, we will learn about the "sieve of Eratosthenes," which is a nonalgebraic method of producing as many primes as one desires.

Comment: Prime Numbers

"Those who are to take part in the highest functions of state must be introduced to approach it (number theory) not in an amateur spirit, but perseveringly, until, by the aid of prime thought, they come to see the real nature of numbers."
—Plato

The study of the properties of whole numbers for their own sake began with Pythagoras and his followers in the fifth century B.C.

Pythagoras is a very important figure in the history of mathematics. He was from the island of Samos, traveled in Egypt and Babylon studying mathematics and religion, and established a secret society or cult which had a mathematical-religious basis. This cult was known as the Pythagorean Brotherhood, although 26 women were members.

The Pythagorean motto is said to have been, "Number rules the Universe," or, "All is Number." Numbers and their properties were intertwined with religious and mystical beliefs. The reader interested in learning more about the Pythagoreans should consult the books by Boyer or Valens listed in the bibliography.

In the previous lab exercises we studied prime numbers. Recall that

1. A number a is a *factor* or *divisor* of b if there is a whole number n such that $n \times a = b$.
2. A whole number p is *prime* if it has exactly two different factors.
3. A whole number other than 0 or 1 that is not prime is called a *composite number*.

Observe that the concepts of "prime" and "composite" are defined such that the numbers 0 and 1 are neither prime nor composite. This is due to the special nature of 0 and 1. All other numbers are either prime or composite, but no number is both.

9. Determine which of the following numbers are prime and which are composite. List *all* the factors for the composite numbers.

(a) 20 _____

(b) 23 _____

(c) 35 _____

(d) 209 _____

(e) 49 _____

(f) 67 _____

■ Around 230 B.C. the Greek mathematician Eratosthenes developed an efficient method for finding all primes less than some given number. Suppose we wish to find all primes less than 50. We list the numbers 1 through 49. Then we cross out 1, because it is not a prime. We circle 2 because it is prime. Now we can eliminate all multiples of 2 (i.e., all numbers that have 2 as a factor). Instead of examining each number in turn to see whether it has 2 as a factor, we can simply cross out every second number. (These numbers are the multiples of 2 because they are obtained by repeatedly adding 2's to the number 2.) Our list looks like

```
 1  ②  3  4̸  5  6̸  7  8̸  9  1̸0̸
11  1̸2̸ 13 1̸4̸ 15 1̸6̸ 17 1̸8̸ 19 2̸0̸
21  2̸2̸ 23 2̸4̸ 25 2̸6̸ 27 2̸8̸ 29 3̸0̸
31  3̸2̸ 33 3̸4̸ 35 3̸6̸ 37 3̸8̸ 39 4̸0̸
41  4̸2̸ 43 4̸4̸ 45 4̸6̸ 47 4̸8̸ 49 5̸0̸
```

Now consider the number 3. It is prime, so circle it and then eliminate all multiples of 3 by mechanically crossing out every third number. (These are multiples of 3 because they are obtained by repeatedly adding 3's to the number 3.) We obtain

```
 1̸  ②  ③  4̸  5  6̸  7  8̸  9̸  1̸0̸
11  1̸2̸ 13 14 15 16 17 18 19 20
2̸1̸ 2̸2̸ 23 2̸4̸ 25 2̸6̸ 2̸7̸ 2̸8̸ 29 3̸0̸
31  3̸2̸ 33 3̸4̸ 35 3̸6̸ 37 3̸8̸ 3̸9̸ 4̸0̸
41  4̸2̸ 43 4̸4̸ 45 4̸6̸ 47 4̸8̸ 49 5̸0̸
```

Continuing this process, we see that 4 is already eliminated. We circle 5 since it is prime and then cross out all multiples of 5 using the method above. The next number not eliminated is 7; we circle it and cross out all multiples of 7. Our chart ends up looking like

```
 1̸  ②  ③  4̸  ⑤  6̸  ⑦  8̸  9̸  1̸0̸
⑪  1̸2̸ ⑬ 1̸4̸ 1̸5̸ 1̸6̸ ⑰ 1̸8̸ ⑲ 2̸0̸
2̸1̸ 2̸2̸ ㉓ 2̸4̸ 2̸5̸ 2̸6̸ 2̸7̸ 2̸8̸ ㉙ 3̸0̸
㉛ 3̸2̸ 3̸3̸ 3̸4̸ 3̸5̸ 3̸6̸ ㊲ 3̸8̸ 3̸9̸ 4̸0̸
㊶ 4̸2̸ ㊸ 4̸4̸ 4̸5̸ 4̸6̸ ㊼ 4̸8̸ 49 5̸0̸
```

The numbers that remain are the prime numbers less than 50.

10. Verify that the remaining numbers are all prime.

■ This procedure is called the *sieve of Eratosthenes* since it "strains out" the composite numbers and "lets through" the primes. Why is it that we had to eliminate only the multiples of 2 through 7 to obtain the primes less than 50? The answer is that if any number less than 50 has a factor larger than 7, then it also must have a factor smaller than 7. In other words, if $n = a \times b$, and both a and b are greater than 7, then n is greater than 49 (since $49 = 7 \times 7$).

11. (a) Use the sieve of Eratosthenes with the chart below to find all primes less than 300.

```
  1   2   3   4   5   6   7   8   9  10
 11  12  13  14  15  16  17  18  19  20
 21  22  23  24  25  26  27  28  29  30
 31  32  33  34  35  36  37  38  39  40
 41  42  43  44  45  46  47  48  49  50
 51  52  53  54  55  56  57  58  59  60
 61  62  63  64  65  66  67  68  69  70
 71  72  73  74  75  76  77  78  79  80
 81  82  83  84  85  86  87  88  89  90
 91  92  93  94  95  96  97  98  99 100
101 102 103 104 105 106 107 108 109 110
111 112 113 114 115 116 117 118 119 120
121 122 123 124 125 126 127 128 129 130
131 132 133 134 135 136 137 138 139 140
141 142 143 144 145 146 147 148 149 150
151 152 153 154 155 156 157 158 159 160
161 162 163 164 165 166 167 168 169 170
171 172 173 174 175 176 177 178 179 180
181 182 183 184 185 186 187 188 189 190
191 192 193 194 195 196 197 198 199 200
201 202 203 204 205 206 207 208 209 210
211 212 213 214 215 216 217 218 219 220
221 222 223 224 225 226 227 228 229 230
231 232 233 234 235 236 237 238 239 240
241 242 243 244 245 246 247 248 249 250
251 252 253 254 255 256 257 258 259 260
261 262 263 264 265 266 267 268 269 270
271 272 273 274 275 276 277 278 279 280
281 282 283 284 285 286 287 288 289 290
291 292 293 294 295 296 297 298 299
```

(b) At what stage were you sure that you had eliminated all the composite numbers less than 300? _____

8/Comment

12. The following are composite numbers. Show that each can be written as a product of primes.

    (a) 21 = _____
    (b) 72 = _____
    (c) 60 = _____
    (d) 120 = _____

■ Have you found a convenient method to factor composite numbers into a product of primes? One way is to make a *factor tree*. For example, to factor 84, write down the number and any two factors as shown below.

Now factor each of these numbers if possible. If not (i.e., if one is a prime) circle it.

Continuing this process, we obtain

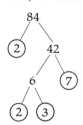

The circled numbers provide the prime factorization, that is, $84 = 2 \times 2 \times 3 \times 7$.

13. Factor each of the following using a factor tree.

    (a) 468   (b) 504

    (c) 5,544   (d) 17,160

■ When writing numbers as products, it is useful to use *exponents*. For example, we write 5×5 as 5^2 and $7 \times 7 \times 7$ as 7^3. Using this notation, you can write the factorization obtained in Exercise 13(b) as $504 = 2 \times 2 \times 2 \times 3 \times 3 \times 7 = 2^3 \times 3^2 \times 7$.

14. Write the factorization for Exercise 13(a), (c), and (d), using exponents.

    _____
    _____
    _____

■ It should be clear that any composite number can be written as a product of primes. We just keep factoring everything in sight until we can't factor any more. What's left are primes.

It is also true that any way we factor, we will end with exactly the same primes. This can be proven but we will not do it here. These two statements are usually combined into what is known as the fundamental theorem of arithmetic.

Fundamental theorem of arithmetic: *Every composite number is expressible (or can be factored) as a product of primes in exactly one way except for the order in which the primes appear.*

15. (a) Factor 5,544 with a different factor tree than the one you used in Exercise 13(c). (Start with two different factors.) Compare the prime factorization obtained with your previous answer.

    (b) Repeat (a) above with Exercise 13(d). _____

16. *Optional.*
    (a) Show that there cannot exist counting numbers m and n such that $m \times m = 2 \times n \times n$.
    (b) Do the same for $m \times m = 3 \times n \times n$.

■ We would now like to determine whether the primes ever stop. If we made a big enough sieve, would we obtain all the primes? The answer is no. There are infinitely many primes. This was proven by Euclid more than 2,000 years ago. To understand his proof, we need more information about divisibility.

Recall that if a and c are whole numbers, a divides c if $\square \times a = c$ has a solution. In other words, a divides c if $c = b \times a$ for some whole number b. If a divides c, we write $a|c$. Observe that $a|c$ means precisely the same as "a is a factor of c."

Question: Can you find numbers a, b, and c such that $a|b$ and $a|(b + c)$, but a doesn't divide c?*

The answer to this question is no. Let us see why. If $a|b$, then there would be some number that solved $b = \square \times a$. So, there is a whole number n such that $b = n \times a$. Similarly if $a|(b + c)$, there would be a solution to $b + c = \square \times a$. Say $b + c = t \times a$. Now $(b + c) - b = c$. Hence $c = (t \times a) - (n \times a) = (t - n) \times a$. But this says that $a|c$, providing $t - n$ is a whole number, which it is.

Fact: If $a|b$ and $a|b + c$, then $a|c$.

17. Compute each of the following.
    (a) $2 \times 3 \times 5 + 1 =$ _____
    (b) $2 \times 3 \times 5 \times 7 + 1 =$ _____
    (c) $2 \times 3 \times 5 \times 7 \times 11 + 1 =$ _____
    (d) $2 \times 3 \times 5 \times 7 \times 11 \times 13 + 1 =$ _____
    (e) Show that 2, 3, 5, does not divide the number in (a).
    (f) Show that 2, 3, 5, 7 does not divide the number in (b).
    (g) Show that 2, 3, 5, 7, 11 does not divide the number in (c).
    (h) Show that 2, 3, 5, 7, 11, 13 does not divide the number in (d).

■ Suppose $p_1, p_2, \ldots, p_k$ are primes. We can use the fact stated above to argue that none of $p_1, p_2, \ldots, p_k$ can divide $(p_1 \times p_2 \times \cdots \times p_k) + 1$.

---

*If you did Exercise 18 in Chapter 7, you should be able to answer this immediately. If $a|(b + c)$ and $a|b$, then a divides their difference $(b + c) - b = c$.

Suppose, for example, that p_1 divides $(p_1 \times p_2 \times \cdots \times p_k) + 1$. Since p_1 certainly divides $p_1 \times p_2 \times \cdots \times p_k$, the fact implies it would divide 1. But that is impossible (even absurd). So p_1 cannot divide $p_1 \times p_2 \times \cdots \times p_k + 1$. Neither can any of the other p's. This means that none of the primes $p_1, \ldots, p_k$ are factors of n. Hence, if there were only a finite number of primes $p_1, \ldots, p_k$, there would be a number $n = p_1 \times \cdots \times p_k + 1$, which cannot be expressed as a product of primes. (Think about this. Let it sink in.) Since this is impossible (it contradicts the fundamental theorem), there must be an infinite number of primes.

18. *Puzzle Problem.* Ann and Barbara, two old college friends who have not seen each other since graduation, meet on the street. They have the following conversation.

What are the ages of the children? _____

Lab Exercises: Set 2

EQUIPMENT: Two containers of C-rods for each group.

In these exercises we will study the concepts of greatest common factor of two numbers and least common multiple of two numbers. Methods are developed to find these numbers using C-rods.

Recall that b is a factor of a if there is a whole number n such that $a = n \times b$.

18. (a) List all the factors of 60. _____
    (b) List all the factors of 42. _____

(c) Let F_{60} be the set of factors of 60 and F_{42} the set of factors of 42. What is $F_{60} \cap F_{42}$?

$F_{60} \cap F_{42} = \{$ _____ $\}$.

(d) $F_{60} \cap F_{42}$ is the set of common factors of 60 and 42. What is the largest number in

$F_{60} \cap F_{42}$? _____

This number is the greatest common factor of 60 and 42.

19. Find the greatest common factor of each of the following pairs of numbers.

   (a) 17, 34 _____

   (b) 10, 35 _____

   (c) 1, 16 _____

   (d) 1, 1 _____

   (e) 26, 26 _____

   (f) 11, 5 _____

   (g) p, q where p and q are prime numbers and

   $p \neq q$ _____

Definition: *We say that d is a* common factor *of a and b if d is a factor of both a and b. The* greatest common factor *of a and b is the largest common factor of a and b.*

For example, 2 is a common factor of 12 and 10, and 3 is a common factor of 27 and 18. The greatest common factor of 12 and 10 is 2. The greatest common factor of 27 and 18 is 9.

Although the method in Exercise 18 will always enable us to find the greatest common factor, it becomes tedious to use when the numbers have many factors. For example, try to find the greatest common factor of 504 and 1,200 using this method.

We will use C-rods to explore two other ways of finding the greatest common factor, and apply these methods to numbers.

20. (a) Represent 336 as a tower of prime rods. (Save the tower.) By examining the tower, find three factors of 336 that are not prime.

   (i) _____

   (ii) _____

   (iii) _____

   (b) Convince yourself that b is a factor of a precisely when the tower that represents a contains within itself a tower identical with the tower that represents b. (One person explain this to the group.)

21. Examine the towers that represent 336 and 360. Find a common factor by extracting from one tower some rods that are contained in the others. Find all common factors in this manner. Record your results below in terms of rod products.

By extracting the maximum possible number of rods, you obtain the train representing the *greatest common factor* (sometimes also referred to as the greatest common divisor). The greatest common factor of $R^3 \times LG^2 \times Y$ and $R^4 \times LG \times Bk$ is $R^3 \times LG$.

22. Find the greatest common factor of each of the following pairs of numbers by forming the rod towers and using the process above.

   (a) 504, 1,200 _____

   (b) 180, 1,400 _____

   (c) 392, 225 _____

23. A number a is said to be a *multiple* of b if b is a factor of a.

   (a) Convince yourselves that a is a multiple of b precisely when the tower representing a contains within itself a tower identical to the tower representing b; see Exercise 20(b). One person explain why to the group.

   (b) Form the tower for 28. Find three towers that represent multiples of 28.

   (i) _____

   (ii) _____

   (iii) _____

   (c) Use towers to check whether 864 is a multiple of 28.

   (d) Use towers to check whether 588 is a multiple of 28.

24. Represent 60 and 21 as towers of primes.

   (a) Find three different towers that represent multiples of both 60 and 21. These towers are called *common* multiples of 60 and 21.

Definition: *The* least common multiple *of a and b is the least nonzero number that is a common multiple of a and b.*

   (b) Use towers to find the least common multiple of 60 and 21. _____

(c) Use towers to find the least common multiple of 336 and 360. _____

(d) Use towers to find the least common multiple of 504 and 1,200. _____

Save the towers in (c) and (d) for later use.

Note: Let GCF(*a*, *b*) denote the *greatest common factor* of *a* and *b*. Let LCM(*a*, *b*) denote the *least common multiple* of *a* and *b*.

25. (a) Consider again the towers for 336 and 360. Form the tower for LCM(336, 360) and the tower for GCF(336, 360). Now form the product tower for LCM(336, 360) × GCF(336, 360). Form also the tower that is the product of the tower for 336 and the tower for 360. Verify by inspecting the towers that LCM(336, 360) × GCF(336, 360) = 336 × 360.

    (b) Repeat the above for 504 and 1,200.

■ What you have discovered is always true: If *a* and *b* are whole numbers, then

$$\text{LCM}(a, b) \times \text{GCF}(a, b) = a \times b$$

There is another procedure for finding the greatest common factor of two numbers. It is based on an old Chinese method that appears in the *Nine Books of Arithmetical Thinking*, which was written about 2,000 years ago. We illustrate it for the trains *t* = **O** + **Bl** + **Bl** and *s* = **Br**.

Step I. Form the two trains with C-rods.

Step II. Subtract the shorter train from the larger train (Figure 8-8). Use the rods.

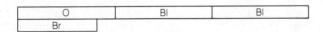

Figure 8-8

Step III. Now take the resulting train, namely **O** + **O**, together with the shorter train of the pair with which you started, namely **Br**.

Step IV. Find the difference of these two trains (Figure 8-9). What is the result? _____

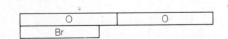

Figure 8-9

Step V. Take the resulting train (**O** + **R**) together with the shorter train from the pair (**Br**). Continue the process of subtracting and forming a new pair from the difference and the shorter train until the difference is zero. The pairs you obtain in this example are:

> **O** + **Bl** + **Bl**, **Br**
> **O** + **O**, **Br**
> **O** + **R**, **Br**
> **P**, **Br**
> **P**, **P**

The last nonzero rod is always the greatest common factor of the original train.

26. (a) Find the lengths of **O** + **Bl** + **Bl** and **Br**, and check the answer above numerically. _____
    _____ _____ _____

    (b) Use the procedure with rods above to find the the GCF for each of the following.

    (i) **O** + **O** + **W** and **O** + **Y** _____

    (ii) **O** + **O** + **O** + **DG** and **O** + **O** + **P**
    _____

    (iii) **O** + **O** + **LG** and **DG** _____

27. Discuss in your group why it works to use the rods to find the greatest common factor in this manner.

■ A complete verification of why the procedure works depends on the following.

Fact: *If a and b are whole numbers with a greater than b, then the common factors of a and b are the same as the common factors of a − b and b.*

For example, the common factors of 60, 36 are 2, 3, 4, 6. The common factors of 24 = 60 − 36 and 36 are also 2, 3, 4, 6.

28. Verify the fact for each of the following.

		Common factors of *a*, *b*	Common factors of *a* − *b*, *b*
(a) *a* = 60,	*b* = 42	_____	_____
(b) *a* = 35,	*b* = 10	_____	_____
(c) *a* = 77,	*b* = 60	_____	_____

■ How can we show that this fact is always valid? All we have to do is to show that if *c* is a factor of both *a* and *b*, it is also a factor of *a* − *b*, and that if *d* is a factor of both $a - b$ and b, it is also a factor of a.

Step I. Suppose c is a factor of a and b. Say $a = c \times t$, $b = c \times s$. Then $a - b = (c \times t) - (c \times s) = c \times (t - s)$. Therefore c is a factor of $a - b$.

Step II. Suppose d is a factor of $a - b$ and b. Say $a - b = d \times u$ and $b = d \times v$. Then $a = (a - b) + b = (d \times u) + (d \times v) = d \times (u + v)$. Therefore d is a factor of a.

29. What property of multiplication are we using in

Step I? _____

In Step II? _____

■ From the fact we stated above, it follows that *the greatest common factor of a and b equals the greatest common factor of $a - b$ and b.* The reason for this is that if the sets of common factors are equal, the greatest one in each set must be the same.

Now we can successively apply this to obtain the GCF of any two whole numbers. For example, to find the GCF (1,200, 504), we compute

$$\begin{aligned} \text{GCF}(1{,}200, 504) &= \text{GCF}(1{,}200 - 504, 504) \\ &= \text{GCF}(696, 504) \\ &= \text{GCF}(696 - 504, 504) \\ &= \text{GCF}(192, 504) \\ &= \text{GCF}(192, 504 - 192) \\ &= \text{GCF}(192, 312) \\ &= \text{GCF}(192, 312 - 192) \\ &= \text{GCF}(192, 120) \\ &= \text{GCF}(192 - 120, 120) \\ &= \text{GCF}(72, 120) \\ &= \text{GCF}(72, 120 - 72) \\ &= \text{GCF}(72, 48) \\ &= \text{GCF}(72 - 48, 48) \\ &= \text{GCF}(24, 48) \\ &= \text{GCF}(24, 48 - 24) \\ &= \text{GCF}(24, 24) = 24 \end{aligned}$$

If you don't make any subtraction errors and eliminate much of the writing above (including the symbol GCF), this method is quite effective, particularly with an abacus or an electronic calculator.

30. Use the old Chinese method to find the GCF for each of the following pairs.

(a) 336, 360 (b) 1,024, 256 (c) 192, 35

31. (a) Someone in the group explain the difference between least common multiple and greatest common factor.

(b) Another person explain why the concept of greatest common multiple would not make sense.

(c) A third person explain why it would not be useful to consider the concept of a least common divisor of two counting numbers.

32. *Optional.* Let a, b, and c be counting numbers. Determine whether or not $\text{LCM}(a, b, c) \times \text{GCF}(a, b, c) = a \times b \times c$.

Comment: The Greatest Common Factor; Odd and Even Numbers

"Mathematics is the queen of Sciences and Number Theory is the queen of Mathematics."—C. F. Gauss

We can adopt the method of Exercise 22 above to find the greatest common factor of two numbers using prime factorization. For example, to find GCF(336, 360), we write $336 = 2^4 \times 3 \times 7$ and $360 = 2^3 \times 3^2 \times 5$. Inspect the primes that appear and for each prime choose the minimum number of times that it appears in either factorization. This gives the greatest common factor. Thus GCF(336, 360) = $2^3 \times 3 = 24$. Compare this process with the procedure you used in Exercise 22.

33. Use the method described above to find the greatest common factor of each of the following pairs.

   (a) $2^3 \times 5^2 \times 7^5 \times 11$, $3^3 \times 5^5 \times 7 \times 13$ _____
   (b) $2^8 \times 3$, $3 \times 5^2 \times 11$ _____
   (c) $2 \times 3^2 \times 7$, $5^3 \times 11 \times 13^2$ _____
   (d) $2^2 \times 3^3 \times 5$, $2^2 \times 3^3 \times 5$ _____
   (e) $2^8 \times 3^3 \times 11^5$, $2^4 \times 3 \times 11^3$ _____

■ This method works exceedingly well when we are given the prime factorization of the numbers involved. However, finding prime factorizations often is quite time-consuming, especially for large numbers.

A very effective method appears in Euclid's *Elements* and is known as the *Euclidean algorithm*. It is essentially a condensation of the old Chinese method that we investigated above. In fact, you might already have noticed that it is sometimes possible to save some steps when applying the Chinese method. For example, instead of subtracting **Br** from **O + Bl + Bl** to obtain the pair **O + O, Br** and then again subtracting **Br** from **O + O** to obtain the pair **O + R, Br**, and then again subtracting **Br** from **O + R** to obtain the pair **P, Br**, we could have subtracted three **Br**'s from **O + Bl + Bl**, obtaining **P**, and then continue with the procedure using **Br** and **P**.

Hence three steps would be condensed into one. (If you are confused here, go through it step by step with the rods.)

If we now translate this procedure into numbers, we would find the greatest common factor of 28 and 8 as follows.

$$28 = 3 \times 8 + 4$$
$$8 = 2 \times 4 + 0$$

4 is the greatest common factor.

Again using numbers, find the greatest common factor of 26 and 10.

$$26 = 2 \times 10 + 6$$
$$10 = 1 \times 6 + 4$$
$$6 = 1 \times 4 + 2$$
$$4 = 2 \times 2 + 0$$

The greatest common factor of 26 and 10 is 2.

Euclidean algorithm: *To find the greatest common factor of two whole numbers, divide the larger number by the smaller number to obtain a quotient and remainder. Then divide the divisor by the remainder. Continue the process of successive division by the remainders until a remainder of zero is reached. The last nonzero remainder is the GCF of the two whole numbers.*

EXAMPLE: Find GCF(1,200, 504) using the Euclidean algorithm.

$$1{,}200 = 2 \times 504 + 192$$
$$504 = 2 \times 192 + 120$$
$$192 = 1 \times 120 + 72$$
$$120 = 1 \times 72 + 48$$
$$72 = 1 \times 48 + 24$$
$$48 = 2 \times 24 + 0$$

The GCF is 24. Compare this with the computation using the Chinese method.

EXAMPLE: Find GCF(121, 35).

$$121 = 3 \times 35 + 16$$
$$35 = 2 \times 16 + 3$$
$$16 = 5 \times 3 + 1$$
$$3 = 3 \times 1 + 0$$

So GCF(121, 35) = 1.

34. Find the GCF of each of the following pairs using the Euclidean algorithm:

   (a) 330; 336   (b) 4,108; 492   (c) 504; 1,200

■ One of the most fascinating aspects of the theory of numbers is the incidence of unsolved problems that are easy to state and understand, but very difficult to solve. In the following exercises we will examine two such problems.

35. If two primes differ by 2, they are called *twin primes*. For example, 3 and 5 are twin primes, as are 5 and 7. Use the sieve of Eratosthenes in Exercise 11(a) to list all twin primes less than 300.

It has been conjectured that there are infinitely many twin primes, but no one has ever been able to prove it. Does it surprise you that no one has ever been able to decide whether the statement is true or false?

36. In 1742, Christian Goldbach made the following conjecture in a letter to Leonhard Euler: Every even whole number can be expressed as the sum

of two primes. Verify Goldbach's conjecture for each of the following numbers.

(a) 4 = _____
(b) 6 = _____
(c) 8 = _____
(d) 10 = _____
(e) 20 = _____
(f) 60 = _____
(g) 98 = _____
(h) 300 = _____
(i) 500 = _____

Goldbach's conjecture has been verified for every even number between 4 and 100,000, but it has never been proved.

Odd and Even Numbers

37. For each of the 10 C-rods, decide whether you can form a train of equal length that is composed of two rods of the same color. That is, decide which rods are two times another rod. These are called the even rods. The ones that are not two times another rod are odd. List them below:

Even rods	Odd rods

Show that every odd rod is of the form $2 \times \square + $ W. Complete the following.

(a) LG = $2 \times \square$ + W
(b) Y = $2 \times \square$ + W
(c) Bk = $2 \times \square$ + W
(d) Bl = $2 \times \square$ + W

Definition: *A number is* even *if it is divisible by 2. A number is* odd *if it is not divisible by 2.*

Every number is either even or odd.

38. Find the numbers that solve the following equations.

(a) 273 = 2 × _____ + 1
(b) 342 = 2 × _____
(c) 0 = 2 × _____
(d) 17 = 2 × _____ + 1
(e) 1 = 2 × _____ + 1
(f) 2 = 2 × _____
(g) 123 = 2 × _____ + 1
(h) 64 = 2 × _____

We can see from the definition above that every even number is two times some other number. That is,

Every even number is of the form $2k$ *for some number k.*

If a number is not divisible by 2, it must leave a remainder of 1 after dividing by 2. Therefore,

Every odd number is of the form $2n + 1$ *for some number n.*

Note: $2k$ means $2 \times k$. For readability we omit the multiplication sign.

39. Form an odd train o with C-rods and even train e with C-rods. Verify that $e + o$ is odd.

This illustrates a general statement about numbers.

If o is an odd number and e is an even number, then $o + e$ is odd.

The reason for this is that since o is odd, $o = 2n + 1$, and since e is even, $e = 2k$. Hence $e + o = 2k + 2n + 1 = 2(k + n) + 1$, which is odd.

We can similarly verify that if o_1 and o_2 are odd numbers and e_1 and e_2 are even numbers, then:

$o_1 + o_2$ *is even. The sum of two odd numbers is even.*
$e_1 + e_2$ *is even. The sum of two even numbers is even.*

40. Illustrate the above with trains of C-rods.

41. Illustrate the above with numbers.

42. *Optional.*
   (a) Prove that the sum of any two odd numbers is even.
   (b) Prove that the sum of any two even numbers is even.

The result of multiplying two numbers is summarized by:

The product of two even numbers is even.
The product of two odd numbers is odd.
The product of an even number and an odd number is even.

43. Illustrate the above with C-rods.

44. Illustrate the above with numbers.

45. *Optional.*
    (a) Prove that the product of any two even numbers is even.
    (b) Prove that the product of any two odd numbers is odd.
    (c) Prove that the product of an even number and an odd number is even.

Important Terms and Concepts

Factor = divisor
Prime number
Composite number
Algebraic expression
Sieve of Eratosthenes
Factor tree
Fundamental theorem of arithmetic
Greatest common factor
Least common multiple
Chinese and Euclidean algorithms
Even and odd numbers

Review Exercises

1. Is there a whole number n such that $n \times n = 13 \times 17 \times 19$? _____ Why? _____

2. Suppose n is a whole number. What is the least common multiple of $2 \times n$ and $3 \times n$? _____

3. Suppose m and n are two whole numbers whose LCM is m. What is the GCF of m and n? _____

4. There exists a method for finding the least common multiple of two numbers that is analogous to the one for finding the greatest common factor of two numbers that we used in Exercise 33. To find LCM(a, b), write each number as a product of primes. Inspect the primes that appear; then choose the maximum number of each prime that appears *in either* factorization and multiply these together to get LCM(a, b). For example, to find LCM(5,880, 6,468), we write $5{,}880 = 2^3 \cdot 3 \cdot 5 \cdot 7^2$ and $6{,}468 = 2^2 \cdot 3 \cdot 7^2 \cdot 11$. Then LCM(5,880, 6,468) $= 2^3 \cdot 3 \cdot 5 \cdot 7^2 \cdot 11 = 64{,}680$. Use the above method to find

    (a) LCM(21, 60) _____
    (b) LCM(336, 360) _____
    (c) LCM(924, 490) _____

5. Explain how the above method correlates with the use of C-rod towers to find least common multiples.

6. Using the methods of Exercise 33 and Review Exercise 4 for finding the greatest common factor and least common multiple of two numbers a and b, show that LCM(a, b) $\times$ GCF(a, b) $= a \times b$ for each pair of numbers below.

    (a) 168, 90
    (b) 140, 825
    (c) 260, 84

7. Determine whether each of the following numbers is prime, composite, or neither. When one of the numbers is composite, express it as a product of primes.

    (a) 2 _____
    (b) 9 _____
    (c) 1 _____
    (d) 52 _____
    (e) 0 _____
    (f) 19 _____
    (g) 144 _____
    (h) 23 _____
    (i) 5 _____

8. Let n be any whole number. Each of the following is an algebraic expression that indicates how to compute another number, starting with n. In certain cases, you will be able to determine that the expression always gives an odd number, no matter what n is. In certain other cases, you can show that the expression always gives an even number. In all other cases, you will be able to show that the expression sometimes gives an odd

number and sometimes gives an even number, depending upon the value of n. Label each of the following expressions with *odd, even,* or *depends on n* to indicate into which of the three cases it falls.

(a) $n + n + n$ _____

(b) $n + 1$ _____

(c) $2n$ _____

(d) $2n + 3$ _____

(e) $4n + 2$ _____

(f) $(2n)(2n + 1)$ _____

(g) $(2n + 3)(2n + 1)$ _____

(h) $n(4n + 1)$ _____

9. Decide whether the following statements are true or false. Write T or F in the blank. The letters a, b, c, and d represent whole numbers.

_____ (a) If c is a factor of both a and b, then it is a factor of both $a - b$ and $a + b$.

_____ (b) If c is any common factor of a and b, then GCF(a, b) divides c.

_____ (c) If d is any common multiple of a and b, then LCM(a, b) divides d.

_____ (d) The number 6,974,286 could be a prime number.

_____ (e) The number 1 is prime since it has only two factors, 1 and itself.

_____ (f) If p and q are two different primes, then GCF(p, q) = 1.

10. Find GCF(990, 182) using the Euclidean algorithm.

11. Using Exercise 10 above, show that LCM(990, 182) = $(\frac{1}{2})(990)(182)$.

12. How would the wording of the fundamental theorem need to be changed if 1 were to be considered as a prime number?

13. *Optional.* Show that there are no "triplet primes," that is, there do not exist three primes p, $p + 2$, $p + 4$, each 2 greater than the previous one. (*Hint:* Show that one of p, $p + 2$, $p + 4$ is always divisible by 3.)

14. *Optional.*

(a) Show that if k divides a and k divides b, then k divides $a + b$. (*Hint:* If k divides a, then $a = nk$ for some counting number n, and if k divides b, then $a = mk$ for some counting number m.)

(b) Show that if k divides a and k divides b, and $a - b$ is a counting number, then k divides $a - b$.

Emilie de Breteuil, Marquis du Chatelet (1708–1749), overcame the profound cultural sentiment against the education of women that existed in eighteenth-century France to become a leading mathematical physicist. She translated Newton's *Principia* into French and wrote a textbook on mathematical physics.

9
Topology

"Beauty in mathematics is seeing the truth without effort." —George Polya

Lab Exercises: Set 1

EQUIPMENT: None.

In these exercises we will explore and discover a solution to a famous problem. The goal is to experience the discovery process rather than learn any particular information.

1. Since we will be working with objects that resemble maps or mazes, each one in your group relate a memory of playing with mazes, learning how to read a map, or trying to find his or her way through a fun house.

■ In a memoir presented to the Russian Academy at St. Petersburg in 1735, the eminent Swiss mathematician Leonhard Euler (pronounced "oiler") described a problem he had solved:

The problem, which I understand is quite well known, is stated as follows: In the town of Königsberg in Prussia there is an island called *Kneiphof*, with the two branches of the river (Pregel) flowing around it. There are seven bridges crossing the two branches. The question is whether a person can plan a walk that will cross each of these bridges once but not more than once. I was told that while some denied the possibility of doing this and others were in doubt, there were none who maintained that it was actually possible. On the basis of the above I formulated the following very general problem for myself: Given any configuration of the river and the branches into which it may divide, as well as any number of bridges, to determine whether or not it is possible to cross each bridge exactly once.

2. (a) The situation at Königsberg is shown in Figure 9-1. Everyone in your group try to plan a walk that would cross each bridge once and only once. After everyone has tried for a few minutes, discuss your results.

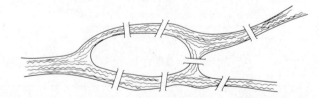

Figure 9-1

(b) Suppose you were allowed to build a new bridge in Königsberg. Each person in the group draw a new bridge on Figure 9-2

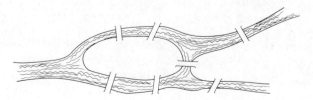

Figure 9-2

that would allow a walk that would cross each bridge once and only once. Discuss your results.

(c) Consider Figure 9-2, with the eighth bridge that you included. Can you start your walk anywhere in the city and still walk across each bridge once and only once? _____

(d) Draw a ninth bridge in Figure 9-2 and answer the question in part (c).

■ As Euler pointed out, we are searching for a solution to the general problem, "Given any configuration of rivers and bridges, when is it possible to walk over each bridge once and only once?" Since it is a little tedious to draw a picture for every instance, let us try to simplify things.

3. (a) In Figure 9-3, put a dot in each of the regions determined by the branches of the river.

There are _____ regions, so there will be _____ dots. Connect each pair of dots by distinct lines so that a line goes over each bridge. Then draw a diagram beneath this figure that shows the lines and dots without the rivers.

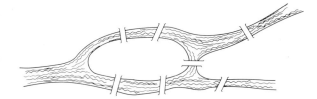

Figure 9-3

Definition: *The resulting configuration of dots and lines is called a* **network**. *The dots are called the* **vertices** *of the network.*

(b) Convince yourself that if you could trace the network (without lifting your pencil from the paper), going over each line once and only once, then you could solve the bridge problem.

4. (a) Try to trace each network in Figure 9-4 without lifting your pencil and without going over any line more than once. (Tracing paper is helpful.) You may start at any vertex (dot) and pass through the vertices in any order.

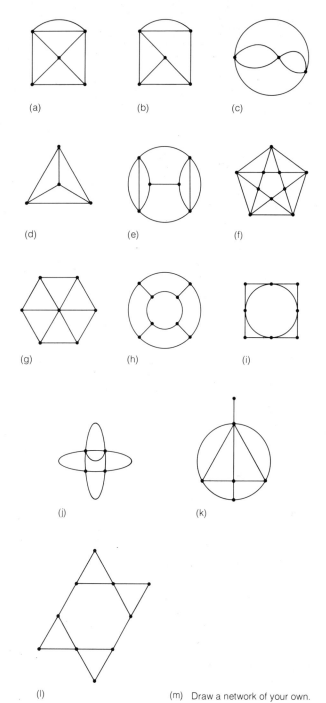

Figure 9-4

(m) Draw a network of your own.

You may find that some networks can be traced only if you start at certain corners, while others cannot be traced at all. Keep a record of which networks can be traced and which cannot.

(b) Each person in your group guess when a network can be traced. Then as a group try to decide whether the guesses may or may not be correct, based on the results in part (a).

■ We need some terminology to proceed further. A *network* is composed of *vertices* and *lines*. Consider the number of lines emanating from a vertex.

Definition: *The number of lines emanating from a vertex is called the* **degree** *of the vertex.*

5. What is the degree of the indicated vertices in the following networks?

    (a) (b)

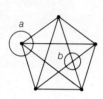 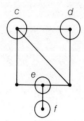

Vertex	Degree
a	
b	
c	
d	
e	
f	

■ If the degree of a vertex is an even number, we say that the vertex is *even*. If the degree of a vertex is an odd number, we say that the vertex is *odd*.

6. (a) Count the number of even vertices and the number of odd vertices for each of the networks in Exercise 4 and complete the table in the next column.
   (b) Using the same table, guess when a network can be traced or when it cannot be traced. Then as a group try to see if the guess could be correct. Make several guesses.

7. Some possible guesses are listed below. Use the table to decide whether the guess may or may not be correct, and indicate your decision by writing *yes* or *no* in the space provided.

    _____ (a) A network can always be traced if it has more even vertices than odd vertices.

    _____ (b) A network cannot be traced if it has more odd vertices than even vertices.

Network	Total number of vertices	Number of even vertices	Number of odd vertices	Can the network be traced?
a				
b				
c				
d				
e				
f				
g				
h				
i				
j				
k				
l				
m				

_____ (c) A network can always be traced if all its vertices are even.

_____ (d) A network can always be traced if it has an even number of even vertices.

_____ (e) A network cannot be traced if it has less than three even vertices.

_____ (f) A network cannot be traced if it has more than two odd vertices.

8. On the basis of network (a) in Figure 9-4 suppose that someone were to guess that networks can be traced if the number of even vertices is one more than the number of odd vertices. How can you show that this guess is incorrect? _____

9. Can a network have an odd number of odd vertices? _____ Try to draw one.

9/Lab Exercises: Set 1

10. **(a)** Consider network (a) in Figure 9-4. Label the vertices *A*, *B*, *C*, *D*, and *E*, and find the degree of each vertex.

Vertex	Degree
A	
B	
C	
D	
E	

    _____ Sum of the degrees

    (b) Add the degrees of the vertices. Now count the number of lines in the network. _____ Verify that the sum of the degrees is two times the number of lines.

    (c) Repeat part (a) for network (l) in Figure 9-4.

Vertex	Degree
A	
B	
C	
D	
E	
F	
G	
H	
I	
J	

    _____ Sum of the degrees

    _____ Number of lines

    (d) Can you convince yourself that this would be true for any network?

Comment: The Königsberg Bridge Problem

"The problem—to cross the seven bridges in a continuous walk without recrossing any of them—was regarded as a small amusement of the Königsberg townfolk. Euler, however, discovered an important scientific principle concealed in the puzzle. . . . Thus began a 'vast and intricate theory,' still young and growing, yet already one of the great forces of modern mathematics."—James Newman

Euler solved the Königsberg bridge problem by showing that it is impossible to walk over each bridge once and only once. He did much more, however. He found conditions that tell us exactly when a network can be traced and when it cannot. This was the beginning of the theory of *topology*, of which the theory of networks is now only a small part.

Let us try to discover what conditions determine whether a network can or cannot be traced. After doing the exercises above, perhaps you have some idea of what they might be. So far we have only a list of examples. What is necessary is a reasonable argument.

Let us consider a simple network (Figure 9-5).

Figure 9-5

11. Verify that the network in Figure 9-5 can be traced if you start at *A* or *D*, but not if you start at *B* or *C*. Verify also that if you start at *A*, you must end at *D*, and vice versa. Conclude that you must either start or end at *A* in order to trace the network. Why is it necessary to start or end at *A*?

■ Suppose that you have started elsewhere. When you pass through *A*, you use up two of the lines emanating from *A*. This leaves one line, which when traveled leaves you back at *A* with no lines left to trace. Hence you must end at *A*. A similar argument would apply if there were five, seven, or any odd number of lines emanating from *A*.

Recall that the vertex is even or odd depending on whether the number of lines emanating from it is even or odd. We conclude the following.

1. *If a network has an odd vertex, to trace it you must begin or end at the odd vertex.*

Now, if you are able to trace the network, there can be only one beginning and one end. Hence we can conclude:

2. *If there are more than two odd vertices, it is impossible to trace the network.*

Verify that this conclusion is in accord with the results of Exercise 4.

This greatly simplifies our problem. There are only three types of networks left to consider—those with no odd vertices, those with one odd vertex, and those with two odd vertices.

In Exercise 9 you tried to draw a network with an odd number of odd vertices. You were unsuccessful since it is is impossible to construct such a network. Why? Suppose we have a network with one odd vertex, labeled O, and even vertices labeled $E_1, \ldots E_k$, where k is some positive integer. Let d be the degree of O, d_1 the degree of E_1, d_2 the degree of E_2, and so on to d_k. Then, as we saw in Exercise 10, $d + d_1 + d_2 + \cdots + d_k = 2 \times l$, where l is the number of lines in the network. But $d_1, d_2, \ldots d_k$ are even; hence $d_1 + d_2 + \cdots + d_k$ is even. But this implies $2 \times l = d + (d_1 + d_2 + \cdots + d_k)$ is odd, since d is odd.

We have arrived at a contradiction: we have shown that the hypothesis of a network having one odd vertex leads us to deduce that $2 \times l$ is an odd number. Since this is absurd, our hypothesis must be incorrect. Therefore,

3. *A network cannot have exactly one odd vertex.*

There are now only two cases left to consider. We omit the arguments and simply state the results.

4. *If a network has all even vertices, then it can be traced. Furthermore, you can start at any vertex and you will always end up at the vertex you start with.*

5. *If a network has exactly two odd vertices, then it can be traced. You must start at one of the odd vertices and you will end at the other one.*

Verify that 4 and 5 are in accord with the results of Exercise 4.

Lab Exercises: Set 2

"The idea that aptitude for mathematics is rarer than aptitude for other subjects is merely an illusion which is caused by belated or neglected beginners."—J. F. Herbart

EQUIPMENT: Geoblocks (optional).*

These exercises continue the study of networks. You will discover a relationship between the number of vertices, the number of paths, and the number of regions of a network.

The networks you investigated in the previous exercises consisted of lines and vertices.

Definition: *A path is a line connecting two different vertices or a vertex to itself.*

Figure 9-6 shows two paths. Vertices are end points of paths. A network, by definition, is a connected set of paths and vertices. By *connected* we mean that you can travel from any vertex to any other vertex along the paths of the network. The drawing in Figure 9-7, for example, is not a network since it is not connected.

Figure 9-6

Figure 9-7

Each network determines a certain number of *regions* in the plane. For example, the network in Figure 9-8a determines 2 planar regions, and the network in Figure 9-8b determines 3 planar regions.

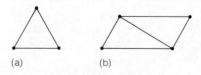

Figure 9-8

Notice that the region *outside* the boundary of the network is always counted.

---

*Information about this equipment may be found on page xvii.

12. How many regions does the network below determine? _____

■ We see that, associated with each network there is a certain number, V, of vertices, a certain number, P, of paths, and a certain number, R, of regions. We will now attempt to discover the relationship between these three numbers.

13. For each of the networks a, b, c, d, f, h, j, and l in Exercise 4, determine the number of vertices, regions, and paths, and record the results below.

	V	R	P
a			
b			
c			
d			
f			
h			
j			
l			

14. Without consulting one another, each person in the group draw a network below with 7 vertices and 10 paths. Count the regions. What do you notice? Can you draw a 7-vertex, 10-path network that has a different number of regions?

15. Without consulting one another, each person in the group draw a network below with 4 vertices and 5 regions. How many paths do you have? _____ Can you draw a 4-vertex, 5-region network that has a different number of paths?

16. Each person draw a 6-path, 5-region network. What do you notice?

17. On the basis of the above you might conclude that there is some connection between V, R, and P, since if any two are given, the third is determined. Can you find a formula that demonstrates the relationship between them? Discuss this in your group for a few minutes before reading on.

18. Let us put some restrictions on $V + R$ to see if that helps us. Remember, V and R are positive integers, so if we want $V + R$ to be some number, there is a limit to the possibilities. What possibilities are there for V and R if $V + R = 3$? List them below.

Draw a network for each of the possibilities and determine P for each network.

What do you notice? _____

19. (a) List all the possibilities for V and R if $V + R = 4$, and draw the networks. Determine P for each network.

What do you notice? _____

(b) Do the same for $V + R = 5$ and $V + R = 6$.

(c) Fill in the following table.

$V + R$	P
3	
4	
5	
6	

Can you now guess the relationship between V, R, and P?

20. What number goes in the box? $V + R - P = \Box$

■ Exercises 21–23 are designed to explore why this formula is always true. First, notice that any network can be constructed by starting with a single path. We can then perform a succession of additions to this simple network in order to obtain the given network. We verify that after every addition the formula still holds.

21. *Optional.*

(a) There are two possible networks that have a single path. Draw them below and verify that the formula $V + R - P = 2$ holds for each network.

First possibility

$V = \underline{\quad}$, $R = \underline{\quad}$, $P = \underline{\quad}$

$V + R - P = \underline{\quad}$

Second possibility

$V = \underline{\quad}$, $R = \underline{\quad}$, $P = \underline{\quad}$

$V + R - P = \underline{\quad}$

(b) One of the possibilities for a 1-path network is shown below. It consists of 1 path, 1 region, and 2 vertices. There are essentially four ways that we can add a path to this network. Can you find them? Draw the four new networks and try to describe how you changed the network in each case.

I   II   III   IV

Description. I: _____

II: _____

III: _____

IV: _____

For each way you found to add a path, answer the following.

(i) How many vertices were added?
(ii) How many regions were added?
(iii) How many paths were added?

	I	II	III	IV
(i)				
(ii)				
(iii)	1	1	1	1

Verify in each case that $V + R - P$ is unchanged.

(c) The other possibility for a 1-path network is shown below. It consists of 1 path, 2 regions, and 1 vertex. Find and draw three ways in which we can add a path to this network. Describe in words how you changed the network.

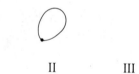

I   II   III

Description. I: _____

II: _____

III: _____

For each new path, answer the following.

(i) How many vertices were added?
(ii) How many regions were added?
(iii) How many paths were added?

	I	II	III
(i)			
(ii)			
(iii)	1	1	1

22. *Optional.* Suppose we have any network and we want to form a new network from it by increasing its number of paths by 1. We do not know what the network may look like, so we simply isolate one part of it and draw some curving lines to symbolize the rest.

(a) One way to increase the number of paths of the original network by 1 is to add the path at one vertex only, as shown below. The dotted path and the ∗ vertex have been added onto the original network.

(i) How many vertices are added? _____

(ii) How many regions are added? _____

(iii) How many paths are added? _____

Convince yourself that $V + R - P$ is unchanged, no matter what the original network may have been.

(b) Another way to increase the number of paths by 1 is to add a new path connecting two vertices of the original network, as shown below.

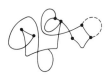

(i) How many vertices are added? _____

(ii) How many regions are added? _____

(iii) How many paths are added? _____

Convince yourself that $V + R - P$ is unchanged, no matter what the original network may have been.

(c) A third way is to add a new vertex in an existing path, as shown below.

(i) How many vertices are added? _____

(ii) How many regions are added? _____

(iii) How many paths are added? _____

Convince yourself that $V + R - P$ is unchanged, no matter what the original network may have been.

(d) A fourth way is to add a closed path to one of the vertices, as shown below.

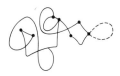

(i) How many vertices are added? _____

(ii) How many regions are added? _____

(iii) How many paths are added? _____

Convince yourself that $V + R - P$ is unchanged, no matter what the original network may have been.

■ The formula

$$V + R - P = 2$$

is called *Euler's formula* and is valid for all networks, since we can construct any network by starting with one path and successively adding paths in one of the four ways investigated above. Any one path network satisfies $V + R - P = 2$, and, as we have seen, at each stage the new network that is obtained satisfies the formula.

23. *Optional (uses geoblocks).* Geoblocks are three-dimensional solids with planar surfaces. The surfaces of the geoblocks are polyhedra (singular = polyhedron). The planar surface of a polyhedron is called a *face*. The line of intersection of two faces is called an *edge* of the polyhedron. The point of intersection of two or more edges is called a *vertex*.

(a) Consider the faces, edges, and vertices of a polyhedron as a three-dimensional network. Can you guess the relationship between the number of vertices, edges, and faces?

(b) Use the geoblocks to tabulate the number of vertices, faces, and edges for four polyhedra. Now can you find a formula? How might you justify your answer?

	Vertices	Faces	Edges
1			
2			
3			
4			

(c) Consider the polyhedron shown below. (If possible, construct it using straws or sticks.) It has an oblong hole through its middle.

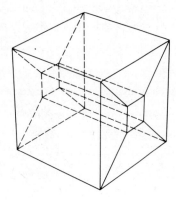

Count the vertices, faces, and edges. Does your formula hold for this polyhedron? _____

Comment: Euler's Formula

"An elegantly executed proof is a poem in all but the form in which it is written."—M. Kline

The investigations in the previous lab exercises have indicated that there is a relationship between the number of vertices, the number of paths, and the number of regions of a network. The discovery of this relationship was made by Leonard Euler and is known as *Euler's formula*. If V = the number of vertices, P = the number of paths, and R = the number of regions, then

$$V + R - P = 2$$

This formula is true for any connected network.

There is a similar relationship between the vertices, edges, and faces of a polyhedron. In Exercise 23 (optional) you observed that for polyhedra the following relationship holds: Vertices + Faces − Edges = 2.

The similarity between this and Euler's formula $V + R - P = 2$ is no coincidence. If we consider the vertices and edges of a polyhedron as a network made of rubber bands, we could stretch and flatten it so it would lie flat on a plane. The vertices would be the same, the number of paths of the network would be the number of edges, and the number of regions of the network would be the number of faces.

The reason that the formula does not hold for the polyhedron of Exercise 23(c) is that it has a hole in the middle and cannot be stretched and flattened to lie on the plane in such a manner that the above correspondence is preserved.

This concludes our study of networks. As we mentioned earlier, this theory is only a small part of the subject of topology. Another aspect of topology is studied in Exploration 3 of Chapter 22.

Important Terms and Concepts

Network   Tracing a network
Vertex   Region
Degree of a vertex   Euler's formula
Path   Connected

Review Exercises

1. The network shown below cannot be traced. Add one path so that the resulting network can be traced.

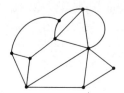

2. The network shown below cannot be traced. Either add a path so that the resulting network can be traced or explain why it is impossible to do so.

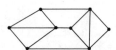

3. If one of the bridges of Königsberg were demolished, would it be possible to walk over each remaining bridge once and only once? _____

4. Below is the floor plan of a house. Can you find a path in this house that would enable you to go through each door once and only once? _____ How can you justify your answer in terms of the discussion in this chapter?

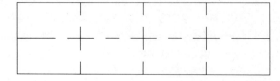

5. *Optional.*
   (a) Show that when a path is deleted from a network, the number of odd vertices either stays the same or changes by 2.
   (b) Use (a) to show that the number of odd vertices in a network is even.

6. If a network has 20 paths and divides the plane into 5 regions, how many vertices must it have? _____

7. Suppose you have a network with 5 vertices and you want to change the network by increasing the number of paths by 4 and the number of regions by 2. How many vertices will the new network have? _____

Maria Gaetano Agnesi (1718–1799) at the age of 30 published *Analytical Institutions*, an influential treatise on the differential and integral calculus. She was elected to the Bologna Academy of Science and appointed an honorary lecturer in mathematics at the University of Bologna.

10
Rational Numbers

"Through and through the world is infested with quantity: To talk sense is to talk quantities. It is no use saying the nation is large—how large? It is no use saying that radium is scarce—how scarce? You cannot evade quantity. You may fly to poetry and music, and quantity and number will face you in your rhythms and your octaves."—Alfred North Whitehead

Lab Exercises: Set 1

EQUIPMENT: Two containers of C-rods for each group.

In this set of exercises you will use diagrams and C-rods to explore the basic concepts of the rational number system.

Definition: *Two regions in a plane are* congruent *if one can be moved so as to coincide exactly with the other.*

Figure 10-1

For example, the squares in Figure 10-2 are congruent squares, and the triangles in Figure 10-3 are congruent triangles.

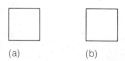

Figure 10-2

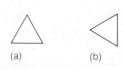

Figure 10-3

1. **(a)** Divide the rectangle below into 2 congruent parts and shade 2 of the parts.

(b) Divide the rectangle below into 3 congruent parts and shade 2 of the parts.

(c) Divide the rectangle below into 4 congruent parts and shade 3 of the parts.

(d) Divide the rectangle below into 6 congruent parts and shade 4 of the parts.

2. (a) Divide the circle below into 3 congruent parts and shade 1 of the parts.

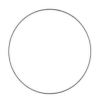

(b) Divide the circle below into 4 congruent parts and shade 2 of the parts.

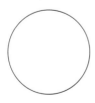

(c) Divide the circle below into 6 congruent parts and shade 2 of the parts.

(d) Divide the circle below into 3 congruent parts and shade 3 of the parts.

3. (a) Circle a subset in the set below that would result from dividing the set into 5 matching subsets and taking 3 of the subsets.

(b) Circle a subset in the set below that would result from dividing the subset into 8 matching subsets and taking 6 of the subsets.

(c) Circle a subset in the set below that would result from dividing the set into 3 matching subsets and taking 1 of the subsets.

■ In Exercises 1 to 3 we chose an amount of something, and then did two things. First, we divided the original amount, which we will call the *unit quantity*, into a certain number (say n) of parts that were equal in some sense (congruent in Exercises 1 and 2, matching in Exercise 3). Then we designated a certain number (say m) of these parts. Notice that it is possible that $m = n$.

We assign the symbol m/n, called a *fraction*, to the result of this process. For example, we assign the fraction 4/6 to the shaded region below, in which the large rectangle is taken to be the *unit region*.

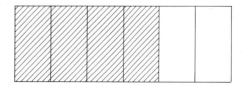

4. Indicate the fraction assigned to the result of each of the processes in Exercises 1, 2, 3: ____ ____ ____ ____ ____ ____ ____ ____ ____ ____

Notice that the choice of the unit quantity will determine the amount represented by a given fraction.

5. Suppose we choose the dark green C-rod as our unit length (i.e., assign **DG** length 1). How many **W**'s are needed to make a train equal in length to **DG**? ____ Thus we assign the fraction 1/6 to **W**. What fraction is assigned to:

   (a) **W** if **Bk** is the unit length? ____

   (b) **W** if **O** is the unit length? ____

   (c) **W** if **Y** is the unit length? ____

   (d) **R** if **Y** is the unit length? ____

   (e) **P** if **DG** is the unit length? ____

6. One way to do Exercise 5(d) is to observe that 5 **W**'s equal **Y** and 2 **W**'s equal **R**. Then, if **Y** is assigned length 1, we can assign 2/5 to **R**. Use this reasoning to assign fractions in each of the following.

   (a) With **Bk** as the unit length, **Y** is assigned the fraction ____.

   (b) With **Bl** as the unit length, **Br** is assigned the fraction ____.

   (c) With **P** as the unit length, **LG** is assigned the fraction ____.

   (d) With **DG** as the unit length, **R** is assigned the fraction ____.

   (e) With **P** as the unit length, **DG** is assigned the fraction ____.

   (f) With **Y** as the unit length, **Bl** is assigned the fraction ____.

7. (a) One person in the group asks a question of the form, "The fraction assigned to the ____ rod is ____. What is the unit?" The other members of the group answer the question. For example, "The fraction assigned to the yellow rod is 5/7. What is the unit?" The other members answer, "Black." Each person take a turn asking a question.

   (b) Now complicate the game a little by asking a two-stage question. For example, "The fraction assigned to yellow is 1/2. What rod has the value 2/5?" Answer: "Purple."

   Each person take a turn asking a two-stage question. Can you formulate a question that does not have a rod as an answer?

8. (a) Shade 1/3 of the region below.

   (b) Shade 2/6 of the region below.

   (c) Shade 4/12 of the region below.

   What do you notice?

Definition: *When two fractions represent the same amount with respect to the same unit quantity, we say that the two fractions are* equivalent. *For example, 1/3 and 2/6 are equivalent fractions.*

9. Use the rectangles below and on the next page to find three fractions equivalent to 1/2 by shading the appropriate regions.

   (a)

10/Lab Exercises: Set 1

(b)

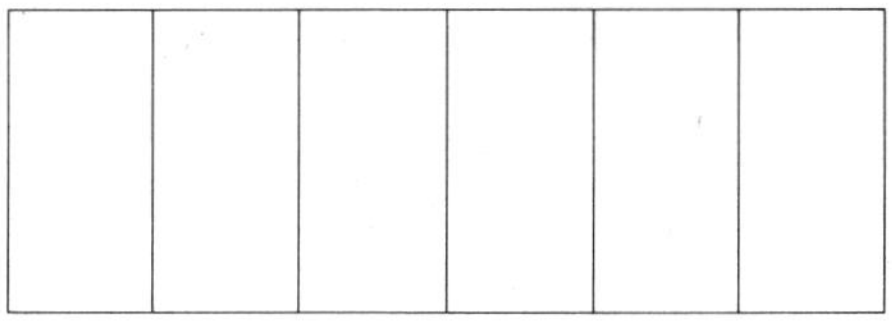

(c)

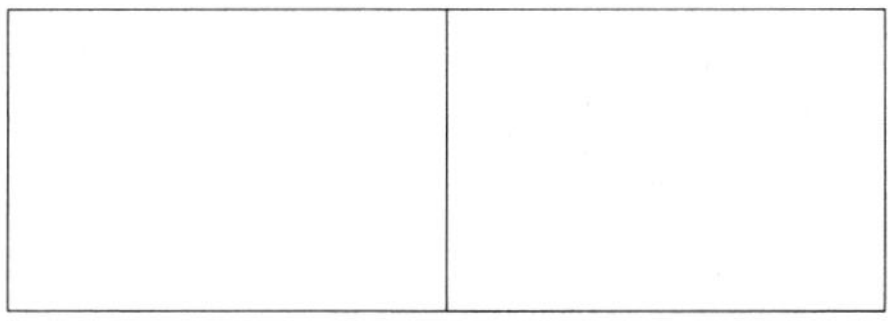

(d)

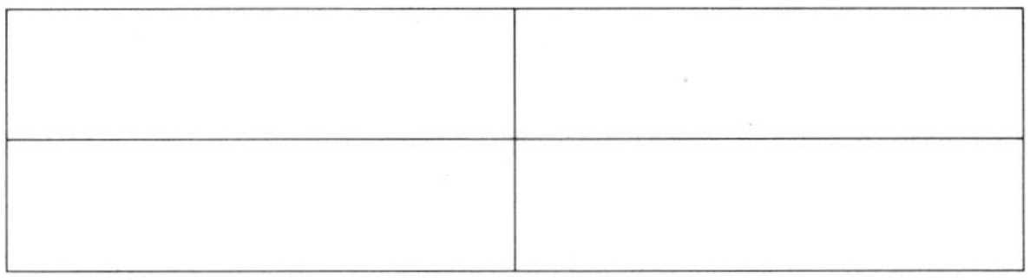

10. Let the length in Figure 10-4 be the unit length. See if you can make a train composed of rods all of the same color equal in length to this unit length. For example, four light green rods will do it. If you find that such a train can be made, carefully mark the dividing line between each rod in the train, and write the fraction name beneath. For example, with the **LG** rod you would have marks at 0/4, 1/4, 2/4, 3/4, and 4/4. Try to do this with each of the ten rods. (It will not be possible with all of them.) As you proceed, certain lengths will have more than one fraction assigned to them. As you find new fractions, write them beneath the previous ones.

11. (a) List all the fractions you found in Exercise 10 that are equivalent to 1/4. _____

    Find two more fractions not on your list that are equivalent to 1/4. _____

    (b) Repeat (a) for the fraction 2/3.

    _____ _____

    (c) Explain how to find as many fractions equivalent to a specified fraction as you want.

12. (a) Shade regions that correspond to the fractions 5/12 and 1/3 on the diagrams (i) and (ii) below.

    (i)

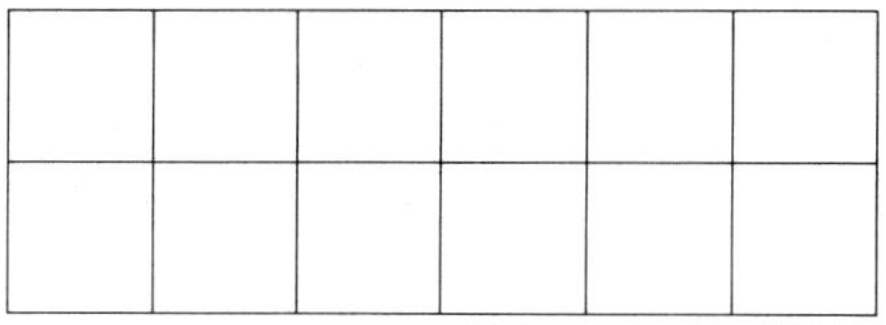

    (ii)

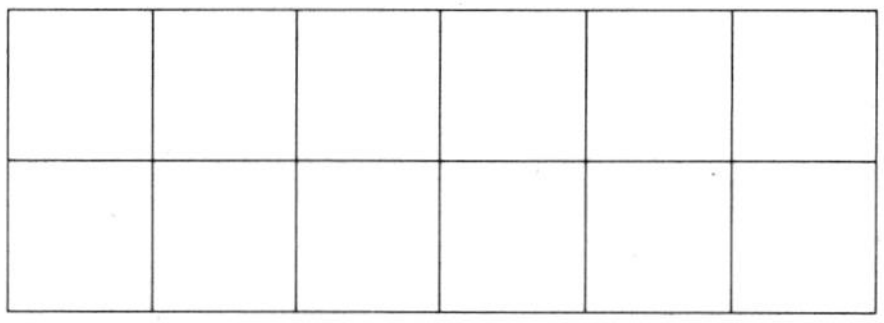

    (b) Which fraction is larger? _____
    (c) How would you demonstrate the same result using the results of Exercise 10?

    (d) Use sets (i) and (ii) below to verify your statement in (b) above.

    (i) * * * *   (ii) * * * *
        * * * *   * * * *
        * * * *   * * * *

Figure 10-4

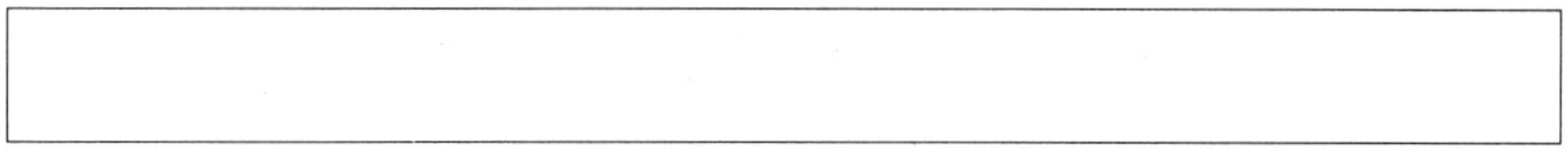

13. **(a)** Shade two rows of the rectangle below. What fraction does it represent? _____

    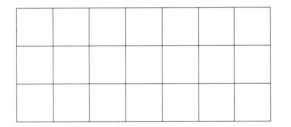

    (b) Shade five columns of the rectangle below. What fraction does it represent? _____

    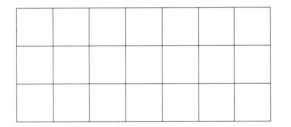

    (c) Which fraction is larger: 2/3 or 5/7? _____
    Why? _____

14. Use the procedure of Exercise 13 to determine whether 3/4 or 4/5 is larger. Shade the appropriate diagrams on the grid paper in Figure 10-5.

15. Suppose you want to determine whether 2/3 or 7/11 is larger by using sets. How many elements should there be in the set? _____ How many elements in the subset corresponding to 2/3? _____ How many elements in the subset corresponding to 7/11? _____ Which fraction is larger? _____

16. Use two different physical models to determine whether 12/5 or 7/3 is larger.

Comment: Fractions and Rational Numbers

"In questions of science the authority of a thousand is not worth the humble reasoning of a single individual."
—Galileo Galilei

Counting numbers are used for counting. We use them when it is possible to identify individual objects that are to be counted. They answer the question, "How many?" But the counting numbers are not always sufficient to deal with the world. For example, we often ask questions like "How much flour is needed in that recipe?" or "How long is that stick?" These questions involve measuring something, which usually does not involve discrete objects. If you are using cups to measure flour, often a recipe calls for an amount of flour between one and two cups. In order to be more precise (most cooks would not appreciate reading "between one and two cups" in their recipe book), we need a number system that allows us to measure parts of a unit.

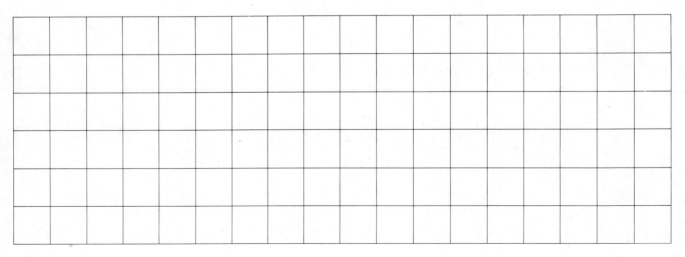

Figure 10-5

Roughly speaking, a *rational number* measures the size of a quantity in relation to the unit we have chosen.

[IF A NUMBER IS RATIONAL, DOES THAT MEAN IT CAN THINK?] *[NO, IT MEANS IT IS THE RATIO OF TWO COUNTING NUMBERS. SOME PEOPLE CALL THEM FRACTIONS.]*

Whatever we measure, we need a unit of measure. Since it is easier to visualize length or area than most of the other things that we measure, such as weight, volume, loudness, brightness, etc., we usually use length of line segments or area of rectangular regions as a model for rational numbers.

To create a model for a particular rational number, let us choose a unit length (Figure 10-6a), divide it into a certain number of segments of equal length, say 3 (Figure 10-6b), and put a certain number of these, say 2, end to end (Figure 10-6c). We say that the resulting segment (which is shaded) has a length *two-thirds* that of the unit length. This process provides one way of thinking of rational numbers. We state it in general terms as follows.

Definition: *A rational number is the concept attached to the result of choosing a unit of measurement (e.g., length), dividing that unit into m parts of equal measure (e.g., congruent segments), and then measuring n of these parts.*

Notice that this concept requires the selection of a unit quantity and is then defined in terms of an operation on the unit quantity, which is specified in terms of a pair of numbers.

We immediately realize that it is possible for different pairs of numbers to give the same result. For example, if we divide a unit length into 6 segments of equal length (Figure 10-7a) and put 4 of them end to end, we obtain Figure 10-7b. We easily see that the resulting segment has the same length as a segment that represents two-thirds (Figure 10-7c).

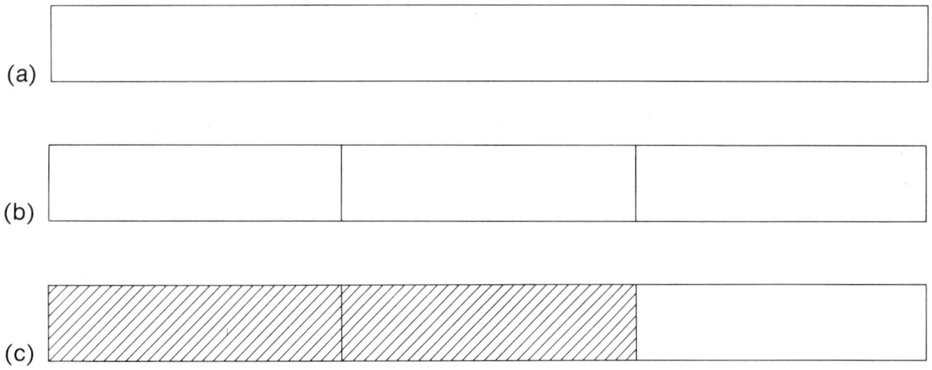

Figure 10-6

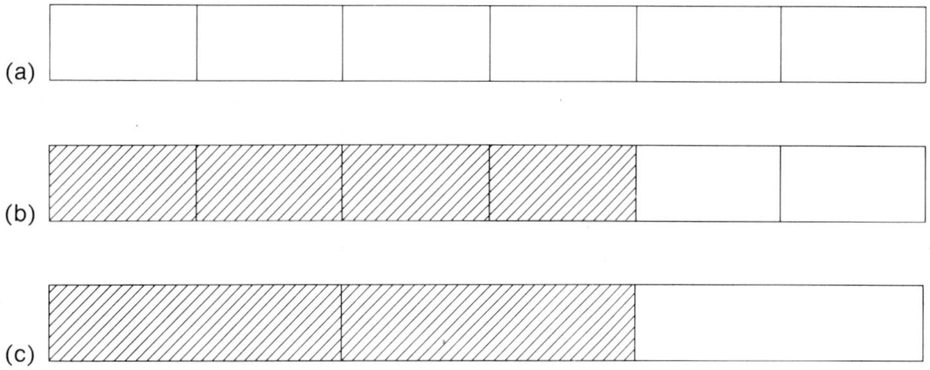

Figure 10-7

Since the results of performing these two operations on the unit quantity have the same length, we agree that they represent the same rational number. Once the unit is chosen, there is only one rational number for any particular length obtained by the type of process specified above.

However, the numbers appearing in the description of the operation are different. We use the symbol 2/3 to represent the operation of dividing the unit into 3 congruent segments and measuring 2 of them, and we use the symbol 4/6 to represent the operation of dividing the unit into 6 congruent segments and measuring 4 of them. The symbols 2/3 and 4/6 are called *fractions*.

Definition: *A fraction is a pair of numbers with an order. The first number is called the* **numerator**; *the second number is called the* **denominator**.

Even if there is only one rational number that specifies a given length, there are many fractions that can describe the same length.

Definition: *If two fractions represent the same rational number, they are said to be* **equivalent**.

When we write, for example, 2/3 = 4/6, we are expressing the fact that these different fractions are equivalent; that is, they represent the same rational number. In our discussion, we will be informal and use fractions to denote rational numbers. Thus, we will talk about the rational numbers 2/3, 5/9, 11/7, m/n, a/b, etc.

17. Show that each of the pairs of fractions in Figure 10-8 are equivalent by shading the region that represents the appropriate part of the rectangle. (See how the shading is done in the example.)

■ Given two fractions, a/b and c/d, we would like to find a way to decide whether they are equivalent, or if they are not, which one represents the larger rational number. To simplify our discussion, we will consider a given quantity, length, throughout this discussion.

This simplest case occurs when the denominators are the same. Consider the fractions, a/n and b/n. They represent the rational number obtained by dividing the unit length into n segments of equal length and putting a of them end to end for a/n, and

Figure 10-8

b of them end to end for b/n. The lengths will be equal precisely when $a = b$. We can thus state:

Fact 1: *If two fractions with the same denominator are equivalent, their numerators are equal.*

Now consider the possibility of the denominators being unequal. You may have observed in Exercise 11(c) that a/b is equivalent to $2a/2b$, $3a/3b$, etc. This is true because dividing the unit length into $2b$ segments and taking $2a$ of them results in a segment equal in length to a segment obtained by dividing the unit into b parts and taking a of them. (For example, see Figure 10-9).

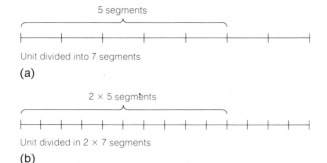

Figure 10-9

This relationship holds similarly for $3a/3b$, etc.

Hence, if a/b is a fraction, $a/b = na/nb$ for any whole number n. In particular,

$$\frac{a}{b} = \frac{da}{db} = \frac{ad}{bd}$$

If we consider c/d, we see that $c/d = bc/bd$. Now, if c/d and a/b are equivalent, then ad/bd and bc/bd are also equivalent.

By Fact 1, it follows that $ad = bc$. Of course, if $ad = bc$, then $ad/bd = bc/bd$ and hence a/b and c/d must be equivalent. Therefore, we can state:

Fact 2: *If a/b and c/d are equivalent fractions, then $ad = bc$ (as whole numbers); and if $ad = bc$, then a/b and c/d are equivalent.*

Put more succinctly,

$$\frac{a}{b} = \frac{c}{d} \Leftrightarrow ad = bc$$

where the $\Leftrightarrow$ means that each statement implies the other.

18. Use Fact 2 to determine whether the fractions are equivalent.

   (a) 2/3, 4/6 _____
   (b) 3/4, 9/12 _____
   (c) 6/10, 9/15 _____
   (d) 7/12, 9/14 _____
   (e) 6/21, 10/35 _____

19. What should p be so that 9/24 is equivalent to $p/16$? _____

20. Write the letter of the equivalent fraction in the blank.

   (a) 15/21   _____ 6/9
   (b) 12/15   _____ 6/8
   (c) 10/15   _____ 10/14
   (d) 9/12   _____ 8/10
   (e) 4/6   _____ 12/16

■ We saw above that a/b and na/nb are *equivalent fractions*. One way of interpreting this is to say that if the numerator and denominator of a fraction have a common factor greater than 1, then the fraction is equivalent to a fraction with a smaller numerator and denominator. For example, 21/35 is equivalent to 3/5 since $21/35 = (7 \times 3)/(7 \times 5)$. Also, 27/15 is equivalent to 9/5 since $27/15 = (3 \times 9)/(3 \times 5)$. We now state the following:

Definition: *A fraction a/b is in reduced form if a and b have no common factor other than 1 (or using the terminology of Chapter 8, $GCF(a, b) = 1$). The reduced form of a fraction is an equivalent fraction that is in reduced form.*

For example, 3/7 is in reduced form because 3 and 7 have no common factor besides 1. The reduced form of 15/35 is 3/7.

21. Find the reduced form of each of the following.

   (a) $\dfrac{220}{154}$ _____

   (b) $\dfrac{7}{1092}$ _____

   (c) $\dfrac{2^4 \times 3 \times 7^4 \times 13}{2 \times 5^2 \times 13^4}$ _____

   (d) $\dfrac{81}{9}$ _____

   (e) $\dfrac{154}{117}$ _____

   (f) $\dfrac{11}{92}$ _____

■ An investigation similar to the one we made for equivalent fractions above can help us decide which of two fractions represents the greater rational number. We know that a/b is greater than c/d if the line segment associated with a/b is longer than the line segment associated with c/d. Considering first the situation when the denominators of the two fractions

Fact 3: $a/n < b/n \Leftrightarrow a < b$

This is true since a/n is associated with the line segment obtained by dividing the unit segment into n segments of equal length and putting a of them end to end, and b/n is associated with putting b of these same segments end to end.

Now consider a/b and c/d. We know that $a/b = ad/bd$ and $c/d = bc/bd$. From Fact 3 we conclude:

Fact 4: $a/b < c/d \Leftrightarrow ad < bc$

We are now ready to formulate the

Trichotomy law for rational numbers: *If r and s are rational numbers, then one and only one of the following holds:*

(a) $r = s$ (b) $r < s$ (c) $s < r$

To determine that the trichotomy law is valid, we find fractions that represent r and s. Suppose $r = a/b$ and $s = c/d$. Now consider ad and bc. These are whole numbers, and by the trichotomy law for whole numbers that we formulated in Chapter 2, either $ad = bc$, $ad < bc$, or $bc < ad$. If $ad = bc$, then $r = s$. If $ad < bc$, then $r < s$. If $bc < ad$, then $s < r$.

22. Decide whether the first fraction of each pair below is equivalent to, less than, or greater than the second fraction of the pair.

   (a) 11/16, 5/7 _____
   (b) 6/9, 34/51 _____
   (c) 20/7, 41/15 _____
   (d) 121/14, 1021/104 _____

■ We have thus created a new number system, which is called the *rational numbers*. However, it is not a system completely detached from the system of counting numbers. First of all, we used the counting numbers to create our new numbers. A pair of counting numbers gives rise to a rational number. Further, there are rational numbers that correspond to counting numbers. If we divide the unit length into 1 segment and take 3 of them, the resulting fraction, 3/1, is associated with 3 copies of the unit length. Similarly $n/1$ is associated with n copies of the unit length. We have the correspondence

$$\frac{1}{1} \leftrightarrow 1$$

$$\frac{2}{1} \leftrightarrow 2$$

$$\frac{3}{1} \leftrightarrow 3$$

This correspondence can be extended to include 0 if we match 0 with 0/1. This makes sense since 0/1 would specify measuring zero segments of length 1. Thus the set of whole numbers can be considered as a subset of the rational numbers.

We can visualize the rational numbers as being stretched out on a "number line." First we indicate the unit length, and mark one end of it 0 and the other end 1/1 (Figure 10-10). Then we mark off 2/1, 3/1, ... (Figure 10-11). And then we mark off the other rational numbers (Figure 10-12).

Figure 10-10

Figure 10-11

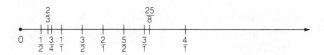

Figure 10-12

By examining the number line, we discover a major difference between rational numbers and whole numbers: we cannot "fill in" any segment on the line. For example, there is no number that immediately follows 2/3. Although the next largest number indicated in Figure 10-12 is 3/4, we can easily find a rational number between 2/3 and 3/4. For 2/3 = 16/24 and 3/4 is 18/24, so 17/24 is between 2/3 and 3/4. (See Figure 10-13.) This is quite different from what happens with whole numbers. Given any whole

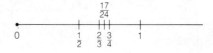

Figure 10-13

number n, there is always a number, namely $n + 1$, that immediately follows it. There are, for example, no whole numbers between 3 and 4. We can state our observation about rational numbers as follows.

Given any two distinct rational numbers r and s, there is always a third rational number between r and s.

Of course, there are many more than just one. Once we find a third, then we can find a fourth between the first and the third, and so on. This property actually permits us to conclude:

Between any two distinct rational numbers, there is an infinite number of other rational numbers.

23. Find a rational number between the two given rational numbers.

   (a) 1/2, 2/3 _____
   (b) 3/4, 7/8 _____
   (c) 11/7, 12/7 _____
   (d) 20/5, 24/6 _____
   (e) 211/83, 191/61 _____

24. Find 5 rational numbers between 1/3 and 1/2.

25. Arrange the following rational numbers in order, with the smallest first.

   19/4, 8/32, 56/14, 11/9, 2/3, 7/28, 9/32, 19/5

   _____

   _____

Lab Exercises: Set 2

EQUIPMENT: Two containers of C-rods for each group.

In these exercises you will explore the addition and subtraction of rational numbers using the physical models we have already developed.

26. Suppose we want to add 3/12 and 5/12.

   (a) Shade 3/12 and 5/12 of the region below, without overlapping the shading. How much of the region is shaded? _____ The sum of 3/12 + 5/12 = _____.

   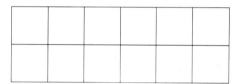

   (b) Circle 2/7 of the set below. Circle 3/7 of the set. What is 2/7 + 3/7? _____ One person explain this procedure to the group.

   ```
         *   *
       *   *   *
         *   *
   ```

   (c) Shade 1/3 of the unit length in Figure 10-14. Then shade 1/4 of the unit length in Figure 10-15. Finally, indicate 1/3 + 1/4 in Figure 10-16.
   The sum of 1/3 + 1/4 = _____.

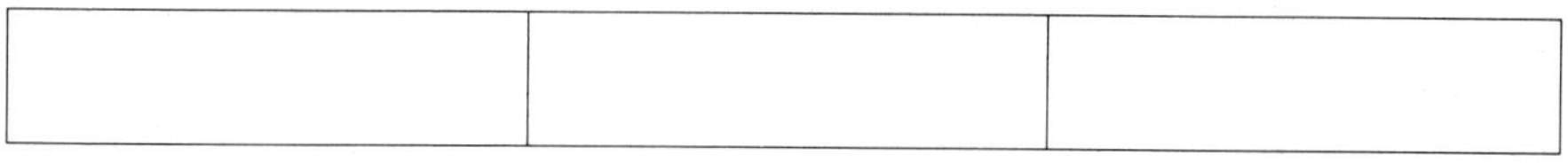

Figure 10-14

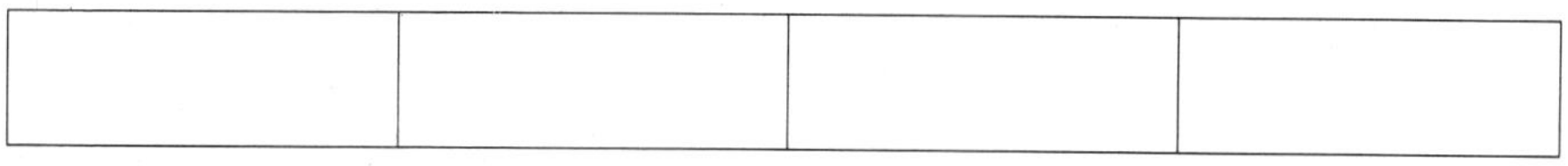

Figure 10-15

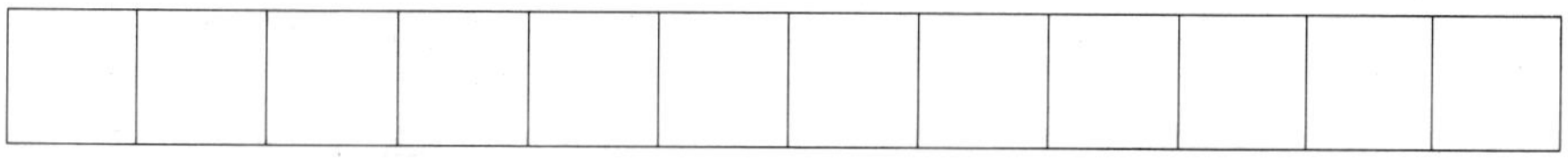

Figure 10-16

27. Use the grid in Figure 10-17 to find the indicated sums and differences.

    EXAMPLE:
    $1/4 + 1/3 = 7/12$

    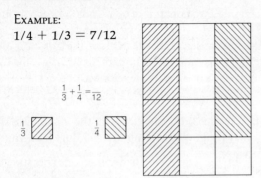

    (a) $1/2 + 1/3 =$ _____
    (b) $2/5 - 1/3 =$ _____
    (c) $3/5 + 1/7 =$ _____
    (d) $5/3 - 1/2 =$ _____
    (e) $2/3 + 2/5 =$ _____
    (f) $3/7 - 2/5 =$ _____

    Each person explain one of the problems to the group.

■ Let us use the C-rods to compute the sum of two fractions. Suppose we want to compute $2/7 + 3/7$. If **Bk** is chosen to be the unit length, **R** represents 2/7 and **LG** represents 3/7, so **Y** = **R** + **LG** represents $2/7 + 3/7$. Since **Y** represents 5/7 when **Bk** is the unit, the sum is 5/7.

28. In each of the following, choose an appropriate unit, find the rod that represents the sum with respect to that unit, and indicate the corresponding fraction. Use only rods; do not use numbers. Each person explain one of the problems to the group.

	Unit rod	Sum	Fraction
EXAMPLE: $2/7 + 3/7$	Bk	Y	5/7
(a) $2/9 + 5/9$			
(b) $3/10 + 7/10$			
(c) $1/7 + 3/7$			
(d) $3/8 + 2/8$			
(e) $1/9 + 2/9 + 5/9$			
(f) $1/5 + 3/10$			

■ Part (f) above required you to express 1/5 using **O** as a unit. The sum of two rational numbers having different denominators can be found using C-rods by choosing a unit length such that each number is represented by a single rod. For example, suppose we wish to find $2/3 + 1/4$. If we choose **LG** as the unit, **R** is 2/3, but there is no rod to represent 1/4. If we choose **P** as the unit, **W** is 1/4, but there is no rod to represent 2/3. If we choose **Y** as the unit, neither 2/3 nor 1/4 is represented by a rod. If we choose **DG** as the unit, **P** represents 2/3, but there is no rod for 1/4. Continuing in this way, we eventually find that if **O** + **R** is the unit, then **Br** represents 2/3 and **LG** represents 1/4. Hence, **Br** + **LG** represents $2/3 + 1/4$, and with **O** + **R** as the unit, **Br** + **LG** is 11/12.

29. In each of the following, choose an appropriate unit length and, *using rods only*, find the train that represents the sum and indicate the corresponding fraction. Each person explain one problem.

	Unit rod	Sum	Fraction
EXAMPLE: $2/3 + 1/4$	O + R	Br + LG	11/12
(a) $1/2 + 1/3$			
(b) $2/7 + 1/2$			
(c) $1/5 + 1/2$			
(d) $2/7 + 1/3$			

30. (a) Can you describe a procedure, *using rods only*, to determine a suitable unit rod without examining each one in turn? Discuss this in your group and then write it down.

    (b) How does this procedure with rods relate to what you know about fractions?

31. Apply procedures similar to the ones in Exercises 28 and 29 to do the following subtraction problems.

	Unit rod	Difference	Fraction
(a) $5/9 - 3/9$			
(b) $7/12 - 3/12$			
(c) $7/12 - 2/6$			
(d) $2/3 - 1/5$			
(e) $3/6 - 1/3$			

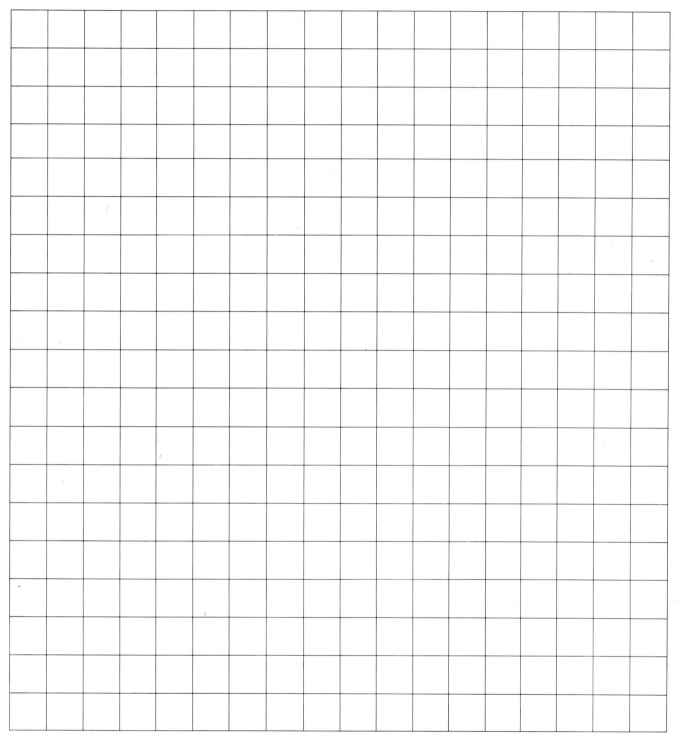

Figure 10-17

Comment: Addition and Subtraction of Rational Numbers

"Mathematicians are like lovers . . . . Grant a mathematician the least principle, and he will draw from it a consequence which you must also grant him, and from this consequence another." —Fontenelle

We have created a new number system, the system of rational numbers. Now we want to define operations with these new numbers. We notice that we can define operations on symbols in many different ways. For example, we could define an operation $*$ on the set of rational numbers by $a/b * c/d = (a + c)/(b + d)$. Then $1/2 * 1/3 = (1 + 1)/(2 + 3) = 2/5$ and $1/5 * 1/2 = (1 + 1)/(5 + 2) = 2/7$.

32. (a) Find $1/2 * 2/3$ ____, and $2/5 * 3/7$ ____.

(b) Show that $*$ is commutative and associative.

■ Now $*$ might look like a reasonable operation (and in fact many people try to use it when they add fractions), but it is not very useful.* For one thing it does not relate to the physical models we have developed for rational numbers. More important, if we use equivalent fractions within the definition of $*$, we get different answers. For example, we have seen that $1/2 * 1/3 = 2/5$. But

$$2/4 * 1/3 = (3 + 1)/(4 + 3) = 3/7$$

In order for an operation on rational numbers to be useful, it must give the same value if we replace the fraction with an equivalent fraction. This is so important that we give it a name.

Definition: *An operation on rational numbers is well defined if the result does not depend on which equivalent fractions are used.*

In other words, if ∘ is any operation, the operation ∘ is well defined if $a/b \circ c/d = u/v \circ c/d$ whenever $a/b = u/v$.

33. In (a), (b), (c), (d) below an operation is defined. Decide whether the given operation is well defined.

(a) $a/b * c/d = (a - c)/(b - d)$ _____

---
*However, a student pointed out that if $a/b < c/d$, then $a/b < (a + c)/(b + d) < c/d$.

(b) $a/b \circ c/d = $ largest of a/b and c/d _____

(c) $a/b \# c/d = (a \times c)/(b \times d)$ _____

(d) $a/b \Diamond c/d = 1$ _____

■ Thus we see that if we are going to define the addition of rational numbers, it must satisfy two criteria: (1) It must relate to the physical model of rational numbers, and (2) it must be well defined.

However, there is a third criterion, which is dictated by our choice of the word "addition." We already have defined an operation called "addition" for counting numbers, and we have seen that counting numbers are contained in a natural way within the rational number system (recall the correspondence $1 \leftrightarrow 1/1$, $2 \leftrightarrow 2/1$, $3 \leftrightarrow 3/1$, etc.). Thus we state the third criterion: (3) The new operation of addition, when applied to the counting numbers, must give the same result as when adding counting numbers according to our previous definition of addition.

34. Verify that if $a/b * c/d = (a + c)/(b + d)$, then $2/1 * 3/1$ does not correspond to $2 + 3$.

■ In the lab exercises above, we have added fractions using rectangular regions, sets, and C-rods. Let us now develop a numerical formulation of that procedure. Suppose that a/b and c/d represent two fractions and we want to compute $a/b + c/d$. If $b = d$, the procedure is clear. For example, to compute $2/7 + 3/7$, we are dividing the unit length into 7 segments of equal length and adding 2 of the sevenths and 3 of the sevenths. Hence $2/7 + 3/7 = (2 + 3)/7 = 5/7$. In general, we see that the definition

$$a/n + b/n = (a + b)/n$$

is in accord with our physical model.

In order to add rational numbers when the denominators of the fractions are unequal, all one needs to do is find equivalent fractions with equal denominators. So, to add $a/b + c/d$, we observe that

$$a/b = ad/bd \quad \text{and} \quad c/d = bc/bd,$$

so $a/b + c/d$ would be the same as $ad/bd + bc/bd$, which is $(ad + bc)/bd$. Thus we have the following:

Definition: $a/b + c/d = (ad + bc)/bd$.

We have seen in the exercises that this definition satisfies our first criterion above, namely that it is in accord with our physical model of rational numbers.

Let us consider the second criterion. Suppose we add $a/b + c/d$ and then we choose a fraction a'/b' equivalent to a/b and a fraction c'/d' equivalent to c/d. If we add $a'/b' + c'/d'$, will the result be equivalent to $a/b + c/d$? The answer is yes—and indeed it must be, since both operations represent adding lengths of line segments. But just for fun (!?), let us verify it algebraically.

We have $a/b = a'/b'$ and $c/d = c'/d'$. Hence

(A) $\qquad ab' = a'b \qquad$ and $\qquad cd' = c'd$

Now $a/b + c/d = (ad + bc)/bd$ and $a'/b' + c'/d' = (a'd' + b'c')/b'd'$. These two will be equivalent if $(ad + bc)b'd' = (a'd' + b'c')bd$. But $(ad + bc)b'd' = adb'd' + bcb'd'$. Using the equations (A) above, this equals $a'bdd' + bb'c'd$. Hence the two resulting fractions are equivalent. This means that our definition $a/b + c/d = (ad + bc)/bd$ is well defined.

It remains only to verify that this definition of addition satisfies the third criterion, namely that it agrees with the definition of addition for counting numbers. For example, if we compute $3/1 + 5/1$, we get $3/1 + 5/1 = (1 \times 3 + 5 \times 1)/(1 \times 1) = (3 + 5)/1 = 8/1$, which is the rational number corresponding to the counting number 8. In general,

$$a/1 + b/1 = (a \times 1 + 1 \times b)/(1 \times 1) = (a + b)/1$$

and therefore the third criterion is satisfied.

Let us use the definition above to investigate briefly the commutative and associative properties. For example, using the definition, we see that $1/5 + 2/3 = (1 \times 3 + 5 \times 2)/5 \times 3 = (3 + 10)/15 = 13/15$ and $2/3 + 1/15 = (2 \times 5 + 3 \times 1)/3 \times 5 = (10 + 3)/15 = 13/15$. In general, we have the following.

Commutative property of addition: $a/b + c/d = c/d + a/b$.

As an example of the associative property, consider $(1/2 + 2/3) + 3/5 = 7/6 + 3/5 = 53/30$ and $1/2 + (2/3 + 3/5) = 1/2 + 19/15 = 53/30$. In general, we have the following.

Associative property of addition: $(a/b + c/d) + e/f = a/b + (c/d + e/f)$.

The operation of subtraction can be considered similarly to the way we considered addition. If we did, we would arrive at the following.

Definition: If $a/b > c/d$, then $a/b - c/d = (ad - bc)/bd$.

35. Verify that $(ad - bc)/bd + c/d = a/b$ by using the definition of addition.

■ Recall that *a fraction is in reduced form if a and b have no common factor other than 1*. Another way of saying this is that a/b is in reduced form if the $GCF(a, b) = 1$.

36. Compute the following sums and differences and put the answers in reduced form.

   (a) $2/3 + 4/9 = $ _____

   (b) $1/8 + 1/7 = $ _____

   (c) $1/5 + 3/10 = $ _____

   (d) $3/10 - 7/24 = $ _____

   (e) $15/14 - 11/56 = $ _____

   (f) $14/9 - 16/15 = $ _____

37. Observe the following:

   $$1/2 - 1/3 = 1/6$$
   $$1/3 - 1/4 = 1/12$$
   $$1/4 - 1/5 = 1/20$$

   (a) What do you notice? _____

   Describe the pattern as a rule. _____

   (b) Compute $1/9 - 1/10$ by your rule. _____

   (c) Compute $1/100 - 1/101$ by your rule. _____

   (d) Write down a justification for the rule.

Lab Exercises: Set 3

EQUIPMENT: Two containers of C-rods for each group.

In these exercises we will investigate multiplication and division of rational numbers using the physical models we have developed.

In light of our earlier discussion, we want to define *multiplication* so that it is in accord with our physical models of rational numbers. If we think of a fraction a/b as representing the result of an operation (divide unit a into b equal parts and consider a of the parts), we can then interpret multiplication as the result of repeating the operation. For example, $2/3 \times 1/2$ would correspond to taking 1/2 of a unit and then taking 2/3 of the result. Let us see what happens.

38. Shade 1/2 of the region below with stripes slanted one way: ▨. Consider this shaded

portion and shade 2/3 of it with stripes slanted the other way: . The doubly-shaded region corresponds to 2/3 × 1/2. How much of the original region is doubly-shaded? ____ Hence 2/3 × 1/2 = ____

39. Calculate each of the following using the procedure in Exercise 38. Each person explain one problem to the group.

   (a) $\frac{1}{2} \times \frac{1}{4}$

   (b) $\frac{1}{6} \times \frac{1}{2}$

   (c) $\frac{1}{2} \times \frac{1}{3}$

   (d) $\frac{1}{2} \times \frac{2}{3}$

40. (a) Circle 3/8 of the set below.

   * * * *
   * * * *
   * * * *
   * * * *

   (b) Circle 1/3 of the subset corresponding to 3/8. What is the fraction corresponding to this set? ____ Hence 1/3 × 3/8 = ____.

   Note: 1/3 × 3/8 means 1/3 of 3/8. That is, we take 3/8 first and then take 1/3 of the result.

41. Draw a set below and use it to calculate 3/8 × 1/3.

42. Draw appropriate sets to calculate the indicated products, as in Exercise 40. Each person explain one problem to the group.

   (a) 3/4 × 1/2 = ____

   (b) 1/3 × 2/7 = ____

   (c) 4/3 × 5/6 = ____

   (d) 5/3 × 3/4 = ____

   (e) 3/4 × 5/3 = ____

■ Let us calculate some products using C-rods. Suppose we want to find 2/3 × 1/2. We must choose a unit such that if we take 1/2 of it and 2/3 of the result, we end up with a whole rod. So let **DG** be the unit. Now, 1/2 of **DG** is **LG**, and 2/3 of **LG** is **R**, and

10/Lab Exercises: Set 3

R corresponds to 1/3 when **DG** is the unit. Why?

43. Use the procedure above to calculate the following. Use the rods and not previously learned rules for multiplication. Each person explain one to to the group.

	Unit	Rod obtained	Fraction
EXAMPLE: 2/3 × 1/2	**DG**	**R**	1/3
(a) 1/2 × 1/4			
(b) 3/2 × 1/4			
(c) 1/2 × 2/3			
(d) 1/3 × 2/3			

■ From this exercise you can see that 2/3 × 1/2 = 1/2 × 2/3. This illustrates the *commutative property* of multiplication for rational numbers. The following exercise illustrates the *associative property*.

44. Use diagram (a) below to indicate 1/4 × 2/3. Now find 1/2 of the resulting region. You have determined 1/2 × (1/4 × 2/3). Now we want to compute (1/2 × 1/4) × 2/3. We know from Exercise 39(a) or 43(a) that 1/2 × 1/4 = 1/8. Hence (1/2 × 1/4) × 2/3 = 1/8 × 2/3. Find 1/8 × 2/3 using diagram (b). Compare with the region in diagram (a). You have verified that 1/2 × (1/4 × 2/3) = (1/2 × 1/4) × 2/3.

(a)
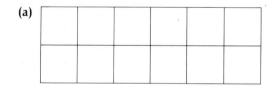

(b)

45. Use the grid in Figure 10-19 (p. 116) to indicate a region corresponding to 2/3 + 1/2. Then indicate 1/2 × (2/3 + 1/2). Now indicate a region corresponding to 1/2 × 2/3 and another corresponding to 1/2 × 1/2. (Be sure to use the same unit region.) Verify that 1/2 × (2/3 + 1/2) = 1/2 × 2/3 + 1/2 × 1/2. This illustrates the *distributive property* of multiplication over addition.

■ One way to interpret the division of rational numbers is to consider, for example, 1/3 ÷ 1/6 as the answer to the question, "How many 1/6's are there in 1/3?" This is analogous to interpreting 8 ÷ 2 as the number of 2's in 8. Suppose you wish to use C-rods to determine how many 1/6's there are in 1/3. What should the unit be? If **Bl** is the unit, then **LG** represents 1/3, but there is no rod representing 1/6. One choice of unit is **DG**. Then **R** = 1/3 and **W** = 1/6, and since there are two **W**'s in **R**, we see that 1/3 ÷ 1/6 = 2.

46. Do the problems in the table at the bottom of the page using C-rods, and fill in the table.

47. Use the grid on page 116 to do each of the problems in Exercise 46 above. (For example, see Figure 10-18.)

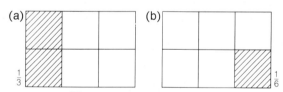

There are two 1/6's in 1/3.

Figure 10-18

	Unit	Representing rods	Explanation
EXAMPLE: 1/3 ÷ 1/6 = 2	**DG**	1/3 is **R**, 1/6 is **W**	There are two **W**'s in **R**.
(a) 1/2 ÷ 1/4		1/2 is ____ 1/4 is ____	
(b) 3 ÷ 1/5		3 is ____ 1/5 is ____	
(c) 1/5 ÷ 1/10		1/5 is ____ 1/10 is ____	
(d) 1/3 ÷ 1/9		1/3 is ____ 1/1 is ____	
(e) 2/3 ÷ 1/6		2/3 is ____ 1/6 is ____	

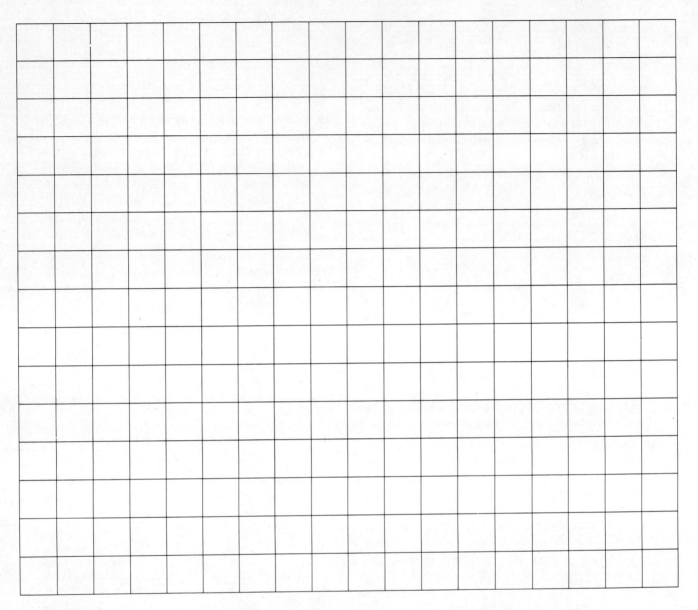

Figure 10-19

Notice that $1/3 \div 1/6 = 2$ and $1/3 \times 6/1 = 2$.

Definition: *If c/d is a fraction, the fraction d/c is called the* reciprocal *of c/d.*

48. (a) Verify for each of the products in Exercise 46 that $a/b \div c/d = a/b \times d/c$.

  (b) Compute each of the following using multiplication by the reciprocal.

   (i) $5/6 \div 2/3$ _____

   (ii) $2/5 \div 1/3$ _____

   (iii) $1/4 \div 5/3$ _____

Comment: Multiplication and Division of Rational Numbers

"The most distinct and beautiful statements of any truth must take at last the mathematical form."—Thoreau

When we wish to define multiplication of rational numbers, we must pay attention to the three criteria we used when we defined addition. Our definition must (1) relate to the physical model of the rational number system, (2) be well defined, and (3) give the same result when applied to 1/1, 2/1, 3/1, etc., as when applied to multiplying the corresponding whole numbers.

How should we interpret multiplication with respect to the physical model? Since $m \times n$ means "n taken m times," we could think of $a/b \times c/d$ as, "c/d taken a/b times." For example, $1/2 \times 3/4$ would be, "$3/4$ of a unit segment taken $1/2$ of a time," or as we ordinarily say it, $1/2 \times 3/4$ would be the result of taking $1/2$ of $3/4$ of a unit.

Let us use the physical model of length of a line segment to determine $1/2$ of $3/4$. A segment of length $3/4$ is obtained by dividing the unit length into 4 segments of equal length and putting 3 of them end to end. Consider the unit length to be equal in length to the brown C-rod. Divide it into 4 congruent pieces (Figure 10-20). Shade a portion to represent $3/4$ (Figure 10-21).

Figure 10-20

Figure 10-21

To obtain $1/2$ of $3/4$, we can divide the shaded portion into 2 equal parts to obtain the dark region in Figure 10-22. The same length will be obtained if we divide each of the original segments into 2 congruent parts (Figure 10-23) and take 3 of the resulting segments (Figure 10-24). In general, to obtain a length corresponding to $1/n$ of a/b, we can divide the unit segment into $n \times b$ congruent segments and put a of them end to end. Hence, $1/n$ of a/b is represented by a segment of length $a/(n \times b)$. Thus, if multiplication is to be in accord with our physical model, we should have $1/n \times a/b = a/(n \times b)$.

Figure 10-22

Figure 10-23

Figure 10-24

What about $m/n \times a/b$? If, for example, we take $3/2$ of $3/4$, we can think of this as first taking $1/2$ of $3/4$, which is equal to $3/(2 \times 4) = 3/8$ (Figure 10-25) and then taking 3 of these $3/8$ (Figure 10-26, at the bottom of the page). We see that this must be

Figure 10-25

$(3 \times 3)/8$. Thus $3/2$ of $3/4$ is $(3 \times 3)/(2 \times 4)$. Although it is not obvious that $3/2$ of $3/4$ would be $(3 \times 3)/(2 \times 4)$ or that $9/11$ of $17/25$ would be $(9 \times 17)/(11 \times 25)$, the exercises and the discussion above should convince you that the following definition is the correct one.

Definition: *If a/b and c/d represent two rational numbers, then $a/b \times c/d = (a \times c)/(b \times d)$.*

From our discussion we can see that this definition is in accord with the physical model, but is it well defined? Suppose $a/b = a'/b'$. Then $a \times b' = a' \times b$. We need to verify that

$$a/b \times c/d \quad \text{and} \quad a'/b' \times c/d$$

are equivalent fractions. Now

$$a/b \times c/d = (a \times c)/(b \times d)$$

and

$$a'/b' \times c/d = (a' \times c)/(b' \times d)$$

and $(a \times c)/(b \times d)$ will be equivalent to $(a' \times c)/$

Figure 10-26

($b' \times d$) precisely when $(a \times c) \times (b' \times d) = (a' \times c) \times (b \times d)$. This equality follows from the fact that $a \times b' = a' \times b$.

The third criterion is that multiplication of fractions extends the operation of multiplication of whole numbers. This is easy to check, since $m/1$ corresponds to m and $n/1$ corresponds to n. We have $m/1 \times n/1 = (m \times n)/1$, which corresponds to $m \times n$.

As with other operations, we wish to investigate the properties of this new operation. By now the following properties should be familiar to you:

Commutative property of multiplication: $a/b \times c/d = c/d \times a/b$.

49. Illustrate the commutative property of multiplication by finding $2/3 \times 1/4$ and $1/4 \times 2/3$ on grid paper.

■ To verify the commutative property for multiplication, we rely on the definition of multiplication:

$$a/b \times c/d = (a \times c)/(b \times d)$$

and

$$c/d \times a/b = (c \times a)/(d \times b)$$

But $a \times c = c \times a$ and $b \times d = d \times b$, since multiplication of counting numbers is commutative. Hence $a/b \times c/d = c/d \times a/b$.

Associative property of multiplication: $(a/b \times c/d) \times c/f = a/b \times (c/d \times e/f)$.

50. Illustrate the associative property of multiplication by finding $2/3 \times (1/2 \times 3/4)$ and $(2/3 \times 1/2) \times 3/4$, using some physical model.

51. Verify the associative property using the definition of multiplication.

Distributive property of multiplication over addition: $a/b \times (c/d + e/f) = a/b \times c/d + a/b \times e/f$.

52. Illustrate the distributive property by finding $1/2 \times (1/3 + 1/2)$ and $(1/2 \times 1/3) + (1/2 \times 1/2)$, using some physical model.

■ We can verify the distributive property by applying the definitions of addition and multiplication. Since it is lengthy, we leave it as an optional exercise.

53. *Optional.* Verify the distributive property of multiplication over addition.

■ It is clear that the rational number 1/1 (which we identify with 1) acts as an identity for multiplication:

Identity for multiplication: $a/b \times 1 = a/b$.

There is a property that the rational numbers have that is not possessed by the whole numbers: If a/b is a rational number, there is another rational number which if multiplied by a/b yields 1. Thus we have the

Reciprocal property: *If a and b are nonzero whole numbers, then $a/b \times b/a = 1$.*

54. Write the reciprocal of the following.

(a) 4/3 _____

(b) 11/29 _____

(c) 5 _____

(d) 1/6 _____

55. Fill in the blanks.

(a) $2/3 \times$ _____ $= 1$

(b) $9 \times$ _____ $= 1$

(c) _____ $\times 13/20 = 1$

(d) $3/5 \times 5/3 =$ _____

■ Multiplication of fractions is often simplified by relying on the procedure for finding the reduced form of a fraction, which was discussed following Exercise 20. Recall that if the numerator and denominator of a fraction have a common factor greater than 1, the fraction is equivalent to a fraction with the common factor removed from both the numerator and denominator. For example, $18/12 = (3 \times 6)/(2 \times 6) = 3/2$. This process is often called *canceling the common factor.* Now if we wish to multiply $2/3 \times 7/5 \times 5/2 \times 2/7 \times 6/5 \times 4/5$, we notice that this product will equal

$$\frac{2 \times 7 \times 5 \times 2 \times 6 \times 4}{3 \times 5 \times 2 \times 7 \times 5 \times 5}$$

Canceling common factors gives

$$\frac{\cancel{2} \times \cancel{7} \times \cancel{5} \times 2 \times 6 \times 4}{3 \times 5 \times \cancel{2} \times \cancel{7} \times \cancel{5} \times 5} = \frac{2 \times 6 \times 4}{3 \times 5 \times 5} = \frac{48}{75}$$

In this last stage, if the 6 had been factored and cancellation then performed, a reduced form of the product 48/75 would have resulted:

$$\frac{2 \times 3 \times 2 \times 4}{3 \times 5 \times 5} = \frac{2 \times \cancel{3} \times 2 \times 4}{\cancel{3} \times 5 \times 5} = \frac{16}{25}$$

56. Find the reduced form of each of the following products.

(a) $3/2 \times 5/7 \times 6/9 \times 2/5$ _____

(b) $18/25 \times 20/9$ _____

(c) $1/2 \times 1/3 \times 1/5 \times 2/1 \times 3/1 \times 5/1$ _____

(d) $4/5 \times 3/2 \times 5/7 \times 11/3$ _____

■ Now let us consider the division of rational numbers. In the lab exercises we noticed that one reasonable physical interpretation of division leads to the observation that $a/b \div c/d$ can be computed by calculating the product of a/b and the reciprocal of c/d, namely $a/b \div c/d = a/b \times d/c = ad/bc$. A simple form of this is used all the time when, for example, we interchange the concept of dividing 12 by 3 with the concept of taking 1/3 of 12. Dividing 12 by 3 is expressed as $12 \div 3$ while taking 1/3 of 12 is expressed as $12 \times 1/3$.

To be sure, there are other ways of looking at division. When we introduced division of whole numbers, we defined division by a number to be the inverse operation of multiplication by that number, that is, it "undid" multiplication by that number. This approach would lead us to the following definition.

Definition: *If a/b represents one rational number and c/d another, where $c/d \neq 0$, then $a/b \div c/d$ is the rational number that solves $\square \times c/d = a/b$. We call $a/b \div c/d$ the* quotient *of a/b and c/d.*

Of course, this definition is not in conflict with our previous interpretation, for if $u/v = a/b \div c/d$, that is, if u/v solves the indicated equation, then $a/b = u/v \times c/d$. If we multiply both sides of this equation by the reciprocal of c/d, we obtain

$$\frac{a}{b} \times \frac{d}{c} = \left(\frac{u}{v} \times \frac{c}{d}\right) \times \frac{d}{c}$$
$$= \frac{u}{v} \times \left(\frac{c}{d} \times \frac{d}{c}\right)$$
$$= \frac{u}{v} \times \left(\frac{c \times d}{d \times c}\right)$$
$$= \frac{u}{v} \times 1$$
$$= \frac{u}{v}$$

Thus, once again $a/b \div c/d = a/b \times d/c$.

Also, in case you are worrying about it . . . BUT IS IT WELL DEFINED? . . . the quotient $a/b \div c/d$ is well defined. That is, if we choose equivalent fractions to a/b and c/d, the new quotient will be equivalent to $(a \times d)/(b \times c)$.

We omit the proof. (Doesn't that make you happy?)

Finally, we should verify that division by a number "undoes" multiplication by that number. We do so as follows.

$$\left(\frac{a}{b} \times \frac{c}{d}\right) \div \frac{c}{d} = \left(\frac{a \times c}{b \times d}\right) \div \frac{c}{d}$$
$$= \frac{a \times c}{b \times d} \times \frac{d}{c}$$
$$= \frac{a \times c \times d}{b \times d \times c}$$
$$= \frac{a}{b} \times \frac{c \times d}{c \times d}$$
$$= \frac{a}{b}$$

We conclude this chapter by noting that for rational numbers, division is always defined, provided that the divisor is not zero. This is quite different from the situation with whole numbers. In fact, since the whole numbers are contained in the rational numbers, we see that, considered as rational numbers, whole numbers can be divided (excluding, of course, division by zero). For example, $3 \div 5$ becomes $3/1 \div 5/1 = 3/1 \times 1/5 = 3/5$.

57. Find the following quotients, considering whole numbers as rational numbers.

(a) $5 \div 7$ _____

(b) $11 \div 4$ _____

(c) $6 \div 3$ _____

(d) $2 \div 4$ _____

58. Show that division of rational numbers is not commutative.

59. Show that division of rational numbers is not associative.

■ Do Exercises 60–62 by thinking about the actual physical situation. Do not resort to memorized techniques from school.

60. It takes Mary $6\frac{1}{2}$ hours to drive from Santa Barbara to San Francisco in her VW. Susan does it in her Porsche in 5/6 of that time. How long does it take Susan? _____

61. One day 2/7 of Tom's third-grade class was

absent. There were 15 students present. How many students are in the class? _____

62. At one elementary school, 58 teachers drive to work. This represents 2/3 of the faculty. How many are on the faculty? _____

Important Terms and Concepts

Fraction
Rational number
Unit quantity
Numerator
Denominator
Equivalent fraction
Ordering of fractions
Trichotomy law
Reduced form
Addition of rational numbers
Subtraction of rational numbers
Multiplication of rational numbers
Division of rational numbers
Reciprocal
Commutative property
Associative property
Distributive property
Well-defined

Review Exercises

1. Two rational numbers a/b and c/d are equivalent if the equation _____ is true.

2. Find two rational numbers equivalent to 0/5. Is 0/5 a rational number? _____ Is 5/0 a rational number? _____

3. Decide whether each of the following is true or false. Write T or F in the blank. Let $n > 0$ be a whole number.

    _____ (a) $an/b = cn/d \Leftrightarrow ad = bc$

    _____ (b) $an/b < cn/d \Leftrightarrow ad < bc$

4. Given two rational numbers a/b and c/d such that $a/b < c/d$, how can you find another rational number between them? (Check your steps using actual numbers.)

5. If $1/a < 1/b$, what must be true of a and b? _____ Use your answer to show that between 0 and any rational number c/d, we can find an infinite number of other rational numbers.

6. We know that 368/1004 cannot be the reduced form of the fraction representing this rational number. Why? _____

7. Find the reduced form of the following fractions.

    (a) 14/24 _____

    (b) 36/8 _____

    (c) 0/5 _____

8. Find the product and express in reduced form: $Bk/Y \times Y/R \times R/Bk$.

9. What fraction p/q will satisfy $2/3 \times 3/5 \times p/q = 1$? _____

10. What is the smallest denominator common to both a/b and c/d that we can use to find $a/b + c/d$? _____

11. Decide whether each of the following is true or false. Write T or F in the blank.

    _____ (a) If a is a prime number, then a/b is in reduced form.

    _____ (b) If b is a prime number, then a/b is in reduced form.

    _____ (c) If both a and b are different prime numbers, then a/b is in reduced form.

12. Find the reduced form of the following products.

    (a) $1/2 \times 5/3 \times 6/10$ _____

    (b) $9/25 \times 3/11$ _____

    (c) $6 \times 2 \times 1/4$ _____

13. Find reciprocals for the following.

    (a) 1/4 _____   (c) 9 _____

    (b) 3/11 _____ (d) 1 _____

14. Calculate the following.

    $(1/2 \times 3/4) \div (5 - 2/3)$ _____

15. Verify that if $a/b < c/d$, then $a/b < (a + c)/(b + d) < c/d$.

16. "But the product of two numbers is always bigger!" exclaimed the student after having seen the equation 1/2 × 1/5 = 1/10. What would you say?

17. Why wasn't it necessary to discuss the concept of being well-defined when we defined operations on the counting numbers?

18. Use one of the physical models for addition to show that

$$\left(\frac{1}{3}+\frac{1}{4}\right)+\frac{1}{6}=\frac{1}{3}+\left(\frac{1}{4}+\frac{1}{6}\right)$$

19. *Optional.* Use the definition of addition to verify the associative property of addition.

This actually happened with a child of one of my students.—J. W.

Caroline Herschel (1750–1848) is most often recognized for her achievements in astronomy, but she was also a competent mathematician. She discovered 14 nebulae, the most notable being the Andromeda nebula in 1783. She is recognized as the first woman to detect a comet, detecting eight in all. Her brother William was a court astronomer to King George III, and they worked together. She was elected an honorary member of the Royal Astronomical Society of London and the Royal Academy of Dublin.

11

Positive and Negative Numbers

"I understand your surprise at my being able to busy myself simultaneously with literature and mathematics. Many who have never had an opportunity of knowing any more about mathematics confound it with arithmetic, and consider it an arid science. In reality, however, it is a science which requires a great amount of imagination. . . . it is impossible to be a mathematician without being a poet in soul. It seems to me that the poet has only to perceive that which others do not perceive, to look deeper than others look. And the mathematician must do the same thing. As for myself, all my life I have been unable to decide for which I had the greater inclination, mathematics or literature."—Sónya Kovalévsky

Lab Exercises: Set 1

EQUIPMENT: None.

In this set of exercises we will extend the system of whole numbers to include negative numbers, and then we will explore the addition and subtraction of positive and negative numbers.

In Chapter 10 we saw that a number line is useful to visualize rational numbers. The number line can also be used to study whole numbers and negative numbers (Figure 11-1).

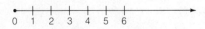

Figure 11-1

1. Discuss how you might use the number line to teach the addition of whole numbers to youngsters.

2. A useful way to explain addition might be to tell about someone who starts at zero, moves 2 units to the right, and then moves 3 more units to the right. The person thus moves ____ units to the right.

3. A variation of this model, shown in the next column, is to regard each of the numbers as addresses. For example, if John's house is 3 units to the right of zero and Mary's house is 4 units to the right of John's house, what is the address of Mary's house? ____

4. An ingenious use of the number line to teach young people about numbers is explored in a forthcoming book.* The idea is that someone is lost on a number line from which all the "addresses" have disappeared. From the given clues the reader is to figure out where the person is located. Find the location of the lost individual in each of the following.

(a)

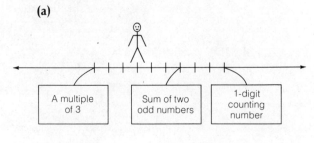

Lost in the Number Line by Margo Nanny and Gene Novak (in preparation).

11/Lab Exercises: Set 1

(b)

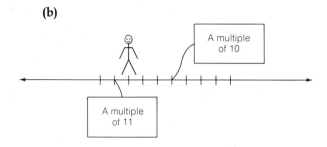

(c)

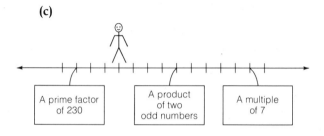

(d) Suppose you are told that one of the clues in (c) has been obliterated by a snowstorm, but there is still enough information to determine the location. Which clue was obliterated? _____
Does it have to be the clue you've chosen, or could it be one of the others? _____
Why or why not?

(e) Make up your own "lost on the number line" problem and present it to the group.

■ Consider the number line in Figure 11-2. The arrow indicates that we should consider the line as extending indefinitely to the right. This is in accord with our knowledge that the counting numbers are infinite.

Figure 11-2

However, the number line above is actually only half a line. A full line extends indefinitely in both directions, so let us extend the line to the left, as in Figure 11-3. If we think of the line as a road with zero

Figure 11-3

as the starting point, we can travel in either of two directions when we leave zero. However, it is necessary to put new "addresses" to the left of zero. We use the symbol $^-1$ to denote the point one unit to the left of zero, $^-2$ to denote the point two units to the left of zero, and so on. The symbol $^-1$ is read "negative one," $^-2$ is read "negative two," etc.

5. Label the points A, B, C, and D on the number line below.

Definition: *The numbers represented by the symbols* $^-1, ^-2, ^-3, \ldots$ *will be called* **negative integers**. *Those represented by* $1, 2, 3, \ldots$ *(which we previously called counting numbers) will be called* **positive integers**. *The entire collection* $\{ \ldots ^-3, ^-2, ^-1, 0, 1, 2, 3, \ldots \}$ *is called the* **integers**.

Note that zero is an integer that is neither negative nor positive.

■ Let us investigate the arithmetical operations for the integers. First we consider *addition*.

6. What do you think $^-2 + ^-3$ should be? _____
Explain why using the number line below.

■ Most people agree that $^-2 + ^-3$ should be $^-5$ because they can see the sum $^-2 + ^-3$ as representing the result of moving 2 units to the left and then 3 units further to the left, ending at $^-5$ (Figure 11-4).

Figure 11-4

7. Calculate each of the following by drawing a similar diagram on the number lines. Take turns explaining each problem to the group.

(a) $^-4 + ^-2 = $ _____

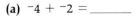

(b) $^-1 + ^-6 = $ _____

(c) $^-2 + ^-4 = $ _____

8. What do you think $^-2 + 5$ should be? _____ Explain using the number line below.

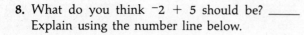

■ If we used the same reasoning as before, we would see this sum as representing a movement of 2 units to the left followed by a movement of 5 units to the right. If we do this, we end up at 3 (Figure 11-5).

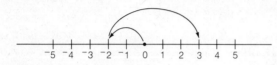

Figure 11-5

Hence we say $^-2 + 5 = 3$.

9. Calculate each of the following by drawing a similar diagram on the adjacent number lines. Take turns explaining each problem to the group.

   (a) $^-3 + 7 =$ _____

   (b) $4 + {^-5} =$ _____

   (c) $5 + {^-2} =$ _____

   (d) $20 + {^-30} =$ _____

10. Discuss in your group whether the commutative property holds for addition of integers. Explain why or why not using the number line.

11. Compute each of the following. What do you notice?

   (a) $(^-3 + 2) + {^-5} =$ _____

   (b) $^-3 + (2 + {^-5}) =$ _____

■ Let us now consider subtraction for integers. Recall that in Chapter 3 when we studied the subtraction of counting numbers, we considered two approaches. One was to remove a subset from a given set. That will not work here, since, for example, no set has $^-3$ elements. The other approach was to say that $a - b$ is the number that solves $b + \square = a$, that is, the number that makes the equation correct when put into the $\square$. We will use this approach to study subtraction of integers. (It might be helpful at this time to review the section on subtraction in Chap-

ter 3.) Thus, to compute $^-5 - {^-3}$, for example, we must find the solution to $\square + {^-3} = {^-5}$. Since $\boxed{^-2} + {^-3} = {^-5}$, we have the solution $^-5 - {^-3} = {^-2}$.

12. Complete each of the following. Check your results with the group.

   (a) $^-8 - {^-5} =$ _____ since $\square + {^-5} = {^-8}$.
   (b) $^-6 - {^-2} =$ _____ since $\square + {^-2} = {^-6}$.
   (c) $^-15 - {^-1} =$ _____ since $\square + {^-1} = {^-15}$.
   (d) $^-3 - {^-7} =$ _____ since $\square + {^-7} = {^-3}$.
   (e) $^-4 - {^-27} =$ _____ since $\square + {^-27} = {^-4}$.
   (f) $5 - {^-3} =$ _____ since $\square + {^-3} = 5$.
   (g) $7 - {^-8} =$ _____ since $\square + {^-8} = 7$.
   (h) $^-7 - 4 =$ _____ since $\square + 4 = {^-7}$.
   (i) $^-5 - 8 =$ _____ since $\square + 8 = {^-5}$.
   (j) $3 - 12 =$ _____ since $\square + 12 = 3$.

13. Compute the following.

   (a) $4 - 17 =$ _____   (c) $^-3 - 7 =$ _____
   (b) $12 - 15 =$ _____   (d) $^-18 - 30 =$ _____

14. Compute the following.

   (a) $4 + {^-17} =$ _____   (c) $^-3 - {^-7} =$ _____
   (b) $12 + {^-15} =$ _____   (d) $^-18 + {^-30} =$ _____

15. (a) Compare 13(a), (b), (c), and (d) with 14(a), (b), (c), and (d). What do you notice? Discuss this in your group.

   (b) Complete the following: The result of subtracting a positive integer from any integer can be found by _____.

16. Is there a similar rule for subtracting negative integers? Investigate on your own to see if you can find one.

17. Compute the following.

   (a) $^-17 - {^-32} =$ _____ $^-17 + 32 =$ _____
   (b) $^-11 - {^-5} =$ _____ $^-11 + 5 =$ _____
   (c) $23 - {^-12} =$ _____ $23 + 12 =$ _____
   (d) $0 - {^-7} =$ _____ $0 + 7 =$ _____

18. Complete the following: The result of subtracting a negative integer from any integer can be found by _____.

11/Comment

19. Discuss how you might explain subtraction by using movements on the number line.

Comment: Addition and Subtraction of Positive and Negative Numbers

"An essential ingredient of the problem is the desire, the will and the resolution to solve it. A problem that you are supposed to do and which you have quite well understood, is not yet your problem. It becomes your problem, you really have it, when you decide to do it, when you desire to solve it."—George Polya

When we discussed the rational numbers above, we noted that they are necessary because counting numbers are not adequate to analyze all situations in, and answer all questions about, the world. For the same reason, we need to develop the integers. In many actual situations we find it necessary to consider the *direction* as well as the *magnitude* of a number, for example:

> Temperatures below or above zero.
> Altitudes below or above sea level.
> Debit balances as well as credit balances.
> Forces acting to the left or to the right.

In each of these cases we could use words to describe the situation, but it is more efficient and concise to choose one specific direction as positive, and then use negative numbers to indicate magnitudes in the opposite direction.

There is also a good mathematical reason for developing a larger number system. If we restrict ourselves to the whole numbers, 0, 1, 2, 3, . . . , there are subtraction problems that would have no solution. This is inconvenient and esthetically displeasing. We desire a system in which we can calculate the difference of *any* two numbers.

Although the laboratory exercises above concerned themselves only with the set of integers, it should be clear that we can consider positive and negative *rational* numbers as well. So, we introduce for each positive rational number, m/n, a negative rational number, which we denote by $^-m/n$. We interpret this number as being represented by the point on the number line that is a distance of m/n to the left of zero. Some of the points on the number line are labeled in Figure 11-6. Pairs of numbers, such as $^-1$ and 1, 3/2 and $^-3/2$, $^-4$ and 4, are called *opposites*.

We say, for example, that $^-4$ is the opposite of 4 and 4 is the opposite of $^-4$.

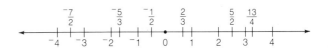

Figure 11-6

20. What is the opposite of each of the following?

(a) 11 _____   (c) $\dfrac{^-3}{2}$ _____   (e) 5 _____

(b) $^-1$ _____   (d) $^-2$ _____   (f) 0 _____

21. What is the opposite of the *opposite* of each of the following?

(a) 4 _____   (b) $^-3$ _____   (c) 0 _____

■ We call the set { . . . , $^-3$, $^-2$, $^-1$, 0, 1, 2, 3, . . . } the *integers*. Note that the dots indicate that the integers go on infinitely in both directions. We will use the term *rational numbers* to refer to the set of both positive and negative rational numbers. What we previously called simply the "rational numbers" will now be called the "positive rational numbers."

Addition

In the exercises above, we interpreted the addition of positive and negative integers by considering them as representing movement on the number line. Negative integers represented movement to the left, and positive integers represented movement to the right. The same interpretation will work for rational numbers. Thus we can determine $^-5/2 + 3/2$ by starting at zero, moving 5/2 units to the left and then 3/2 units to the right (Figure 11-7). We see that $^-5/2 + 3/2 = ^-1$.

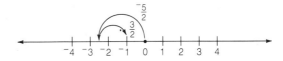

Figure 11-7

22. Do each of the following on the number line.

(a) $2/3 + ^-2/3 =$ _____

(b) $^-3/5 + 1 =$ _____

(c) $2/3 + ^-7/3 =$ _____

(d) $5 + ^-3/2 =$ _____

(e) $4/3 + ^-5/2 =$ _____

■ We saw in the lab exercises that addition of integers is commutative. This is true because it does not matter in which order you do the specified movement on the number line. Addition of integers is also associative: $(a + b) + c = a + (b + c)$. This follows because both sides of the equation indicate the result of doing the three movements associated with a, b, and c. We will see in the following set of lab exercises that another model for the integers makes the commutative and associative properties much more clear.

Subtraction

In the lab exercises we considered subtraction as the inverse operation of addition. If we wish to find the difference $^-5 - 4$, for example, we seek a number n such that $4 + n = {^-5}$. How do we find the number n? Trial and error will work, or we can look at the number line. We seek a number n such that if we move 4 units to the right and then do "n," we end up 5 units to the left (Figure 11-8). By examining the number line, we see that n is $^-9$.

Figure 11-8

Subtraction can also be considered in terms of direct movement on the number line. In general, to find a difference, we move as indicated by the first number and then reverse the direction of the movement indicated by the number being subtracted. For example, to find $^-4 - {^-3}$, we move 4 units to the left and then move 3 units to the right (Figure 11-9). The $^-3$ by itself would denote 3 units to the left, but the minus sign indicates that we change the direction.

Figure 11-9

23. Do each of the problems below using this method.

(a) $^-2 - {^-3} =$ _____

(b) $2 - {^-2} =$ _____

(c) $3 - 5 =$ _____

(d) $^-5 - 4 =$ _____

Of course, the same methods work for rational numbers.

24. Do each of the following in both of the ways discussed above.

(a) $\dfrac{5}{2} - \dfrac{^-3}{2} =$ _____

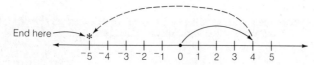

(b) $\dfrac{^-2}{7} - \dfrac{6}{7} =$ _____

(c) $\dfrac{^-3}{5} - \dfrac{^-7}{5} =$ _____

■ Consider again the second method we discussed above. The process of reversing the direction of the movement corresponds to taking the opposite of the number. We can therefore say that *subtracting a number is equivalent to adding the opposite of that number.* That is, if r and s are any rational numbers, $r - s = r +$ opposite of s.

25. Write each of the following as an addition problem using the above rule.

(a) $3 - \dfrac{5}{2}$ _____

(b) $^-2 - 3$ _____

(c) $\dfrac{6}{7} - \dfrac{^-3}{2}$ _____

(d) $\dfrac{^-4}{3} - \dfrac{^-2}{5}$ _____

(e) $6 - {^-11}$ _____

(f) $^-8 - {^-2}$ _____

■ We have achieved our goal of creating a new set of numbers in which every subtraction problem has a solution. Next we see whether multiplication of these numbers can be defined in a reasonable manner.

Lab Exercises: Set 2

EQUIPMENT: For each group, 20 red and 20 black squares cut from construction paper, or any similar collection of red and black counters.

In these exercises we will develop a model for the multiplication of integers. Learning and teaching how to multiply positive and negative integers are often quite challenging. Why, for example, should a negative number times a negative number be positive? And how do you explain it?

We wish to find a physical model to help explain the rules that so many students simply memorize. Although the number line is a useful model for explaining addition and subtraction, it does not do much to clarify multiplication.

We will represent positive integers by black counters and negative integers by red counters. (It is standard bookkeeping procedure to use red ink for debits and black for credits.) For example, 5 red counters will represent $^-5$ (Figure 11-10) and 3 black counters will represent 3 (Figure 11-11).

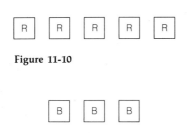

Figure 11-10

Figure 11-11

26. What number is represented by each of the following collections?

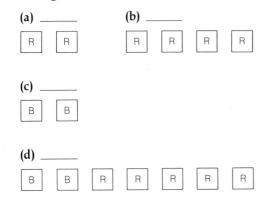

27. Exercise 26(d) poses a problem. If we have a collection that contains both red and black counters, we need a rule to determine what number the collection should represent. What do you think the rule might be? Discuss with your group.

■ One sensible rule—and the one we will adopt—is to agree that red and black counters "cancel each other out." That is, as far as possible, we match the set of red and black counters, remove the matched pairs, and count the counters of one color that remain.

Applying this procedure to Exercise 26(d), we have Figure 11-12. Hence the collection represents $^-3$.

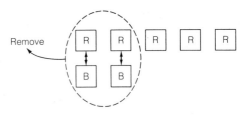

Figure 11-12

28. Use the above rule to determine the number represented by each of the following collections.

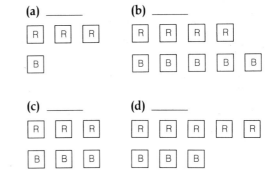

29. (a) How would you interpret the addition of integers using the red and black counters?

   (b) Each member of your group pair off with another member and teach the partner how to compute one of the following. For this exercise, pretend that the partner is a sixth-grader.

   (i) $^-5 + 3 =$ _____

   (ii) $^-7 + {}^-4 =$ _____

   (c) One person in the group show the rest how one might use the counters to justify the commutative property of addition for the integers.

30. (a) Using the counters, compute each of the following.

   (i) $(^-5 + 7) + {}^-4 =$ _____

   (ii) $^-5 + (7 + {}^-4) =$ _____

   (b) Someone in the group explain to the rest how one might use the counters to justify the associative property of addition for integers.

■ You have probably observed that different collections can represent the same number. The collections in Exercises 28(a) and 28(d), for example, both represent ⁻2.

31. Indicate another collection of counters that represents ⁻2.

32. Use the counters to find 3 different representations for each of the following integers.

   (a) 2
   (b) 15
   (c) ⁻7
   (d) ⁻5
   (e) 0

■ In order to define the multiplication of integers, we recall that multiplication of positive integers (counting numbers) may be considered as repeated addition. Thus $3 \times 2 = 2 + 2 + 2$. Let us extend this idea to the set of integers, using the model of the counters. Observe that if we represent 2 as [B] [B], we can find 3×2 by taking the union of 3 sets of 2 black counters (Figure 11-13). In order to find $3 \times {}^-4$, represent ⁻4 by [R] [R] [R] [R] and take the union of 3 sets of 4 red counters. We obtain Figure 11-14, which represents ⁻12. Hence $3 \times {}^-4 = {}^-12$.

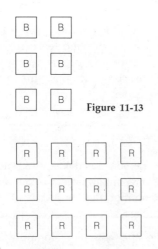

Figure 11-13

Figure 11-14

33. Find each of the following using the above procedure. Take turns explaining each problem to your group.

   (a) $2 \times {}^-5$ _____
   (b) $4 \times {}^-1$ _____
   (c) $3 \times {}^-2$ _____

34. Represent ⁻2 by

Then take the union of three of these collections.

   (a) Describe the collection you obtain. _____
   (b) What number does it represent? _____
   (c) Should it agree with the answer to Exercise 33(c)? _____
   (d) Why or why not?

35. Compute $3 \times {}^-2$ using

as a representation for ⁻2. Verify that your answer agrees with the previous answers.

ALL THIS IS EASY. BUT WHAT ABOUT MULTIPLYING BY A NEGATIVE NUMBER?

I'm glad you asked that. There are two cases to consider: multiplying a positive number by a negative number, and multiplying a negative number by a negative number.

36. What do you think $^-3 \times 2$ should be? _____
   Why? _____
   How would you explain it using red and black counters? Discuss this for a few minutes before reading on.

■ The difficulty of computing $^-3 \times 2$ is that you cannot take the union of ⁻3 sets of [B] [B]. If you say that $^-3 \times 2$ should be ⁻6 because multiplication should be commutative, that is an ingenious answer, but it won't help us deal with $^-3 \times {}^-2$. We need a different interpretation.

The solution is to consider again the situation in which we were interpreting 3×2 with the counters.

Instead of thinking of 3 × 2 as 3 sets of B B, we could have considered it as the result of adding 3 sets of B B to a representation of zero—namely, the representation consisting of zero blacks and zero reds. The next step is to look upon 3 × 2 as adding 3 sets of B B to *any* representation of zero. For example, if zero is represented as shown in Figure 11-15, then adding 3 sets of 2 blacks yields Figure 11-16 and the resulting set represents 6.

Figure 11-15

Figure 11-16

37. Verify that 3 × ⁻2 can be obtained by adding 3 sets of 2 reds

   (a) to the representation of zero given by

   (b) to the representation of zero given by

■ Of course, you would not complicate things in this way when multiplying by a positive number, but the idea is useful as we interpret ⁻3 × 2. We define ⁻3 × 2 as the result of subtracting (removing) 3 sets of 2 blacks from a suitable representation of zero. You must choose a representation of zero that will allow you to remove the required counters.

For example, to compute ⁻3 × 2, we represent zero as shown in Figure 11-17. We subtract one set of 2 black counters to obtain Figure 11-18. Then we subtract a second set (Figure 11-19). After subtracting a third set of 2 black counters, we are left with Figure 11-20, which represents the number ⁻6.

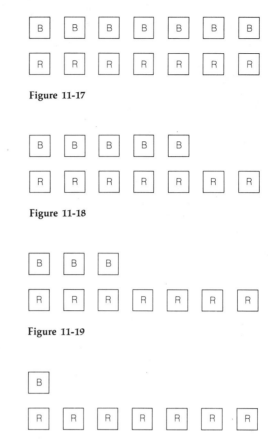

Figure 11-17

Figure 11-18

Figure 11-19

Figure 11-20

38. Represent zero as

   and compute ⁻3 × 2 using the above procedure. What do you obtain? _____

39. Compute each of the following by selecting a representation of zero and using the above procedure. Fill in the table.

	Representation of zero selected	Counters left	Answer
EXAMPLE: ⁻3 × 2	7B, 7R	1B, 7R	⁻6
(a) ⁻3 × 3			
(b) ⁻5 × 1			
(c) ⁻4 × 2			
(d) ⁻1 × 1			

40. Do Exercise 39 again, choosing a different representation of zero. Fill in the table below.

	Representation of zero	Counters left	Answer
Example: $^-3 \times 2$	6B, 6R	6R	$^-6$
(a) $^-3 \times 3$			
(b) $^-5 \times 1$			
(c) $^-4 \times 2$			
(d) $^-1 \times 1$			

41. Discuss how you might interpret $^-3 \times {}^-2$ using our model. Think of a procedure you would follow to compute $^-3 \times {}^-2$ using red and black counters.

■ The solution to the above is to consider $^-3$ as indicating that we should subtract 3 sets of "something" from a representation of zero. This "something" we should subtract is some representation of $^-2$, the simplest being 2 red counters. If we represent zero as shown in Figure 11-21 and subtract 3 sets of 2 red counters, we obtain Figure 11-22, which represents 6. Hence $^-3 \times {}^-2 = 6$.

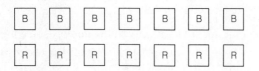

Figure 11-21

Figure 11-22

42. Calculate each of the following using the above procedure, and fill in the table in the next column.

	Representation of zero	Counters left	Answer
Example: $^-3 \times {}^-2$	7B, 7R	7B, 1R	6
(a) $^-2 \times {}^-4$			
(b) $^-2 \times {}^-1$			
(c) $^-1 \times {}^-4$			
(d) $^-2 \times {}^-3$			

43. (a) Compute $^-4 \times {}^-2$ using the counters. Compare with Exercise 33(c).

   (b) Compute $^-2 \times 3$ using the counters. Compare with Exercise 33(c).

   (c) Discuss in your group how you could use the counters to argue that multiplication of integers is commutative. (This is more challenging than it seems. In considering the product $a \times b$, it may be that one or both of a, b are negative.)

44. Optional.
   (a) Compute $^-3 + {}^-4$ using the counters. _____ Now use the counters to compute $2 \times ({}^-3 + {}^-4)$. _____
   (b) Compute $2 \times {}^-3$ and $2 \times {}^-4$ using the counters. _____ Now compute $(2 \times {}^-3) + (2 \times {}^-4)$ using the counters. _____ Compare with your answer in (a).

45. Optional.
   (a) Compute $^-2 \times ({}^-3 + {}^-4)$ using the counters. (Choose an appropriate representation of zero.)
   (b) Compute $^-2 \times {}^-3$ and $^-2 \times {}^-4$ and then $({}^-2 \times {}^-3) + ({}^-2 \times {}^-4)$ using the counters. Compare with your answer in (a).

46. Optional. Verify that $^-2 \times ({}^-3 + 5) = ({}^-2 \times {}^-3) + ({}^-2 \times 5)$ using the counters.

47. Optional. By now you have probably noticed that for integers, multiplication distributes over addition. That is,
$$a \times (b + c) = (a \times b) + (a \times c)$$
for all integers a, b, c.

Discuss in your group how you might use counters to argue that the distributive property is valid.

Comment: Multiplication and Division of Positive and Negative Numbers

"What you have been obliged to discover by yourself leaves a path in your mind which you can use again when the need arises." —G. Lichtenberg

We have been developing a physical model that justifies the algebraic laws for multiplying integers. We can state these laws concisely by first observing that every negative integer is the opposite of a positive integer. Therefore, all possible products of two integers where one or more factors are negative are included in the following.

Definition: *If p_1 and p_2 are nonnegative integers, then*

$$p_1 \times {}^-p_2 = {}^-(p_1 \times p_2)$$
$${}^-p_1 \times p_2 = {}^-(p_1 \times p_2)$$
$${}^-p_1 \times {}^-p_2 = p_1 \times p_2$$

The multiplication on the right side of each equation is the usual multiplication of whole (nonnegative) numbers.

This definition is in accord with the model developed in the lab exercises. In addition, there are other ways to justify these rules. Suppose you have one of those bank accounts that allow you to write checks without having any money deposited. (Of course, you have to pay it back.) If you start with a zero balance and write 7 checks of $10 each, you will owe the bank $70, or have a balance of $^-70$ dollars. This corresponds to the computation $7 \times {}^-10 = {}^-70$.

Suppose a friend has deposited checks for $8 three days in a row and now her balance is zero. Before she deposited the checks, she therefore must have owed $24; that is, her balance 3 days ago was $^-24$ dollars. This corresponds to the computation $^-3 \times 8 = {}^-24$. Now suppose another friend tells you that he had written 5 checks for $7 each to pay some bills, and that now his balance is zero. He then discovers that the bills were his roommate's and destroys the checks. You know that his balance is $35. This corresponds to the calculation $^-5 \times {}^-7 = 35$.

48. Make up similar examples for each of the following products.

    (a) $3 \times 7 =$ _____   (c) $2 \times {}^-3 =$ _____

    (b) $^-4 \times 6 =$ _____   (d) $^-6 \times {}^-8 =$ _____

49. Imagine a movie film in which each frame is labeled by an integer. Answer each of the following and state what multiplication problem corresponds to the physical situation.

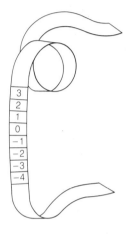

   (a) The film is running backward at the rate of 3 frames per second. What frame will be shown 4 seconds after the 0 frame? _____

   (b) The film is running forward at the rate of 3 frames per second. What frame will be shown 4 seconds after the 0 frame? _____

   (c) The film is running forward at the rate of 3 frames per second. What frame was shown 4 seconds before the 0 frame? _____

   (d) The film is running backward at the rate of 3 frames per second. What frame was shown 4 seconds before the 0 frame? _____

50. Make up a similar problem for each of the products in Exercise 48.

■ Once we accept the definition of multiplication given above, we can ask whether multiplication according to that definition is commutative, associative, and distributive. To verify the commutative and associative properties, it is useful to introduce the concept of *absolute value*.

Definition: *The absolute value of an integer measures the magnitude of the integer, regardless of whether the integer is positive or negative.*

The absolute value of an integer is always positive. Thus the absolute value of $^-3$ is 3, and the absolute value of 4 is 4. If you think of the integers on the number line, the absolute value indicates the distance from 0.

51. Write the absolute value of each of the following.

    (a) $^-4$ _____
    (b) $^-2$ _____
    (c) 15 _____
    (d) 0 _____
    (e) $^-127$ _____
    (f) 923 _____

Note: To indicate the absolute value of a number, we enclose it between two vertical lines. Thus we write $|^-3| = 3$ for the statement "the absolute value of $^-3$ equals 3."

52. Complete the following.

    (a) $|^-10| = $ _____
    (b) $|10| = $ _____
    (c) $|^-6| = $ _____
    (d) $|6| = $ _____
    (e) $|0| = $ _____
    (f) $|^-2 \times 3| = $ _____
    (g) $|^-2| \times |3| = $ _____
    (h) $|^-5 \times ^-7| = $ _____
    (i) $|^-5| \times |^-7| = $ _____

■ If we now look back at our definition of multiplication, we see that the product of any two integers can be obtained by multiplying their absolute values, and then attaching a negative sign if one of the factors is negative and the other is positive. In other words, if a and b are any integers

$$a \times b = \begin{cases} |a| \times |b| \text{ if } a \text{ and } b \text{ are both positive} \\ \quad \text{or both negative.} \\ -(|a| \times |b|) \text{ if one of } a, b \text{ is positive and} \\ \quad \text{the other is negative.} \end{cases}$$

53. Verify the above for each of the following.

    (a) $^-7 \times ^-3$ _____
    (b) $^-5 \times 7$ _____
    (c) $9 \times ^-3$ _____
    (d) 4×6 _____

■ The commutative property of multiplication follows immediately from the above, since $|a|$, $|b|$ are counting numbers and therefore $|a| \times |b| = |b| \times |a|$ because we know the commutative property is valid for counting numbers. We will omit a somewhat lengthier argument that would verify the associative property.

54. Verify each of the following.

    (a) $(^-2 \times 5) \times ^-7 = ^-2 \times (5 \times ^-7)$ _____
    (b) $(^-3 \times ^-2) \times ^-4 = ^-3 \times (^-2 \times ^-4)$ _____
    (c) $(2 \times ^-8) \times 5 = 2 \times (^-8 \times 5)$ _____
    (d) $(^-3 \times 7) \times ^-4 = ^-3 \times (7 \times ^-4)$ _____

■ A verification of the distributive property of multiplication over addition would also require considerable algebraic manipulation. However, we can argue informally using the red and black counters.

To compute $a \times (b + c)$ using the counters, we first compute $b + c$. If b and c are both negative, we can represent $b + c$ as $|b| + |c|$ red squares. To compute $a \times (b + c)$, we either add or subtract $|a|$ (depending on whether a is positive or negative) of these collections to or from a representation of zero. Thus we are either adding or subtracting the collection shown in Figure 11-23. In this diagram, $b = ^-5$, $c = ^-2$, and $|a| = 3$. It should be clear that the result is the same as adding or subtracting the collection of $|a| \times |b|$ red counters and then the collection of $|a| \times |c|$ red counters. That is, $a \times (b + c) = (a \times b) + (a \times c)$.

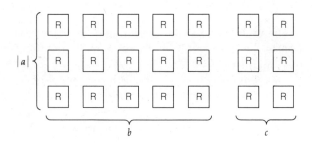

Figure 11-23

A similar argument works if b and c are both positive. In that case, all the squares are black. Now, suppose b is positive and c is negative. To compute $b + c$, we must form a collection of b black counters and c red counters and either add or subtract $|a|$ of these collections to a representation of zero, as shown in Figure 11-24. Again, the result will be the same as adding or subtracting the collection of $|a| \times |b|$ black counters and $|a| \times |c|$ red counters. That is, $a \times (b + c) = (a \times b) + (a \times c)$. A similar argument works if b is negative and c is positive.

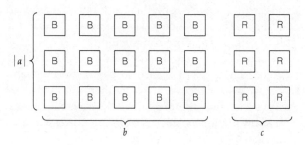

Figure 11-24

55. Verify each of the following, first using counters and then by computation.

(a) $^-2 \times (5 + ^-3) = ^-2 \times 5 + ^-2 \times ^-3$

(b) $2 \times (^-2 + ^-1) = (2 \times ^-2) + (2 \times ^-1)$

■ There is another way of understanding the multiplication of positive and negative integers. Suppose that after we had defined addition of integers, we decided to define multiplication so that our new system would satisfy the following properties:

(a) Multiplication is commutative.
(b) The product of zero and any number is zero.
(c) Multiplication distributes over addition.

Recall that any number plus its opposite is zero and the opposite of the number is the only number that has that property. In symbols, $a + {^-a} = 0$, and if $a + b = 0$, then $b = {^-a}$. For example, $^-3 + 3 = 0$, and if $^-3 + b = 0$, we must have $b = 3$.

We can use this fact and the above three properties to compute any product. For example, to compute $3 \times ^-4$, we write

$(3 \times 4) + (3 \times ^-4) = 3 \times (4 + ^-4)$ by (c) above
$ = 3 \times 0$
$ = 0$ by (b) above

So we have $3 \times ^-4 = ^-(3 \times 4) = ^-12$.

Now consider a negative times a negative, for example, $^-4 \times ^-3$. We write

$(^-4 \times ^-3) + (^-4 \times 3) = ^-4 \times (^-3 + 3)$ by (c) above
$ = ^-4 \times 0$
$ = 0$ by (b) above

Therefore, $^-4 \times ^-3 = ^-(^-4 \times 3)$. But $^-4 \times 3 = 3 \times ^-4$ by property (a) above, and we have just seen that $3 \times ^-4 = ^-12$. Hence, $^-4 \times 3 = ^-12$. Hence $^-4 \times ^-3$ is the opposite of $^-12$, or $^-4 \times ^-3 = 12$.

A similar argument would work for any two integers. One advantage of this type of argument is that it easily extends to rational numbers. If we went through the necessary justification, we would end with the following rules for multiplication of rationals.

Suppose r and s are positive rationals. Then $r \times s$ is positive (computed as in Chapter 10) and the following hold:

$$^-r \times s = ^-(r \times s)$$
$$r \times ^-s = ^-(r \times s)$$
$$^-r \times ^-s = r \times s$$

56. Compute each of the following.

(a) $2/3 \times ^-3/5$ _____

(b) $^-4/7 \times 1/2$ _____

(c) $^-3/2 \times ^-1/6$ _____

(d) $^-3/5 \times 0$ _____

(e) $^-4/3 \times 6$ _____

(f) $^-3 \times ^-5/6$ _____

■ Division of rational numbers is defined in terms of multiplication just as in Chapter 10. That is, $r \div s$ is defined to be that number which solves $\square \times s = r$. For example, $2/3 \div ^-1/2$ is the solution to $\square \times ^-1/2 = 2/3$. Since $^-4/3 \times ^-1/2 = 2/3$, we have $2/3 \div ^-1/2 = ^-4/3$. It is easy to verify that if r and s are any positive or negative rational numbers with $s \neq 0$, then $r \div s$ can also be found by multiplying r by the reciprocal of s. For example, suppose we want to compute $3/4 \div ^-2/5$. We observe that the reciprocal of $^-2/5$ is $^-5/2$, since $^-2/5 \times ^-5/2 = 2/5 \times 5/2 = 1$. Hence $3/4 \div ^-2/5 = 3/4 \times ^-5/2 = ^-15/8$.

57. Compute the following.

(a) $\dfrac{^-2}{3} \div \dfrac{^-3}{5} =$ _____

(b) $\dfrac{3}{4} \div \dfrac{^-3}{2} =$ _____

(c) $\dfrac{3}{7} \div \dfrac{^-3}{7} =$ _____

(d) $\dfrac{^-5}{4} \div \dfrac{^-5}{4} =$ _____

(e) $\dfrac{^-2}{7} \div \dfrac{2}{7} =$ _____

(f) $0 \div \dfrac{^-4}{3} =$ _____

(g) $4 \div \dfrac{^-2}{5} =$ _____

(h) $\dfrac{^-3}{2} \div \dfrac{^-2}{3} =$ _____

Important Terms and Concepts

Number line   Opposites
Negative integers   Addition of integers
Positive integers   Subtraction of integers

Multiplication of integers
Division of integers
Addition of rational numbers
Subtraction of rational numbers
Multiplication of rational numbers
Division of rational numbers

Review Exercises

1. Illustrate each of the following on a number line.
   (a) $^-2 + 3 = 3 + {}^-2$
   (b) $^-2 + {}^-4 = {}^-4 + {}^-2$
   (c) $(5 + {}^-2) + {}^-1 = 5 + ({}^-2 + {}^-1)$

2. Solve each of the following.
   (a) $^-3/5 + 4/5 = $ _____
   (b) $2/3 + {}^-3/2 = $ _____
   (c) $^-3/5 + 5/3 = $ _____
   (d) $8 - {}^-3 = $ _____
   (e) $^-3/2 - {}^-5/2 = $ _____
   (f) $5/7 - 2 = $ _____

3. Solve each of the following.
   (a) $3/5 \times {}^-2/3 = $ _____
   (b) $^-4/7 \times {}^-3/2 = $ _____
   (c) $^-4 \times {}^-7 = $ _____
   (d) $2/3 - {}^-2/3 = $ _____
   (e) $^-3/2 - {}^-5/2 = $ _____
   (f) $^-2/7 \times 0 = $ _____

4. In each of the following, decide which of the two distances on the number line is greater, or if the two distances are equal.
   (a) The distance between $^-3$ and 6, or between 3 and 6.
   (b) The distance between $^-5$ and $^-2$, or $^-2$ and 5.
   (c) The distance between $^-4$ and 5, or $^-5$ and 4.

5. In four downs a football team loses 5 yards, gains 7 yards, loses 12 yards, and gains 20 yards. Write an equation using positive and negative integers to describe the situation, and compute the total change in yards.

6. *Optional.* If x and y are integers, we say that $x > y$ (x is greater than y) if there is a positive integer w such that $x = y + w$. If $x > y$, then we also write $y < x$ (y is less than x). Put either $<$ or $>$ in each of the blanks below.
   (a) 3 ___ 8
   (b) $^-2$ ___ 4
   (c) $^-3$ ___ $^-5$
   (d) 0 ___ $^-2$
   (e) $^-2$ ___ $^-1$
   (f) $^-11$ ___ $^-7$
   (g) If $x > y$, then ^-x ___ ^-y.

7. Suppose you get a part-time job in which you will earn between $450 and $520 next month. You plan on buying a stereo that will cost between $490 and $540. Determine the largest amount you may owe and the largest amount you may have left over at the end of the month.

8. Write an argument that the multiplication of integers distributes over the subtraction of integers.

9. (a) Find two different representations of $^-3$ with the red and black counters.
   (b) If m red counters and n black counters represent the same integer as k reds and l blacks (m, n, k, l counting numbers), then $m + l = $ ___.

10. Verify each of the following using counters, and then verify by computation.
    (a) $^-7 \times {}^-2 = {}^-2 \times {}^-7$
    (b) $^-2 \times ({}^-3 + {}^-4) = ({}^-2 \times {}^-3) + ({}^-2 \times {}^-4)$

11. Verify that if a and b are integers, then $a \times b$ can be represented as counters in the following manner.
    (a) Represent b as a set of red or black counters.
    (b) Take the union of $|a|$ of these sets.
    (c) If a is negative, change the color of the resulting collection; otherwise leave it alone.

You have learned a lot of mathematics so far. You can be proud of your accomplishments. Here is a present. It is a game that was popular in French military circles during the Franco-Prussian War and is described in Martin Gardner's "Mathematical Games" column in *Scientific American* (October 1973) as well as in his *Sixth Book of Mathematical Games*, pp. 39–40. Any young people you know will enjoy playing it with you.

Player A places a penny on each of the three shaded circles. Player B places a nickel (or some other object) on the center circle. Player A moves first, then B, then A, etc. Player B can move the nickel in any direction from one circle to any neighboring circle. Player A moves the pennies similarly, but is restricted to moving left, right, or forward (straight ahead or diagonally). The pennies cannot be moved backward. A wins if he or she can trap B's nickel so that it cannot move. B wins if he or she slips behind "enemy lines," so that the nickel cannot be trapped, or if a situation occurs in which the same moves are repeated endlessly. I hope you enjoy the game.

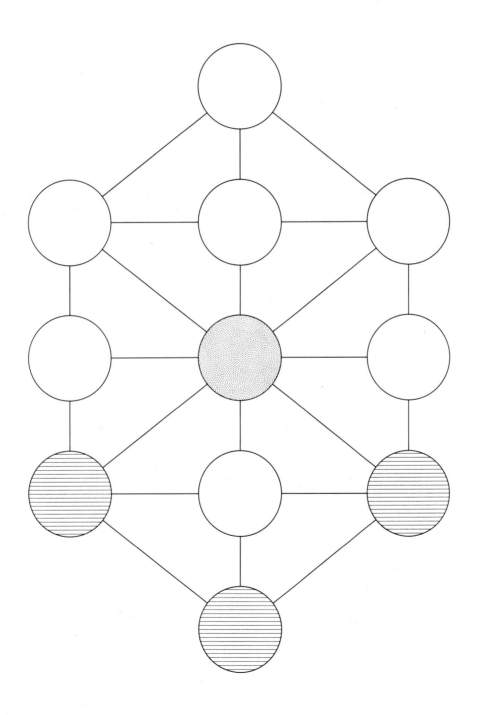

12
Geometric Concepts

"Let no one ignorant of geometry enter here."—Inscribed above the door of Plato's Academy

Lab Exercises: Set 1

EQUIPMENT: One set of geoblocks* for each three groups of students; one set of tangrams* for each student (can be constructed from cardboard). (If there are not enough geoblocks available for your group, do Exercises 5-13 first and then return to Exercises 1-4.)

This set of exercises is divided into three groups. Exercises 1-4 explore configurations of three-dimensional solids. Exercises 5-9 investigate two-dimensional shapes. These first nine exercises are designed to provide physical experiences with geometry rather than communicate any specific information. Exercises 10-13 are concerned with points, lines, planes, and their relationships.

1. Play with the geoblocks for a few minutes. Discuss the following questions in your group. What can you do with them? What would children do? What kinds of questions could you ask children when they get bored with building with the blocks? How might you use the blocks in other subject areas?

---
*Information about this equipment may be found on pages xvii-xviii.

2. Choose eight to ten shapes and put two or three blocks of each shape in a pile on the table. Two people, *A* and *B*, close their eyes. Someone place a block in *A*'s hands. *A* describes the block to *B*. *B* (with eyes still closed) tries to find a block in the pile that is the same size and shape.

3. There is a set of yellow activity cards that comes with the geoblocks. Each card has outline drawings on it of one of the geoblocks. Each drawing shows a different side of the block. If the block looks the same on two sides, there will be a "2" in that drawing, similarly for three sides, etc. If there is no number, it means *only* one side has that outline. Take a yellow card and find the block that matches the card. Do it with four more cards.

4. (a) Read the directions for the white activity cards that come with the geoblocks. Work in pairs and build the constructions indicated by the five photographs. Be careful not to look at the answer card or it will spoil your fun. Do another construction.

   (b) There is also a set of blue activity cards. Read the direction card and, working in pairs, choose a card and find the construction. Be careful not to look at the answer card (the

12/Lab Exercises: Set 1

one with the photograph) or it will spoil your fun. Do one or two more.

■ Exercises 5–9 use *tangrams*. The tangram is an old Chinese puzzle consisting of seven pieces that can be cut from a square, as illustrated in Figure 12-3.

5. The object of classical tangram puzzles was to form silhouettes of various objects. If commercially manufactured tangrams are not available to you, use Figure 12-3 to make a set of tangram pieces from heavy construction paper or (preferably) cardboard. Make your set accurate and durable, since we will be using them again. Use your tangram pieces to form the duck, candle, and cat pictured in Figures 12-4, 12-5, and 12-6, respectively.

Figure 12-1

Figure 12-2

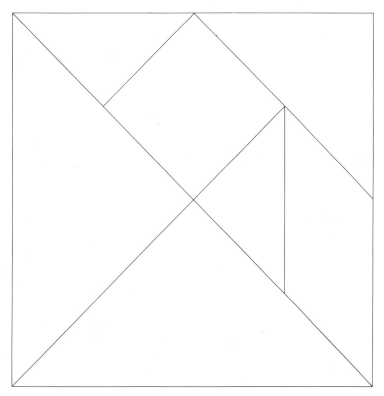

Figure 12-3

Figure 12-4

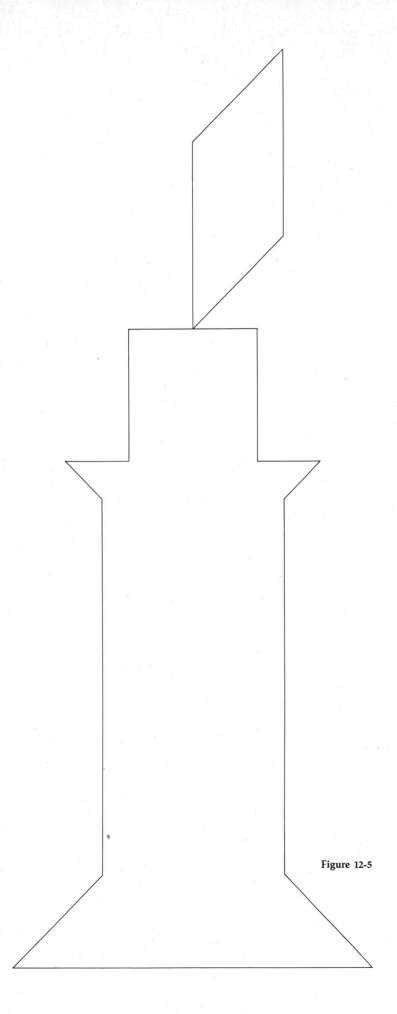

Figure 12-5

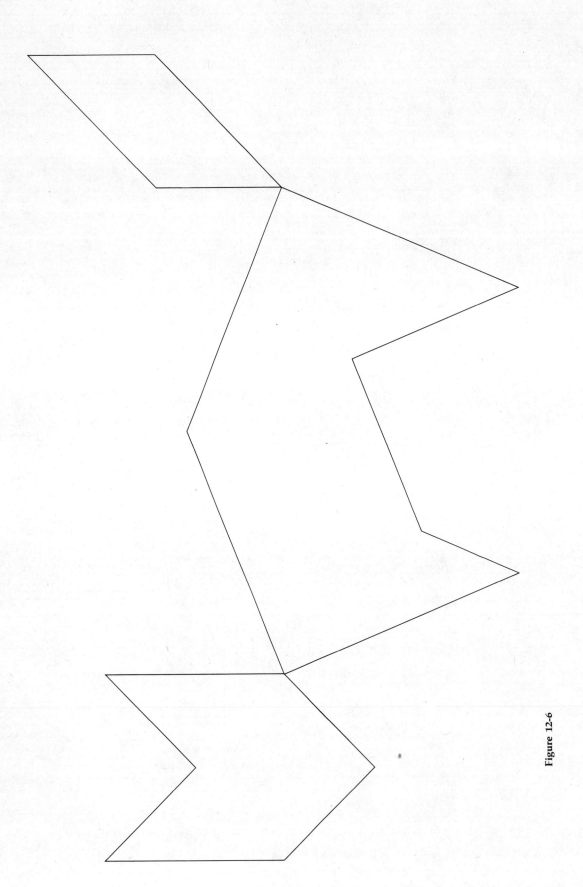

Figure 12-6

6. Form the following shapes with *all seven* tangram pieces: rectangle, triangle, parallelogram (not a rectangle), trapezoid.

7. Make some of your own shapes (people, animals, boats, geometric shapes, etc.). See Figure 12-7a and b for some examples.

Figure 12-7a

Figure 12-7b

8. Form a square using the five smallest tangram pieces.

9. Can you form a square using any six of the tangram pieces?

■ A good reference for tangrams is *Tangrams—330 Puzzles* by R. C. Read (New York: Dover Publications, 1965). In recent years, tangrams have been used to teach various geometric concepts. Two references for the use of tangrams as teaching aids are *Tangramath* by Dale Seymour (Creative Publications, P.O. Box 1038, Palo Alto, CA) and *Math Experiments with Tangrams* by John Ginther (Midwest Publications, Inc., P.O. Box 129, Troy, MI).

■ Exercises 10-13 are concerned with points, lines, and planes. Discuss Exercises 11-13 in your group before writing down your answers.

10. Look around you. Do you have a notion of space? Each person take a minute and try to define what space is. Is it hard or easy to do?

11. Make a list of four different ways to enclose space (e.g., the inside surface of the four walls plus the surfaces of the ceiling and floor enclose the space within your classroom; the surface of your skin encloses the space occupied by your body).

(a) _____

(b) _____

(c) _____

(d) _____

■ The boundaries that separate parts of space from other parts of space are called *surfaces*. Flat surfaces are called *planes*.

12. (a) List six planes you can see.

   (i) _____ (iv) _____

   (ii) _____ (v) _____

   (iii) _____ (vi) _____

(b) List three surfaces you can see that are not planes.

   (i) _____

   (ii) _____

   (iii) _____

(c) Imagine enclosing space with a single sheet of paper. What must you do with the paper?

_____

(d) How many pieces of paper would you need to enclose a space if the pieces were to stay flat? (You can use scissors and tape.) _____

(e) What is the minimum number of plane surfaces that will enclose a space? _____ Draw a picture of the figure.

13. (a) Find some place in your room where two planes intersect. What two planes are you thinking of? _____
_____
What is the intersection of these two planes? _____ Describe two planes that do not intersect. _____
_____
Complete the following: If two different planes intersect, their intersection is _____
_____.

(b) Find some place where a line and a plane intersect. _____ What is the intersection? _____
Complete the following: If a plane and a line *not in the plane* intersect, their intersection is _____. Why are the italicized words necessary; i.e., what is the intersection of a plane and a line in the plane? _____

(c) Find some place where two lines intersect. _____ What is their intersection? _____ Complete the following: If two different lines intersect, their intersection is _____. Find two lines that do not intersect. If two lines in the same plane do not intersect, they are called *parallel*. If two lines are not in the same plane, they are called *skew*. Find two skew lines in your room. Find two parallel lines in your room.

(d) Below are two points. How many straight lines can you draw through both points? _____

•   •

Is the same true for any two points in space? _____ Complete: Through any two points in space there is exactly _____ line. The two points, of course, are *in the plane* of the paper. Where is the line? _____.
Complete the following: If two distinct points lie in a plane, the line determined by the points lies _____.
Use a piece of cardboard or paper to indicate another plane through these two points. How many planes are there? _____
Complete the following: Through two points in space there are _____ planes.

(e) Hold three pencils in one hand, or arrange them in such a way that allows you to place a piece of cardboard so it touches each of the three points of the pencils. Can you find a different plane through all three of these points? _____ Arrange four pencils so that there is no plane containing all four of their points. Complete the following: Any _____ points not on the same line determine one and only one plane. Why do we need the condition that the points are "not on the same line"?

Comment: Points, Lines, and Planes

"Sesostris . . . made a division of the land of Egypt among the inhabitants . . . . If the river carried away any portion of a man's lot . . . the king sent persons to examine, and determine by measurement the exact extent of the loss . . . . From this practice, I think, geometry first came to be known in Eqypt, whence it passed into Greece."—Herodotus

The word *geometry* is derived from the Greek words *geos* for "earth" and *metros* for "measure." Thus, geometry literally means the measurement of the earth. In Herodotus' view geometry originated from the practical problems of surveying and measuring the earth. It should be noted that Aristotle disagreed with this view, claiming that the priestly leisure class in Egypt was responsible for the beginning of geometry. The question has never been resolved, but whether geometry originated for practical needs or through an aesthetic sense for design and order, the Greeks developed the subject into a truly impressive body of mathematical knowledge. You no doubt studied geometry in a high school course. If so, you probably remember a sequence of axioms (assumptions), definitions, theorems, and proofs.

As you probably surmised from Exercises 1–13 above, we will use a different approach. The goal is for you to have a good intuitive understanding of the fundamental concepts of geometry, and for you to acquire the ability to think about geometric situations.

The basic concepts of geometry are point, line, plane, and space. We will not attempt to define these concepts. Rather, we assume an intuitive understanding of them and an agreement on what they are. This procedure is necessary for any investigation. If you try to define every concept in a system, you eventually must define one or more concepts in terms of another.

14. Look up *point* in a dictionary and write down a key word in the definition. Now look up that word and write down a key word in its definition. Continue until some word appears twice on your list.

■ By *space*, we mean the three-dimensional space of our everyday experience. Space is the set of all points. A *line* is a subset of space. We think of lines as being straight and extending infinitely in both directions.

Two *points* determine a line. If A and B are points, then the line determined by A and B is denoted $\overleftrightarrow{AB}$ (Figure 12-8). The arrows indicate that the line extends infinitely in both directions. We refer to the points between A and B together with the points A and B as the *line segment* $\overline{AB}$. Line segment $\overline{AB}$ is a subset of line $\overleftrightarrow{AB}$ (Figure 12-9). Sometimes we want to refer to half of a line. This is called a *ray*. The ray $\overrightarrow{AB}$ is a subset of the line $\overleftrightarrow{AB}$, consisting of A and all the points on the line on the same side of A as B (Figure 12-10). A *plane* is also a subset of space. We

think of a plane as being flat and extending infinitely in all directions.

In Exercise 13 we explored some properties of points, lines, and planes. Let us review them and make some additional comments.

1. If two different planes intersect, their intersection is a line (Figure 12-11). Of course, two planes may not intersect at all. Two planes that do not intersect are called *parallel* (Figure 12-12).

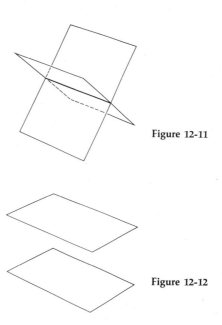

Figure 12-11

Figure 12-12

2. If a plane and a line not in the plane intersect, their intersection is a *point* (Figure 12-13). It is possible, of course, for a line not to intersect a plane at all (Figure 12-14). Further, if a line does lie in a plane, their intersection is the *line* (Figure 12-15).

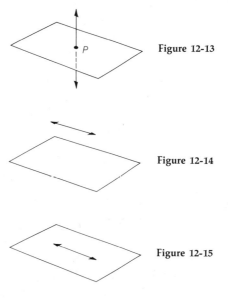

Figure 12-13

Figure 12-14

Figure 12-15

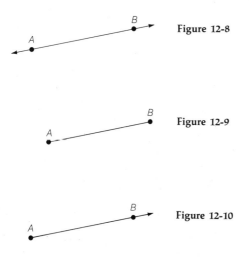

Figure 12-8

Figure 12-9

Figure 12-10

3. If two different lines intersect, their intersection is a point (Figure 12-16). There are two distinct possibilities for lines that do not intersect. The two lines may lie in the same plane (Figure 12-17). In this case they are said to be *parallel*. Or, the two lines may not lie in the same plane (Figure 12-18). In this case they are said to be *skew*.

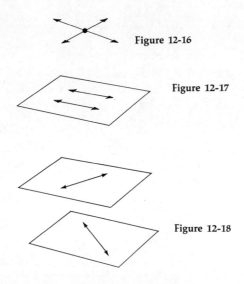

Figure 12-16

Figure 12-17

Figure 12-18

4. Given any two points in space, there is exactly one line containing both points.
5. Given any two points in space, there are many planes containing both points.
6. A line containing two points is contained in every plane that contains the two points.
7. Given three points in space not on the same line, there is exactly one plane containing all three points.
15. Show that it is possible for four points, no three of which lie on a line, to lie in one plane. Show that it is also possible to have four points that do not lie in the same plane. Draw pictures.

16. Find three points that do not determine exactly one plane.

Lab Exercises: Set 2

EQUIPMENT: At least one geoboard per group.

In these exercises you will learn about curves drawn in a plane. You will also learn about a special class of curves named *polygons*. The concept "congruent" is also introduced.

17. (a) Put your pencil on point A in the figure below, and move it along the paper without lifting it and without retracing, so that you end up at B. You have drawn a curve. Did anyone in the group draw the line segment $\overline{AB}$? (It is permitted that a curve be straight or contain portions that are straight and still be called a curve.)

   B
   •

   A
   •

   (b) Draw a curve that starts at P and ends at P, without any retracing. This is called a *closed curve*.

   P
   •

   (c) Indicate which of the following are closed curves. Remember that a closed curve must start and end at the same point and retracing is not permitted, but the curve may intersect itself.

   (i) _____   (iv) _____

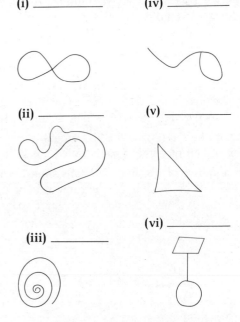

   (ii) _____   (v) _____

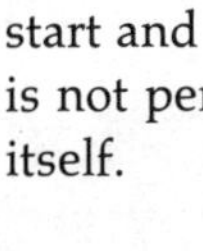

   (iii) _____

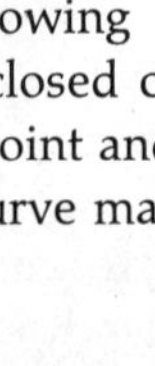

   (d) In (b) above you may have drawn a *simple closed curve*. If a closed curve does not intersect itself, then it is a simple closed curve. Which of the closed curves in (c) above are simple closed curves? _____

Definition: *A region is a set of points in a plane such that any two points in the set can be connected by a curve in the set.*

12/Lab Exercises: Set 2

18. **(a)** In the space below draw a closed curve that separates the plane of the paper into two regions. Everyone in the group compare their drawings. Discuss whether such a curve must be a simple closed curve.

Figure 12-20

(b) How many nonintersecting simple closed curves must be drawn to divide a plane into

3 regions? _____ 4 regions? _____

10 regions? _____ n regions? _____

19. A *sphere* is a surface in space consisting of the set of points that are all equidistant from a given point. Discuss Exercise 18(b) for the surface of a sphere and write the answers below.

3 regions _____ 4 regions _____

10 regions _____ n regions _____

20. Consider the surface of a torus. A *torus* is a surface that has the shape of an inner tube or a doughnut (Figure 12-19).

Figure 12-19

Discuss the following questions. Does every circle on a torus separate the surface into two regions? _____ How many nonintersecting simple closed curves must you draw to be *sure* that the surface is divided into 2 regions? _____ 3 regions? _____ n regions? _____ (Be careful here! It might help to look at an inner tube or doughnut. We are asking how many curves you need in order to be *sure* that the surface is divided into 2, 3, or n regions.)

21. **(a)** Draw a complicated simple closed curve in the space below Figure 12-20 and choose a point that may or may not be inside the curve. For example, see Figure 12-20. The problem is to decide whether point A is inside or outside the curve. Try to stump your group.

(b) Now mark some points that are clearly outside the curve, for example, points X and Y in Figure 12-21.

Figure 12-21

Draw the straight line segments $\overline{AX}$ and $\overline{AY}$. How many times does $\overline{AX}$ cross the curve? _____ How many times does $\overline{AY}$ cross the curve? _____ Answer the same questions for $\overline{BX}$ and $\overline{BY}$. _____, _____ Do the same for the curves you drew in (a). Do you notice anything? Try to formulate your conclusion before reading part (c) below.

(c) Complete the following sentence: If a straight line segment between a point outside the curve and another point crosses the curve an even number of times, then the other point is

_____ the curve; if it crosses an odd

number of times, the point is _____ the curve.

Figure 12-22

(d) Decide whether point X in Figure 12-22 is inside or outside the curve. Select some other points and determine whether they are inside or outside the curve.

Definition: *A polygon is a simple closed curve that is the union of three or more line segments.*

22. Use rubber bands to represent the following shapes on the geoboard.
    (a) A polygon with three sides (triangle).
    (b) A polygon with four sides (quadrilateral).
    (c) Another quadrilateral that looks different from the one above.
    (d) A polygon with five sides (pentagon).
    (e) A polygon with six sides (hexagon).
    (f) A polygon with seven sides (septagon).

23. *Game: Changing Shapes.* This game starts with a polygon on the geoboard. A move consists of moving one corner of the figure to obtain a new figure. For example, start with the shape in Figure 12-23. Find a sequence of moves that will make each of the following shapes in turn. Use only one move to go from one shape to the next.
    (a) *PQRS* as shown.
    (b) A triangle.

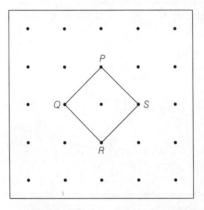

Figure 12-23

(c) A trapezoid (four sides with two sides parallel).
(d) A trapezium (four sides, no sides parallel).
(e) Another trapezium.
(f) A rectangle.
(g) A pentagon.
(h) A hexagon.

Can you find a way of doing it with shape (e) eliminated, i.e., going directly from a trapezium to a rectangle?

Note: You may have noticed by now that there are two interpretations of the term "move." Suppose we have the shape shown in Figure 12-24. Moving corner C to pin P results in either of two

12/Lab Exercises: Set 2

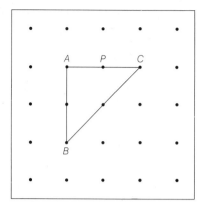

Figure 12-24

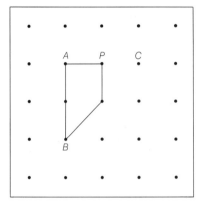

(a)

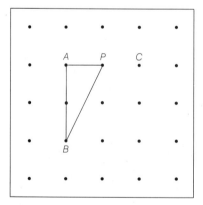

(b)

Figure 12-25

possibilities: Figure 12-25a or b. Do both on your geoboard. For simplicity, we allow both possibilities as moves in this game.

24. *Optional.* Everyone in the group make up a "Changing Shapes" puzzle for the other members of the group to solve. Some additional shapes you might use in your puzzles are an isosceles triangle, a square, a parallelogram, a rhombus (four sides of equal length with opposite sides parallel), a quadrilateral, a rectangle, and an octagon.

Figure 12-26

Definition: *A polygon is* convex *if, whenever A and B are two points on the polygon, then all the points on the line segment AB lie on the polygon or in the interior of the polygon.*

For example, the polygon in Figure 12-27 is convex, but that in Figure 12-28 is not.

Figure 12-27

Figure 12-28

25. **(a)** Use the bands on the geoboard to make a convex quadrilateral, a nonconvex quadrilateral, a convex pentagon, a nonconvex pentagon, a convex hexagon, a nonconvex hexagon, a convex septagon, and a nonconvex septagon.

    Draw your example of a convex septagon and a nonconvex quadrilateral on the diagrams below and on the following page.

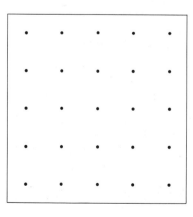

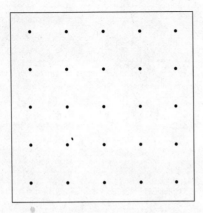

(b) What is the greatest number of sides possible for a convex polygon made with one band on the geoboard? _____ Draw a convex polygon with that number of sides.

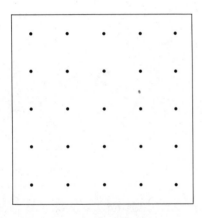

Definition: *Two figures (line segments, curves, polygons, etc.) are* congruent *if one could be moved so as to exactly coincide with the other. We shall use the word "congruent" rather than "equal," reserving "equal" to mean that the two figures are in fact identical (are the same set of points).*

26. (a) Form two line segments on the geoboard that are congruent but not equal.

    (b) Form two triangles on the geoboard that are congruent but not equal.

    (c) How many noncongruent line segments can you make on the 5-pin × 5-pin geoboard?

    _____ Draw your results on the diagram in the next column. What method did you use to find them all?

27. *Optional.* In Exercise 26 you determined the number of noncongruent line segments that can be constructed on the 5-pin × 5-pin geoboard. Do the same for 4-pin × 4-pin, 3-pin × 3-pin, etc., and fill in the table in the next column.

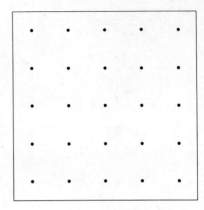

Geoboard	Number of different line segments
EXAMPLE: 1-pin × 1-pin	0
(a) 2-pin × 2-pin	
(b) 3-pin × 3-pin	
(c) 4-pin × 4-pin	
(d) 5-pin × 5-pin	

Comment: Curves

"Nothing puzzles me more than time and space; and yet nothing troubles me less, as I never think about them."—Charles Lamb

If you place the tip of a pencil at a point on a plane and move it without lifting the tip from the plane, you will have traced a *plane curve*. Another way of thinking of a curve is as the result of continuously deforming a line segment. Thus the segment $\overline{AB}$ in Figure 12-29 could be deformed into each of the parts in Figure 12-30. However, the segment could not be deformed into either Figure 12-31a or 12-31b. Similarly, neither of these could be drawn without lifting the pencil from the paper.

Figure 12-29

Note that according to the above description a curve may be straight or contain straight-line segments. A *closed curve* is a curve that can be drawn without retracing so that the tip of the pencil starts and ends

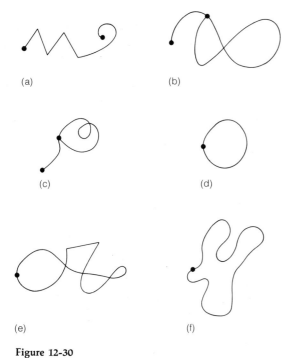

(a) (b) (c) (d) (e) (f)

Figure 12-30

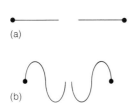

(a)

(b)

Figure 12-31

at the same point. Equivalently, a curve is closed if after deforming the line segment $\overline{AB}$, the points A and B coincide.

28. Which of the curves in Figure 12-30 are closed curves? _____

Definition: *A simple closed curve is a closed curve that does not intersect itself.*

29. Which of the curves in Figure 12-30 are simple closed curves? _____

30. Construct some curves on the geoboard that are not closed. Construct some simple closed curves. Construct a curve that is closed but not simple.

■ A simple closed curve separates the plane in which it is drawn into two regions, the *interior* of the curve and the *exterior* of the curve. Observe that the interior of the curve is a *connected* region, in the sense that, if A and B are any two points of the region, there is a curve joining A and B that lies completely in the region. For example, see Figure 12-32. Similarly, the exterior of a simple closed curve is also connected (Figure 12-33).

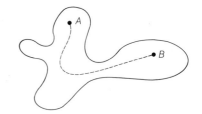

Figure 12-32

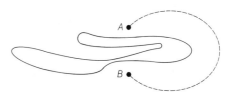

Figure 12-33

If a curve joins a point in the interior of a simple closed curve to a point in the exterior, the joining curve must always intersect the given simple closed curve (Figure 12-34). This describes the situation on plane surfaces. As you noticed in Exercise 4, this statement is not valid for all surfaces. On the torus, you can draw a simple closed curve that does not separate the surface into two regions. If we draw the simple closed curve in Figure 12-35, any two points not on the curve can be joined by a curve that does not intersect the simple closed curve.

Figure 12-34

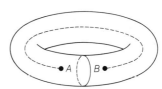

Figure 12-35

Polygons are one of the most familiar types of simple closed curves. Review the definition preceding Exercise 22.

31. Indicate which of the curves below are polygons and which of the polygons are convex. (We shall discuss polygons in greater detail in Chapter 17.)

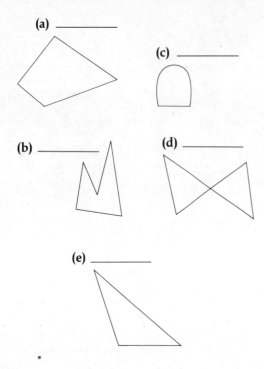

Figure 12-36

$\overrightarrow{CB}$. Which pairs of rays are equal; that is, which pairs of rays represent the same set of points?

2. Indicate the rays $\overrightarrow{AB}$, $\overrightarrow{BC}$, $\overrightarrow{CA}$ below.

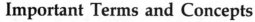

Important Terms and Concepts

Surface	Closed curve
Plane	Simple closed curve
Line	Region
Point	Connected
Line segment	Polygon
Ray	Convex polygon
Parallel lines	Torus
Skew lines	Congruent
Curve	

Review Exercises

1. Consider the line below with the three points A, B, C. Indicate the rays $\overrightarrow{AB}$, $\overrightarrow{BA}$, $\overrightarrow{AC}$, $\overrightarrow{CA}$, $\overrightarrow{BC}$,

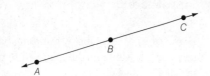

3. Next to each phrase on the following page put the letter of the phrase below that matches it (the phrases can be used more than once).

   (a) Parallel
   (b) A line
   (c) Three points that are not on the same line
   (d) Four points (no three of which are on the same line)
   (e) Two points
   (f) One point
   (g) Angle
   (h) Skew
   (i) Polygon
   (j) Perpendicular
   (k) Infinite

12/Review Exercises

_____ A simple closed curve made up of the union of three or more line segments.

_____ The intersection of two different lines in a plane.

_____ The intersection of two different planes.

_____ That which determines a plane.

_____ That which determines a line.

_____ The intersection of a plane and a line not in the plane.

_____ The union of two rays with the same end point.

_____ Two nonintersecting lines in a given plane.

_____ Two nonintersecting lines that are not in in the same plane.

_____ The number of planes that contain a given line.

4. Indicate which phrase best describes each picture below.

(a) Curve   (c) Simple closed curve
(b) Closed curve   (d) None of these

(i) _____

(iii) _____

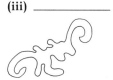

(ii) _____

(iv) _____

(v) _____

By permission of Johnny Hart and Field Enterprises, Inc.

Figure 12-37

5. How many nonintersecting simple closed curves must be drawn to divide a plane into seven regions? ____

   (a) 6   (b) 5   (c) 4   (d) None of these

6. Four points can determine at most ____ lines.

   (a) 6   (b) 3   (c) 4   (d) None of these

7. You need at least ____ planes to enclose space.

   (a) 3   (b) 4   (c) 5   (d) None of these

8. In a certain fairy tale, a princess P, a dragon D, and a knight K are in the maze determined by the closed curve in Figure 12-37. The castle is at C. The fairy tale ends "And so the princess killed the dragon and went to live in the castle. The knight, unable to join the princess, went off to study mathematics at the University." Based on the curve in Figure 12-37, which parts (if any) of the ending are possible and which are not? Justify your answer.

9. If a line connecting points A and B crosses a simple closed curve an even number of times, then _____.

   (a) One of A or B is inside and one is outside the simple closed curve.
   (b) Both A and B are outside.
   (c) Both A and B are inside.
   (d) Both A and B are either inside or outside.

■ In Exercises 10–14, decide whether the statement is true or false. Write T or F in the blank.

____ 10. All three-sided figures are convex.

____ 11. All four-sided figures are convex.

____ 12. If two planes have three points in common, the planes must be identical.

____ 13. Suppose A is a point not on line $\overleftrightarrow{CD}$. There is only one plane that contains both the point A and the line $\overleftrightarrow{CD}$.

____ 14. If a triangle has two sides congruent to two sides of another triangle, then the triangles are congruent.

13

Lines, Angles, and Triangles

"[T]he universe cannot be read until we have learned the language. It is written in mathematical language, and its characteristics are triangles, circles, and other geometric figures, without which it is humanly impossible to comprehend a single word." —Galileo Galilei

Lab Exercises: Set 1

EQUIPMENT: One Mira* per student or pair of students; one compass per group.

In these exercises you will learn how to do some basic geometric constructions using the Mira.

Figure 13-1

---

*Information about this equipment may be found on page xviii.

To use the Mira it should be held so that you can read the words at the top corners. Then the recessed edge will be facing you. The drawing edge is recessed in order to compensate for the thickness of the Mira. Be sure to keep the page flat when you use the Mira.

1. Place the edge of the Mira on the line $\overleftrightarrow{m}$ below and read the hidden message. What you were reading (the reflection of the words) is called the *image* of the words.

$$\xleftarrow{\hspace{6cm}} m \rightarrow$$

Hello! What you are reading is the reflection of the printing below.

2. Place your Mira so that the image of the circle below lies inside the square. The *image* of any point P is the point obtained by reflecting P in the Mira.

3. Place your Mira so that the image of the point A coincides with the point A' (read A-prime). Without moving the Mira, indicate the image of the points B, C, D with the letters B', C', D'. Draw a line along the edge of the Mira. (Be sure you are drawing along the recessed edge.)

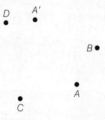

The line is called the *Mira line*. It should be clear that, given a large enough Mira, one could find an image for every point on the plane. We say that *A' is the image of A through the Mira line*. If Q is a point on the Mira line, where is the image of Q? _____

■ Go over each of the exercises that follow with the group before proceeding to the next exercise.

4. Place the Mira on the line $\overleftrightarrow{m}$ below, and find the images of the points A, B. Draw the image of the line segment $\overline{AB}$.

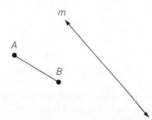

5. Place the Mira on the line $\overleftrightarrow{m}$ below. Choose two points A and B and find their images A' and B'.

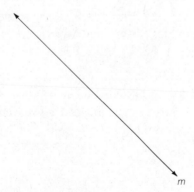

Discuss the following questions in your group.

(a) How does the distance between A and B compare with the distance between A' and B'? Use a ruler, a compass, or two marks on the edge of a piece of paper to measure the distances.

(b) Would your answer hold if A and B were on different sides of the line $\overleftrightarrow{m}$? _____ Try it and see.

(c) What if one of A and B lies on $\overleftrightarrow{m}$? _____

■ We say that reflection through the Mira is *distance-preserving* because it does not change *distances*. Another way of saying the same thing is to say that a line segment and its image under reflection are *congruent*.

6. Find a point P on the segment $\overline{AB}$ below so that $\overline{AP}$ and $\overline{BP}$ are congruent.*

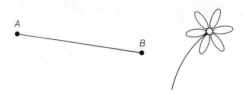

We say that P is the *midpoint* of $\overline{AB}$ or that P *bisects* $\overline{AB}$.

■ In a geometric construction it is often convenient to indicate congruent segments with one, two, three, etc., dashes, as in Figure 13-2a or 13-2b. This will be particularly helpful when you refer to the construction at a later time. Indicate the congruent segments in Exercise 6.

(a)   (b)

Figure 13-2

Definition: *An angle is the union of two rays that have the same endpoint, but do not lie on a straight line.*

For example, see Figure 13-3. We will often denote this angle by $\angle AXB$ or $\angle X$.

Figure 13-3

---

*A flower will be used to indicate those constructions that will be referred to later.

13/Lab Exercises: Set 1

Definition: *A ray from X is said to* **bisect** *∠AXB if it cuts the angle exactly in half, resulting in two angles that are congruent, i.e., one could fit exactly on top of the other (Figure 13-4).*

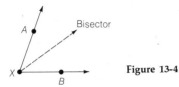

Figure 13-4

7. Use the Mira to find a ray that bisects the angle below.

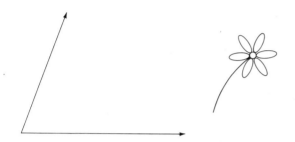

■ When two angles in a construction are congruent, we use the notation shown in Figure 13-5a or 13-5b.

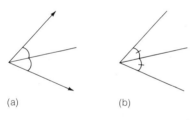

Figure 13-5

8. Indicate the congruent angles in the diagram of Exercise 7.

9. Find the image of A through the Mira line m below, and indicate it by A'.

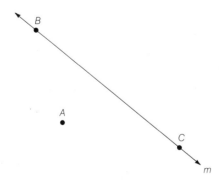

(a) Draw the line segment $\overline{AA'}$ and let $P = \overline{AA'} \cap m$. How does the distance between A and P compare with the distance between A' and P?

(b) What do you notice about the angles ∠APB and ∠APC? _____

(c) Place your Mira on $\overline{AA'}$ and verify that the angles ∠APB and ∠APC are the same size and shape.

10. In the diagram below use the Mira to find point D on line $\overleftrightarrow{AB}$ so that ∠CDA is congruent to ∠CDB. Draw the line $\overleftrightarrow{CD}$. You have constructed the *perpendicular* to $\overleftrightarrow{AB}$ through C.

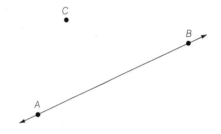

Definition: *We say that two intersecting lines are* **perpendicular** *if the adjacent angles formed at their point of intersection are congruent.*

We indicate that lines are perpendicular with a ⌐ or a ⌐, as shown in Figure 13-6a or 13-6b.

Figure 13-6

11. (a) Use the Mira to construct a perpendicular to the line $\overleftrightarrow{AB}$ at P in the drawing below. Discuss this procedure in the group.

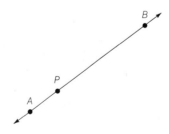

(b) Find the midpoint of $\overline{AB}$ above and label it X. Construct a perpendicular to $\overleftrightarrow{AB}$ at X. This line is called the *perpendicular bisector* of the segment $\overline{AB}$.

■ Recall that two lines in the same plane are said to be *parallel* if they do not intersect.

12. Use the Mira to construct a line through point P parallel to line $\overleftrightarrow{l}$ in the drawing below. (*Hint:* First construct a perpendicular from P to $\overleftrightarrow{l}$. Make sure the perpendicular extends past P.) Indicate any perpendicular lines on the diagram. Discuss in your group and write down in words what you did. What have we assumed in this construction?

(c) Use the Mira to construct a perpendicular to $\overleftrightarrow{l}$ to $\overline{A_5B}$.

(d) Now construct perpendiculars to $\overleftrightarrow{l}$ through $A_4, A_3, A_2,$ and A_1.

(e) What do you know about these lines? _____

(f) Label the points where these perpendiculars intersect $\overline{AB}$ X_4, X_3, X_2, X_1. The line segments $\overline{AX_1}, \overline{X_1X_2}, \overline{X_2X_3}, \overline{X_3X_4},$ and $\overline{X_4B}$ are all _____.

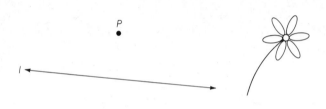

Comment: Angles

"[T]he investigation of mathematical truths accustoms the mind to method and correctness in reasoning, and is an employment peculiarly worthy of rational beings. In a cloudy state of existence, where so many things appear precarious to the bewildered research, it is here that the rational faculties find a firm foundation to rest upon. From the high ground of mathematical and philosophical demonstration, we are insensibly led to far nobler speculations and sublime meditations."—George Washington

A simple and important geometric figure is the angle.

Definition: *An angle is the union of two rays that have the same endpoint but that do not form a line.*

If $\overrightarrow{AB}$ is one ray and $\overrightarrow{BC}$ is another, then the angle formed by their union is written $\angle ABC$. The common endpoint of the two rays is called the *vertex*. For example, see Figure 13-7. The vertex of $\angle ABC$ is B, and the vertex of $\angle XYZ$ is Y.

Fact: *Lines perpendicular to a given line are parallel.*

13. We can use the above fact to divide a line segment into any given number of congruent pieces. Suppose $\overline{AB}$ shown below is to be divided into 5 congruent segments. We draw a ray through A, as shown.

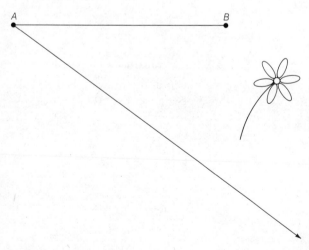

(a) Use a compass or a marked piece of paper to measure off 5 congruent segments of any length on the lower ray. Label the points obtained A_1, A_2, A_3, A_4, A_5.

(b) Draw $\overline{A_5B}$.

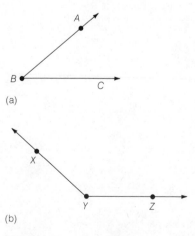

Figure 13-7

Two angles are congruent if one can be moved so as to exactly fit on top of the other. Thus, ∠PQR in Figure 13-8 is congruent to ∠XYZ in Figure 13-7.

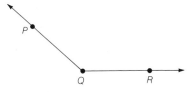

Figure 13-8

Recall from Exercise 10 that two lines are said to be perpendicular if the adjacent angles formed by the intersections are congruent. These angles are said to be *right angles*.

In order to measure the size of angles, a unit must be chosen. Just as with length, the choice of unit is arbitrary. Any angle may be chosen as the unit, and all other angles are then compared to this unit. The most commonly used unit is called a *degree*. The choice of this unit originates with the Babylonians. An angle of one degree is defined to be 1/90 of a right angle. This means that 360 angles of one degree comprise a complete rotation of the plane, or circle. The Babylonian choice of the degree as the standard unit for measuring angles was undoubtedly influenced by their calendar, which had 360 days to a year, and by their use of a Base 60 numeration system.

Measuring angles is best done with the use of a protractor. The angle shown on the protractor in Figure 13-9 measures 35°.

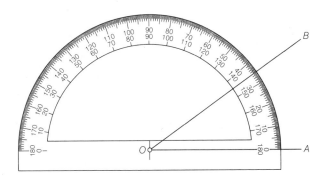

Figure 13-9

14. Use a protractor to measure the angles in (a) and (b).

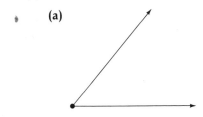

(a)

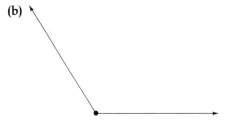

(b)

15. (a) Use a protractor to draw a triangle in which one angle measures 40° and another 36°.

Measure the third angle. ____
(b) Repeat (a) with 108° and 35°.

(c) Calculate the sum of the angles for the triangle in (a) and in (b).

16. Without a protractor, try to draw angles of 30°, 45°, 60°. Check your results by measuring them with a protractor.

Definition: *Angles that measure less than a right angle, i.e., angles whose measure is less than 90°, are called* acute *angles. Angles whose measure is greater than 90° are called* obtuse *angles.*

Lab Exercises: Set 2

EQUIPMENT: One Mira for each student or each pair of students; one compass, two 5 × 8 file cards (or the equivalent amount of cardboard), scissors, one washer or other weight, and 20 centimeters of string or thread for each group.

In these exercises you will investigate the side-angle-side congruence axiom using the Mira. You will then use this axiom to prove a famous theorem about isosceles triangles. The study of triangles continues with a study of the *centroid* and the *orthocenter* of a triangle.

17. Use the Mira to construct an angle at A' congruent to $\angle A$ and one at X' congruent to $\angle X$.

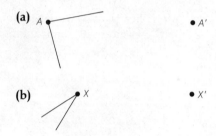

18. (a) Use the procedure in Exercise 17 to construct an angle at B' congruent to $\angle ABC$. (Do not draw a triangle yet—just draw the angle.)

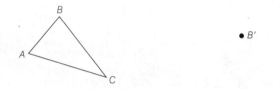

Now, on one ray of $\angle B'$, use a compass to mark off segment $B'C'$ congruent to BC. On the other ray mark off $B'A'$ congruent to BA. Draw the triangle $A'B'C'$. What do you notice?

(b) Draw the angle as above, but this time use a compass to mark off $B'C'$ and $B'A'$ on different rays from those on which they were drawn in part (a).

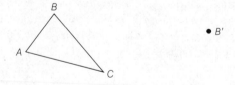

(c) Do the two triangles still have the same size and shape? _____

Definition: *Two triangles are said to be* congruent *if they have the same size and the same shape.*

19. (a) Suppose you are given an angle and two line segments and want to construct a triangle with two sides congruent to the given line segments, and the angle *between* the two sides congruent to the given angle. Construct such a triangle.

(b) How many noncongruent triangles can you draw satisfying those conditions? _____

■ After doing Exercises 18 and 19, you should agree that once two sides and the included angle of one triangle are drawn congruent to two sides and the included angle of another triangle, then the two triangles are congruent. We state this as an *axiom*.

Side-angle-side congruence axiom: *If two sides and the included angle of one triangle are congruent respectively to the corresponding parts of another triangle, then the two triangles are congruent.*

20. (a) Look at the line segment $\overline{AB}$ below. Which of the points C, D, E, F, G, H are equidistant (i.e., the same distance) from A and B? _____
_____.

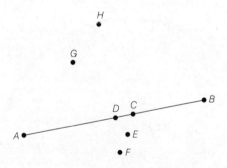

(b) Find two other points that are equidistant from A and B. What do you notice about all the points that are equidistant from A and B? _____

(c) Draw the perpendicular bisector of $\overline{AB}$ above. Complete the following: If a point P is equidistant from A and B, then it lies on the _____ $\overline{AB}$.

21. (a) With the Mira, construct the perpendicular bisector of segment $\overline{AB}$ on the following page.

(b) Choose a point C on the perpendicular bisector and draw $\overline{AC}$ and $\overline{BC}$. Verify with the Mira that $\overline{AC}$ and $\overline{BC}$ are congruent.

(c) Complete the following: If point P lies on the perpendicular bisector of $\overline{AB}$, then it is _____ from A and B.

Definition: *A triangle with two congruent sides is called an* isosceles triangle. *The nonequal side is called the* base.

(d) The triangle $\triangle ABC$ that you drew in part (b) is an isosceles triangle. Draw two more isosceles triangles with base $\overline{AB}$.

22. Use the Mira to construct an equilateral triangle with one side $\overline{PQ}$.

23. (a) In the diagram below, P is chosen to be a point on one ray of the angle O. Use the Mira to find Q on the other ray so that OP and OQ are congruent. Now construct $\triangle OPQ$. It is an isosceles triangle.

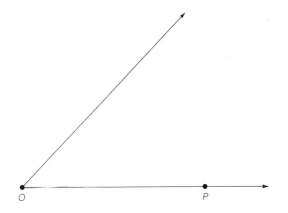

(b) Use the Mira to draw the line that bisects $\angle POQ$. Call X the point of intersection of this line and PQ. Then $\angle XOQ$ and $\angle XOP$ are congruent. Why? _____

(c) We can conclude that $\triangle XOQ$ is congruent to $\triangle XOP$. Why? _____

Now, if $\triangle XOQ$ and $\triangle XOP$ are congruent, then $\angle P$ is congruent to $\angle Q$. We have shown the following theorem.

Theorem: *The base angles of an isosceles triangle are congruent.*

24. Draw a triangle on a piece of paper and cut it out. Cut off the angles and put them together so that the three vertices are at a point.

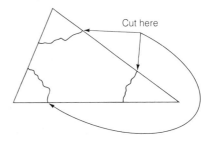

What do you notice? _____

Let us verify that this fact is true for every triangle without having to cut up the triangle.

25. Consider the triangle ABC below.

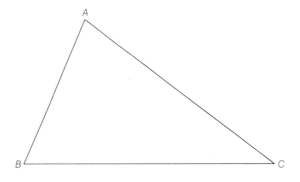

(a) With the Mira, find the midpoint M of AB and the midpoint N of AC.

(b) Place the Mira so that the Mira line would be MN. Where does the image of A lie? _____

(c) Let A' be the image of A. then MA' is congruent to MA, and NA' is congruent to NA.

Why? (*Hint:* See Exercise 5.) _____

_____

(d) Can you conclude that △*MBA'* and △*NCA'* are both isosceles triangles? _____ Why?

_____

(e) From Exercise 23, what can you conclude about ∠*MA'B* and ∠*B*? _____

About ∠*NA'C* and ∠*C*? _____
This shows that if the three angles ∠*A*, ∠*B*, ∠*C* were placed adjacent to one another so that *A*, *B*, *C* were at the same point, the two outer rays would lie on a straight line. Sometimes a straight line is referred to as a "straight angle" (although to be precise it is not even an angle). The above result is often stated as:

The sum of the angles of the triangle is a straight angle.

Since a "straight angle" would measure 180°, the above could be stated:

The sum of the angles of a triangle is 180°.

26. Draw two triangles on a file card or a piece of cardboard. Draw one triangle with all acute angles and one triangle with an obtuse angle. (The exercise works best if the triangles are large.) Cut out the two triangular regions and do the following with each:
    Tie a washer or some other weight to one end of a piece of thread and make a knot about one inch from the other end. Make a small hole near one vertex of the triangle, and put the thread through the hole so that the weighted string acts as a plumb line, as shown below. Mark on the triangular region the line determined by the plumb line. Repeat for the other two corners. What do you notice about the three lines? _____

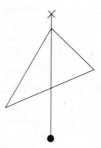

Verify that each plumb line bisects a side of the triangle. Try to balance the triangular region on a pencil point. Where is the balance point?

_____

■ The point you have found is called the *centroid* (or center of gravity) of the triangle. The centroid can be defined as follows.

Definition: *A* median *is a line connecting a vertex of a triangle with the midpoint of the opposite side. The three medians intersect in a single point called the* centroid.

27. (a) Use the Mira to find the centroid of this triangle.

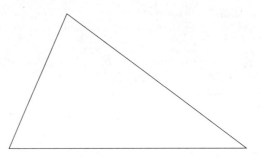

(b) Can you explain why a triangular region balances at the centroid of the triangle? _____

_____

(c) Can you draw a triangle where the centroid is on the edge of the triangle? _____

28. (a) Use the Mira with the triangle below to construct a line from *A* perpendicular to the opposite side. This line is an *altitude line* for the triangle. Draw the altitude lines from the other two vertices. What do you notice?

_____

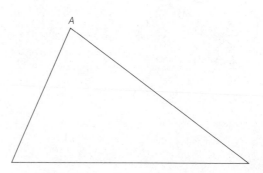

(b) Use the Mira with the triangle on the following page to construct an altitude line from *A*. Notice that you first have to extend the opposite side. Why? _____

Draw the other altitude lines. What do you notice? _____

Definition: *The intersection of the three altitude lines is called the* orthocenter *of the triangle.*

29. What other words do you know that use the prefix *ortho*? If you cannot think of any, consult a dictionary. What do these words have in common?

_____

30. (a) Can you find a triangle with the orthocenter on a side?

Describe the triangle. _____

(b) If the orthocenter is outside of the triangle, what do you know about one of the angles?

_____

Definition: *An altitude of a triangle is the line segment determined by a vertex and the point of intersection of the altitude line with the line determined by the opposite side.*

31. (a) Explain the difference between an altitude and an altitude line.

(b) Indicate the altitudes for the two triangles in Exercise 28. Draw a triangle in the space provided below that has two altitudes that are the same length. Draw a triangle that has three altitudes that are the same length. Describe the triangles.

Comment: Triangles—Congruence Axioms, Centroid, Orthocenter

"A triangle is the most economical and natural path a structure can take. Triangle is structure and structure is triangle. Period. It's what the whole Universe is based on."—Buckminster Fuller

The triangle is undoubtedly the most studied of geometric figures. There is evidence that the Egyptians were aware of some of its properties, and as the above quotation from Buckminster Fuller indicates, the triangle is still attracting attention today (see Figure 13-10).

The triangle is the simplest of polygons because it is determined by three points that are not on a straight line. Perhaps it is surprising that so much can be said about so simple a figure.

The first concept you investigated in this chapter was congruence. Two triangles are congruent if they have the same size and the same shape, i.e., if one could be moved so as to fit exactly on the other. This means that two triangles are congruent when their three sides and three angles are congruent. But as you discovered in the lab exercises, you can conclude that two triangles are congruent from much less information. You discovered in Exercise 19 that if two sides and the included angle of a triangle are specified, then the triangle is determined. This is an example of a case in which partial information about a situation leads to a much more specific conclusion. However, since our conclusion was based on physical experience and could not be proven from previous assumptions about points, lines, and planes, we say it is an *assumption* or an *axiom*.

Side-angle-side congruence axiom: *If two sides and the included angle of one triangle are congruent respectively to the corresponding parts of another triangle, then the two triangles are congruent.*

Let us investigate some of the other assumptions about two triangles being congruent.

Figure 13-10 Geodesic dome at Expo 67. Courtesy Buckminster Fuller Archives.

32. In (a) and (b) below, use a compass to construct a triangle with sides congruent to the indicated segments. Can you construct noncongruent triangles when the three sides are specified? _____

    (a) _____
    _____
    _____

    (b) _____
    _____
    _____

■ Exercise 32 should have convinced you that the size and shape of a triangle is completely determined by its three sides. We state this as an axiom.

Side-side-side congruence axiom: *If three sides of one triangle are congruent to three sides of another triangle, then the triangles are congruent.*

Let us now investigate whether there is an angle-side-angle axiom.

33. In (a) and (b) below, construct a triangle that has two angles and the included side congruent to the two angles and line segment indicated. How many noncongruent triangles can you construct when two angles and the included side are specified?

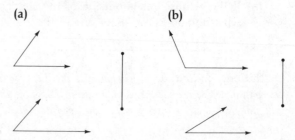

■ Exercise 33 should have convinced you of the validity of the following axiom.

Angle-side-angle congruence axiom: *If two triangles and the included side of one triangle are congruent respectively to the corresponding parts of another triangle, then the two triangles are congruent.*

34. Show that there is no side-side-angle congruence axiom by drawing two triangles that have two sides and the nonincluded angle congruent, but are drawn in such a way that the two triangles are not congruent.

■ The congruence axioms are very useful when we wish to prove geometrical theorems. For example, we were able to prove that the base angles of an isosceles triangle were congruent using the side-angle-side congruence axiom. Using this result and the Mira, we were then able to show in Exercise 25 that the sum of the angles of a triangle is 180° (a straight angle).

Exercises 25–29 were concerned with determining special points that are associated with a triangle. Let us summarize the information.

A *median* of a triangle is a line that connects one vertex with the midpoint of the opposite side. Do you find it surprising that for any triangle the three medians are *concurrent* (i.e., they intersect in a single point)? This point is called the *centroid* of the triangle.

The centroid of a triangle is always in the interior of the triangle. It is the point where the triangular region would balance if you tried to balance it on the tip of a finger. The reason it balances at the centroid is that each median divides the triangular region into two smaller regions each of which has the same "weight." In fact, it can be shown (Review Exercise 10 in Chapter 14) that the two regions have the same area.

An *altitude line* of a triangle is a line through one vertex perpendicular to the line determined by the other two vertices. The three altitude lines are always concurrent. Their point of intersection is called the *orthocenter*. The orthocenter may lie in the interior, on a side, or in the exterior of the triangle.

So far we have investigated two special points, the centroid and the orthocenter, of a triangle. We will investigate two more in the following set of lab exercises.

Lab Exercises: Set 3

EQUIPMENT: One Mira and one compass for each group.

We continue our study of the triangle by learning about the *circumcenter* and the *incenter*, and a relationship between the four "centers."

35. Suppose you want to construct a circle that passes through all three vertices of the triangle below. Where do you think its center, C, would be? Using a compass, try to find the center of this circle by trial and error. Try to explain the process you used to the group before proceeding to the next exercise.

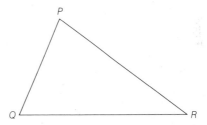

36. Did you find that you were trying to put the center C so that the lengths of $\overline{PC}$, $\overline{QC}$, and $\overline{RC}$ were equal, i.e., so that C was equidistant from each of the vertices? If you did, you were on the right track. Since a circle is the set of points all equidistant from a given point (called the center), the center of the desired circle must be equidistant from P, Q, and R.

(a) Use your Mira to find all points equidistant from P and Q in the triangle below. (If you do not see how to do this, review Exercise 20.)

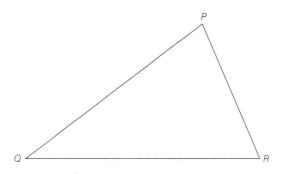

(b) What line did you construct? _____

(c) Now draw the line of points equidistant from Q and R.

(d) Draw the line of points equidistant from P and R.

(e) What do you notice? _____

37. In Exercise 36, you drew the perpendicular bisectors of each side of the triangle and discovered that they intersected in a point. Put the point of a compass on the point of intersection and the compass's pencil point on one of the vertices. Draw a circle. It should pass through the other two vertices. Explain why.

Definition: *The point of intersection of the perpendicular bisector of the sides of a triangle is called the* **circumcenter** *of the triangle.*

38. Why do you think it is given this name? _____

_____

What does the prefix *circum* mean? _____

39. (a) When would the circumcenter of a triangle lie outside the triangle? _____

(b) Draw a triangle below with the circumcenter on a side. What is special about this triangle?

_____

40. For each of the triangles below, construct the circumcenter and label it C. Construct the centroid and label it G (for gravity). Construct the orthocenter and label it O. What do you notice about O, C, and G?

(a)
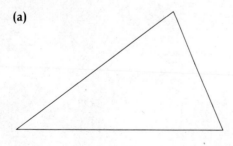

(b)
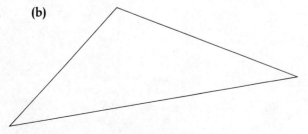

41. Draw a triangle where O, C, and G coincide. Describe the triangle.

42. In Exercise 40, you discovered that O, C, and G were *collinear* (i.e., lie on the same line) and that G was between O and C. Can you find a triangle in which G is not between O and C?

43. For the triangles of Exercise 40, how do the lengths of $\overline{CG}$ and $\overline{GO}$ compare? _____
Draw another triangle below and see if the same relationship holds.

44. Suppose you want to draw a circle inside the triangle below so that the circle touches each side at exactly one point (in the language of geometry, so that it is *tangent* to each side). Where do you think its center would be? Using a compass, try to find the center of this circle by trial and error. Try to explain your thinking to the group as you do this, before proceeding to the next exercise.

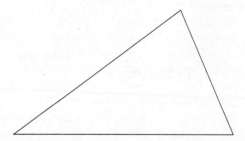

45. Were you trying to put the center of the circle so that it was the same distance from each of the sides? Since a circle is the set of all points equidistant from its center, if the circle is to touch each side at one point, then the perpendicular distance from the center to each side must be the same.

(a) In the triangle on the following page, use the

Mira to draw the line of points that are equidistant from $\overleftrightarrow{PQ}$ and $\overleftrightarrow{QR}$.

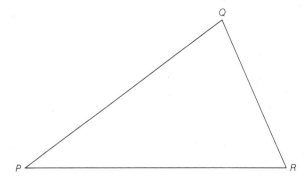

- (b) What line did you construct? _____
- (c) Draw the line of points equidistant from $\overleftrightarrow{PR}$ and $\overleftrightarrow{QR}$.
- (d) Draw the line of points equidistant from $\overleftrightarrow{QP}$ and $\overleftrightarrow{PR}$.
- (e) What do you notice? _____

46. You have found that the lines that bisect each angle intersect at a point. Put the point of a compass on this point. You should be able to draw a circle that touches each side at exactly one point.

Definition: *The* incenter *of a triangle is the intersection point of the angle bisectors.*

47. (a) How could you find the incenter of a triangle without a Mira and without a compass or straightedge? (*Hint:* It would help to cut out the triangular region.)

    (b) Can the incenter of a triangle lie outside the triangle? _____ On an edge? _____

48. When would the incenter *I* lie on the same line as *O*, *G*, and *C*? _____ When would *I*, *O*, *G*, and *C* all coincide? _____

Comment: Triangles—Circumcenter and Incenter

"The moving power of mathematical invention is not reasoning but imagination."—Augustus De Morgan

The perpendicular bisectors of each of the sides of a triangle are concurrent. Their intersection is called the *circumcenter* of the triangle. Since the circumcenter is on the perpendicular bisector of every side, and every point on a perpendicular bisector of a line segment is equidistant from the endpoints, the circumcenter must be equidistant from each vertex. Hence it is the center of a circle that will pass through each of the three vertices. This circle is called *the circle circumscribed about the triangle* (Figure 13-10).

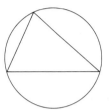

Figure 13-10

In any triangle, the centroid, orthocenter, and circumcenter are always collinear (i.e., they lie on the same line). The centroid is always between the orthocenter and the circumcenter, and the distance of the centroid to the orthocenter is twice the distance from the circumcenter.

The lines that bisect the three angles of a triangle are also concurrent. The point of intersection is called the *incenter* of the triangle. It is the center of a circle that just touches, or is tangent to, each side. This circle is said to be *inscribed* in the triangle (Figure 13-11).

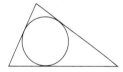

Figure 13-11

49. Consider the isosceles triangle below.

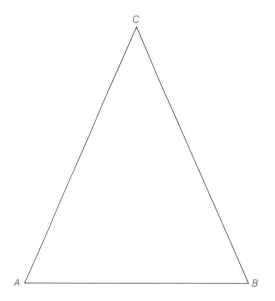

Use your Mira to construct each of the following.

(a) The median to side $\overline{AB}$
(b) The altitude to side $\overline{AB}$
(c) The perpendicular bisectors of $\overline{AB}$
(d) The bisector of angle C

What do you notice?

■ Notice that the centroid lies on the median to $\overline{AB}$, the orthocenter lies on the altitude to $\overline{AB}$, the circumcenter lies on the perpendicular bisector of $\overline{AB}$, and the incenter lies on the bisector of angle C. Since for an isosceles triangle ABC with base AB all of these lines are equal, the four points are all collinear. That is, for an isosceles triangle the incenter lies on the same line as the orthocenter, centroid, and circumcenter. If the triangle is equilateral, the four "centers" will coincide.

50. Construct an equilateral triangle and verify with your Mira that I, O, G, and C coincide.

Important Terms and Concepts

Image	Side-side-side axiom
Mira line	Angle-side-angle axiom
Mapping	Isosceles triangle
Midpoint	Equilateral triangle
Bisects	Median
Perpendicular	Centroid
Perpendicular bisector	Altitude line
Parallel lines	Altitude
Angle	Orthocenter
Right angle	Circumcenter
Degree	Incenter
Side-angle-side axiom	

Review Exercises

1. Use your Mira to find the perpendicular bisector to the line segment $\overline{AB}$.

2. Use the Mira to bisect $\angle A$.

3. Use the Mira to construct a perpendicular to $\overline{CD}$ at X. Describe the process in words.

4. Use the Mira to divide $\overline{AB}$ into three congruent line segments.

5. Use the Mira to find the line that bisects $\overline{AB}$ and that is parallel to BC.

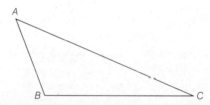

6. (a) Draw a circle inside the triangle below so the circle touches each side at only one point. What point must you find in order to use a compass to draw this circle? _____

(b) On the same triangle draw a circle that touches each vertex. What point must you find in order to draw this circle with a compass? _____

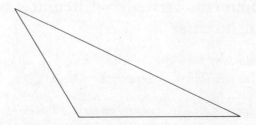

7. Draw a triangle with an obtuse angle and find O, C, and G (orthocenter, circumcenter, and centroid, respectively). Show that they lie on the same line (are collinear).

8. Suppose you are given isosceles △ABC with $\overline{AX}$ bisecting angle A. Explain why △ABX is congruent to △ACX.

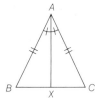

9. In triangle ABC below, M is the midpoint of $\overline{AB}$, N is the midpoint of $\overline{AC}$, and A' is the image of A in the Mira line $\overline{MN}$.

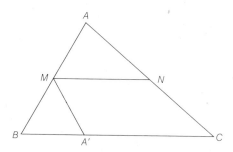

The best reason for concluding that $\overline{MA}$ is congruent to $\overline{MA'}$ is _____.

(a) Base angles of an isosceles triangle are equal.
(b) If the base angles of a triangle are equal, the opposite sides are equal.
(c) Reflection in a Mira line preserves distance.
(d) If two triangles are such that two sides and the included angle of one is congruent respectively to two sides and the included angle of another, then they are congruent.
(e) There is no reason, since $\overline{MA}$ is not congruent to $\overline{MA'}$.

10. How would you tell someone how to find the image of a point A through the line m if no Mira were available? (Assume the person can construct perpendiculars and measure distances.)

In Exercises 11–14, write the letter of the correct answer in the blank.

11. If the triangle ABC below has its orthocenter at point X, then the orthocenter of the triangle ABX will be _____.
    (a) Point X
    (b) Point A
    (c) Point B
    (d) Point C
    (e) Unknown from this information

12. Which of the following sets of points will *always* be inside the triangle? _____
    (a) Centroid, circumcenter, and incenter
    (b) Orthocenter and centroid
    (c) Centroid and incenter
    (d) Circumcenter, incenter, orthocenter, and centroid

13. For an obtuse triangle, the orthocenter is always _____ the triangle.
    (a) Outside
    (b) Inside
    (c) On

14. If a circle, when reflected in the Mira line m, coincides with its image, then _____
    (a) The circumference of the circle touches the Mira line m in only one place.
    (b) The center of the circle lies on the Mira line.
    (c) A mistake has been made.
    (d) The circle does not intersect the Mira line.

In Exercises 15–19, decide whether the statement is true or false. Write T or F in the blank.

_____ 15. Each point on the perpendicular bisector of a line segment is equidistant from the endpoints of the line segment.

_____ 16. The point on an angle bisector is equidistant from the sides of the angle it bisects.

_____ 17. If two lines are each perpendicular to a given line, then the two lines must be parallel.

_____ 18. All triangles must have at least two acute angles.

_____ 19. In an equilateral triangle, the centroid, orthocenter, circumcenter, and incenter are all the same point.

By permission of Johnny Hart and Field Enterprises, Inc.

> **Sophie Germain** (1776–1831) overcame strong parental opposition in order to study mathematics. Her work in number theory earned the praise of Gauss, the most eminent number theorist of all time. She also investigated the mathematical laws of elastic surfaces and won the grand prize of the French Academy of Sciences for her *Memoir on the Vibrations of Elastic Plates*.

14

Area and the Geoboard

"If you don't read poetry how the hell can you solve equations?"—Harvey Jackins

Lab Exercises: Set 1

EQUIPMENT: One geoboard for each pair of students.

In these exercises we will explore the concept of area. We will determine formulas for the area of familiar geometric figures and develop a method for finding the areas of regions on the geoboard.

Definition: *The area of a plane region is a number that measures the size of the region.*

Let us investigate the concept of area on the geoboard by assigning an area of 1 to the square region indicated in Figure 14-1.

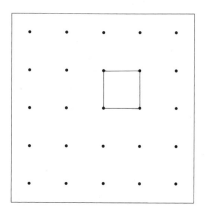

Figure 14-1

1. Determine the area of each of the following regions.

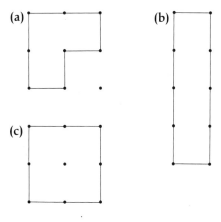

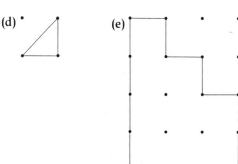

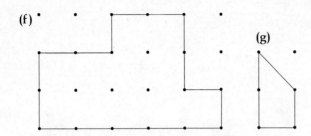

2. (a) Construct rectangles of areas 2, 3, 6, 8, and 12 on your geoboard.

   (b) Verify that the area of each rectangle is the product of the *base* and the *height*.

■ We use A to denote the area of a region and write

$$A \text{ (rectangle)} = \text{base} \times \text{height}$$
$$A = b \times h$$

3. On your geoboard construct the parallelogram shown by the solid line below. Discuss in your group how you would explain to a young person that the area of the parallelogram is the same as the area of the rectangle (dotted line). What is the area of this parallelogram? Construct a parallelogram of area 6 and verify its area by indicating a rectangle of equal area. (Be sure you construct the figures on the geoboard.)

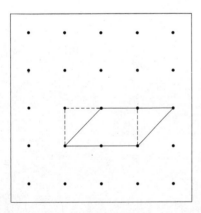

Are you convinced that the area of any parallelogram is the product of its base and height? If not, do some more examples. Be sure to note that the height of a parallelogram is the distance between the two sides (Figure 14-2).

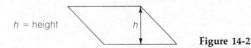

Figure 14-2

4. (a) Discuss in your group how to find the area of a triangle on the geoboard. For example, how would you determine the area of triangles (i), (ii), and (iii) below?

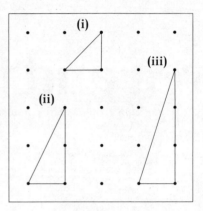

(b) How would you find the area of triangles (iv) and (v) below?

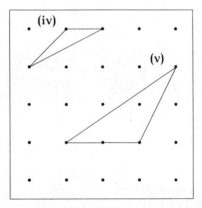

■ A simple method for finding the area of the above triangles is to notice that for each there is a parallelogram that has an area twice the area of the triangle. If you did not discover this method, do Exercise 4 again.

5. Construct another triangle on the geoboard, and have a partner determine the area using this method. Then reverse roles and repeat.

■ Since the area of a triangle is always $\frac{1}{2}$ the area of its associated parallelogram, we know that the area of a triangle is $\frac{1}{2}(b \times h)$, where b is the length of the base and h is the height of the triangle.

6. Construct triangles on the geoboard having areas 3, $4\frac{1}{2}$, 6, and 8.

7. Areas of more complicated regions can be calculated by partitioning the region into triangular and rectangular regions of known area and adding. Discuss this method in your group and use it to find the area of each of the following regions.

14/Lab Exercises: Set 1

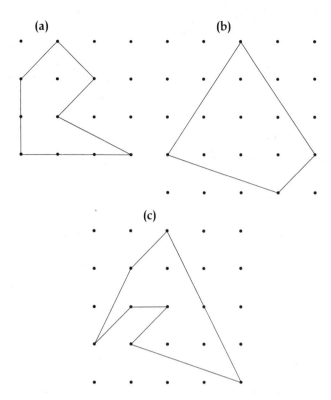

■ In (c) above, finding a workable partition is challenging. It is usually easier to enclose such a figure in a rectangle and subtract the appropriate numbers from the area of the rectangle. For example, to find the area of the region in Figure 14-3, enclose it in a rectangle (dotted line) and calculate $6 - 2\frac{1}{2} - 1 - 1 = 1\frac{1}{2}$, which is the area.

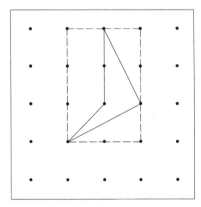

Figure 14-3

8. Find the area of the figures in Exercise 7 using this method.

9. Find the area of each of the following figures using this method. In (c) you will have to apply the method a second time in order to find the area of one of the figures you want to subtract.

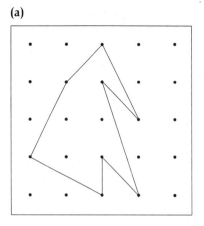

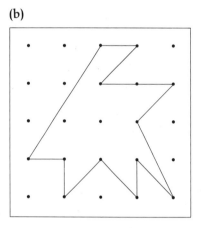

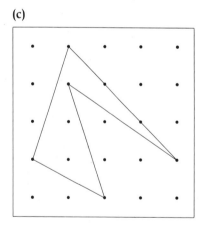

10. Will the method in Exercise 8 work for finding the area of any figure on the geoboard? _____ If it does, you can conclude that any figure on the geoboard made with one rubber band that does not cross itself has area equal to an integer or an integer plus $\frac{1}{2}$. Why is this statement true?

_____

_____

■ In Chapter 22 we will discover another method of finding the area of such regions, called *Pick's formula*,

172   14/Area and the Geoboard

which also verifies this statement. Let us now determine the area of a fairly common geometrical figure—the trapezoid.

Definition: *A trapezoid is a quadrilateral with two parallel sides.*

11. Which of the following are trapezoids? _____

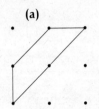
(a)

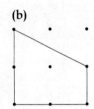
(b)

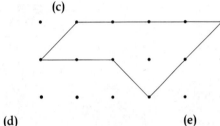
(c)

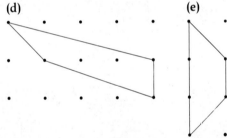
(d)   (e)

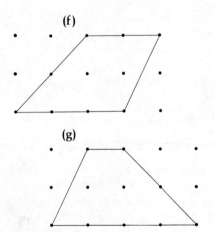
(f)

(g)

12. Find the area of each of the trapezoids in Exercise 11.

■ On your geoboard, construct trapezoid (g). Now construct a congruent trapezoid so that the two trapezoids together form a parallelogram. Your geoboard should look like Figure 14-4a or b.

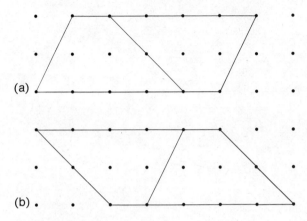
(a)

(b)

Figure 14-4

13. (a) What is the area of either of these parallelograms? _____ Hence the area of the trapezoid is _____.

(b) Find the area of trapezoid (d) of Exercise 11, using the above method. _____

■ Suppose now we take any trapezoid with height h and parallel sides of length c and d (Figure 14-5). We can make a parallelogram having twice the area of this trapezoid (Figure 14-6).

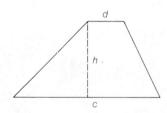

Figure 14-5

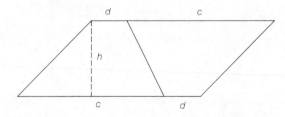

Figure 14-6

14. What is the area of this parallelogram? _____ Hence the area of the trapezoid is _____.

15. *Puzzle Problem—Optional.* Can you divide the non-hatched area on the geoboard on the following page into four parts, each having the same shape and the same area?

14/Comment

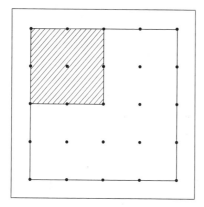

Comment: The Concept of Area and Its Properties

"[B]ut for harmony beautiful to contemplate, science would not be worth following."—Henri Poincaré

Area is a concept associated with regions in a plane. The area of a region measures the size or extent of the region.

16. Use the tangrams you made in Chapter 12 to show that regions (a), (b), and (c) in Figure 14-7 have the same area.

■ Exercise 16 illustrates a fundamental property of area:

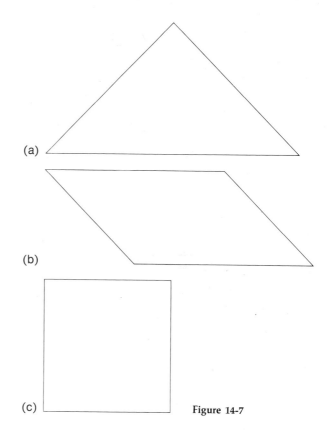

Figure 14-7

Property: *If a region can be divided and reassembled to form another region, then the two regions have the same area.*

17. Do regions (a) and (b) in Figure 14-8 have the same area? _____ Why or why not?

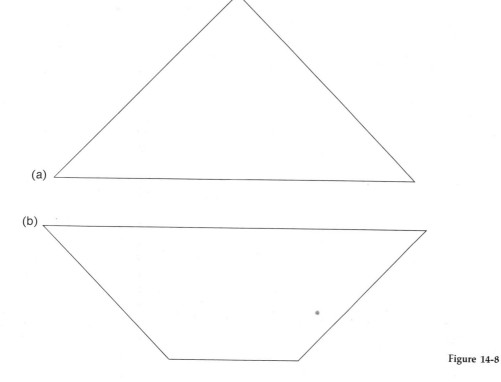

Figure 14-8

18. Use the tangrams to determine which of the regions in Figure 14-9 have the same area.

■ Observe that if a region can be divided and reassembled so as to fit inside another region, then the first region has a smaller area than the second region.

19. Guess which of the regions in Figure 14-10 has the larger area. Then verify by using tangrams.

■ You should have been able to answer Exercises 16–19 without determining the area of any of the figures drawn. If you did not do it this way, try to do so now. Exercise 18 is somewhat challenging

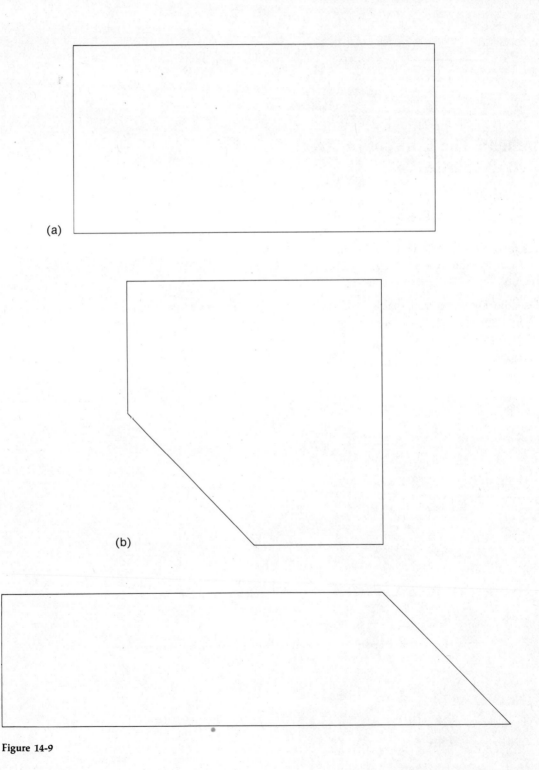

Figure 14-9

14/Comment

in this regard and requires exchanging tangrams of equal area.

Thus, it is clear that although it is possible to compare the size of regions without knowing their specific areas, it is both theoretically and practically useful to be able to assign a number to the area of a region—a number that describes its size in relation to a unit region, which is assigned area 1. The unit region can be chosen arbitrarily.

20. (a) Suppose the small triangle in your tangram set is assigned area 1. Find the area of each of the regions in Figure 14-11 on the following page.

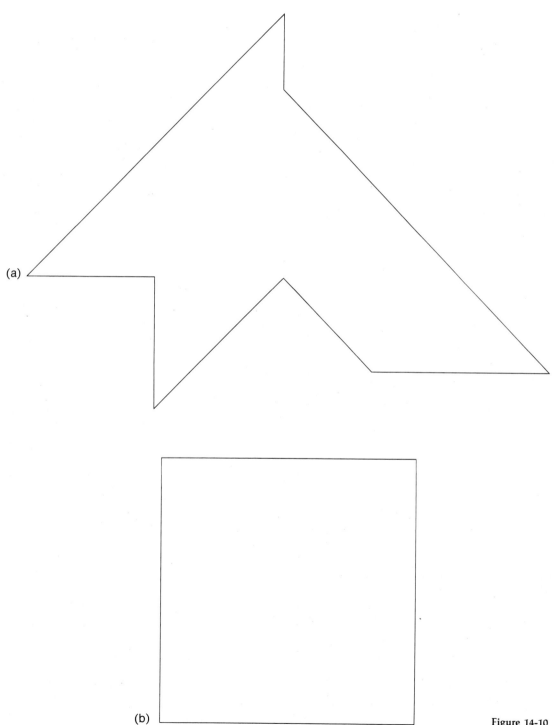

Figure 14-10

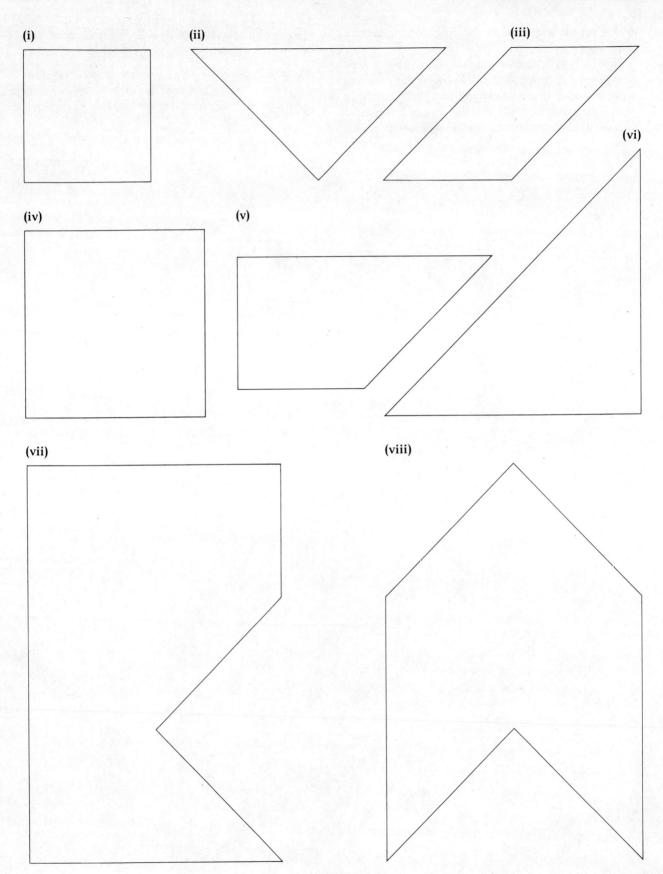

Figure 14-11

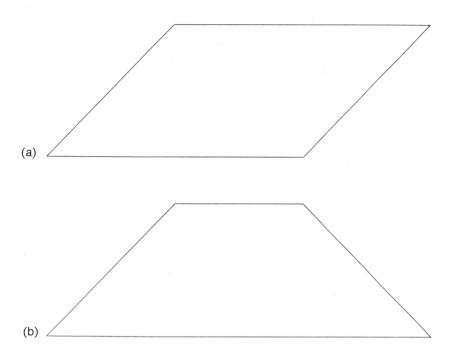

Figure 14-12

(b) If the square (i) is assigned area 1, what is the area of each figure? (i) ____ (ii) ____ (iii) ____ (iv) ____ (v) ____ (vi) ____ (vii) ____

21. Suppose the tangram parallelogram is assigned area 1. Determine the area of each of the other tangram pieces.

22. With the parallelogram assigned area 1, determine the area of the regions in Figure 14-12.

    (a) ____ (b) ____

■ Notice that the area of every region is not as easy to measure as the above examples. For example, we might wish to determine the area of the region in Figure 14-13 with respect to a unit area ▢. The investigation of this problem requires knowledge of the subject known as calculus and is beyond the scope of this book. We will restrict ourselves to very simple figures and provide an intuitive justification for some of the formulas used to calculate the areas of rectangles, parallelograms, and trapezoids.

Figure 14-13

For the rest of this chapter we will assume that the unit region is chosen to be a square having sides of length 1.

The Rectangle

In Exercise 2, you constructed some rectangles on the geoboard and verified that the area was always the product of the numbers that measured the base and the height. These rectangles were special in that the lengths of their sides were whole numbers. If the base and the height of a rectangle are measured in whole numbers, then the definition of multiplication immediately gives the area as the product of the base and the height. For example, consider the rectangle in Figure 14-14, base 7 and height 3. Since the rectangle can be divided into a three-by-seven array of unit squares, the area of the rectangle is $7 \times 3 = 21$. But how could we determine the area of a rectangle with base 7/3 and height 3/2?

Figure 14-14

Suppose we use the same unit length as above, namely ____ so that the rectangle looks like Figure 14-15. We solve the problem by forming a large rectangle using duplicates of the given rectangle, so that the new rectangle has whole-number dimen-

Figure 14-15

sions. For example, we can put together six of our original rectangles to obtain Figure 14-16. We determine that the dimensions of the rectangle are 7 units by 3 units. Why? Hence the area of this rectangle, as we saw above, is 21. Since it takes six of our original rectangles to form the large rectangle, we can conclude that the area of the original rectangle is 21/6, which is the product of 7/3 and 3/2.

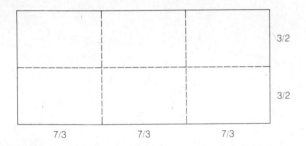

Figure 14-16

A similar procedure will work if the base and height of the rectangle are any rational numbers.

23. Use an analogous procedure to the one outlined above to show that the area of a rectangle with dimensions 7/4 and 5/3 is 35/12.

■ It seems as if we are justified in accepting the formula A (rectangle) = base × height. However, there is one catch. As we shall see later in this chapter, there are line segments whose lengths cannot be expressed as rational numbers. Even then, it is possible to verify the formula; but the verification is difficult, so let us accept it as a definition:

Definition: *The area of a rectangle is the number obtained as the product of the numbers representing the base and the height.*

The Parallelogram

Once we know the area of a rectangle, it is easy to find the area of a parallelogram. Consider the parallelogram in Figure 14-17. Any side can be designated as the *base* of a parallelogram. The *height* is the distance between the base and its opposite side. To see that the area of the parallelogram is the product of the length b of the base and the height h, consider Figure 14-18. Since the triangular regions EAD and FBC have the same area, the parallelogram $ABCD$ has the same area as the rectangle $EFCD$, which is $b \times h$.

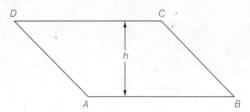

Figure 14-17

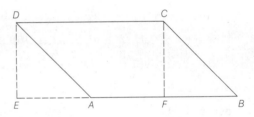

Figure 14-18

Note we are using a fundamental property of area in the argument:

Property: *If two regions are congruent, they have the same area.*

Area of a Triangle

To find the formula for the area of a triangle, we observe that, given any triangle ABC, we can draw a parallelogram $ABCD$ such that the area of $\triangle ABC$ is exactly one half the area of the parallelogram $ABCD$ (Figure 14-19).

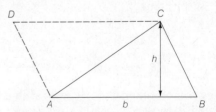

Figure 14-19

Hence $b \times h = A(ABCD)$
$= A(\triangle ABC) + A(\triangle BDC)$
$= 2A(\triangle ABC)$

Therefore $A(\triangle ABC) = \frac{1}{2} b \times h$

We used the following property in the above argument:

Property: *If R and S are two planar regions with $R \cap S = \{\ \}$, then $A(R \cup S) = A(R) + A(S)$.*

We used the same property when we determined the area of a trapezoid. Recall that, given any trapezoid (Figure 14-20), two copies can be joined to form a parallelogram (Figure 14-21).

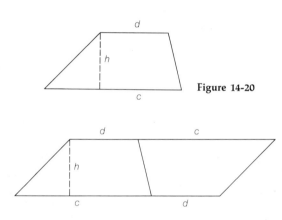

Figure 14-20

Figure 14-21

Using this property, we see that

$A(\text{parallelogram}) = A(\text{trapezoid}) + A(\text{trapezoid})$

Therefore

$A(\text{trapezoid}) = \frac{1}{2} A(\text{parallelogram}) = \frac{1}{2} h(c + d)$

Lab Exercises: Set 2

EQUIPMENT: Four geoboards and one set of tangrams for each group.

One of the most famous discoveries of Greek mathematics is the *Pythagorean theorem*. In this set of exercises you will use the geoboard and tangrams to study this theorem and its consequences.

24. Form a 10 × 10-pin geoboard by putting four geoboards together. On this geoboard, construct a right triangle with legs (the sides of a right triangle perpendicular to each other are called *legs*) of lengths 1 and 2. Then construct a square on each side of the triangle.

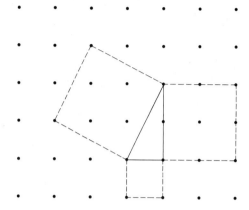

Compute the area of the square you constructed on the *hypotenuse* (the side opposite the right angle) using the method of Exercise 7. Verify that the area of the square on the hypotenuse is the sum of the areas of the squares on the legs.

25. Construct a right triangle on the geoboard with legs of lengths 1 and 3 and again verify that the area of the square on the hypotenuse equals the sum of the areas of the squares on the legs.

Theorem: *For any right triangle, the area of the square on the hypotenuse equals the sum of the areas of the squares on the other two sides.*

■ This statement is usually referred to as the *Pythagorean theorem*, although historians are not in agreement

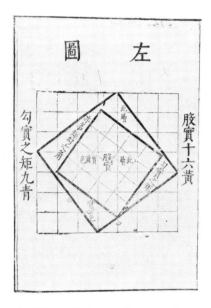

Figure 14-22 Chinese block print illustrating the Pythagorean theorem, from the *Chou pei suan ching*, an ancient mathematical text. Copyright: The British Library.

as to when and where its validity was first demonstrated. Certainly the principle for specific triangles was well known by the Egyptians and Babylonians before the time of Pythagoras (circa 540 B.C.), and some historians contend that the first proof of the theorem is due to the Chinese.

There have been hundreds of different proofs of the Pythagorean theorem. We shall investigate a few of them.

Notice that the theorem has been stated in geometric terms. We can also state it algebraically by using the fact that the area of a square with sides of length x is $x^2 = x \cdot x$. The Pythagorean theorem states that if the lengths of the two legs and hypotenuse are, respectively, a, b, and c, then $a^2 + b^2 = c^2$ (Figure 14-23).

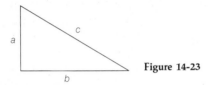

Figure 14-23

It is important to realize that although this shortened expression of the theorem is available to us, it was not available to the Greek geometers, since algebraic notation was not in use at the time. Their interpretation of the theorem was exclusively in terms of the areas of certain squares.

Exercises 26–28 indicate how one might prove the Pythagorean theorem. Two alternative proofs will be discussed later.

26. Consider the right triangle with legs having lengths of 1 and 3 units. Arrange four such triangles on one geoboard as shown below.

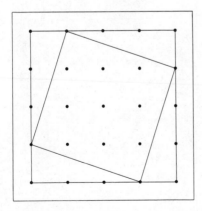

(a) Is the figure in the center a square? _____
To prove that it is, you must show that its sides are congruent and its angles are right angles.

(b) Why are the sides congruent? (*Hint:* Use the side-angle-side congruence axiom.)

(c) Why are the angles right angles? (*Hint:* Use the fact that the sum of the angles of a triangle is 180°.)

(d) Now, the figure in the center is the square on the hypotenuse of the triangle under consideration. Without calculating the area of the center square, conclude that the area of the square on the hypotenuse plus four times the area of the triangle equals the area of the large square.

(e) On another geoboard, arrange the four triangles within the large square, as shown below.

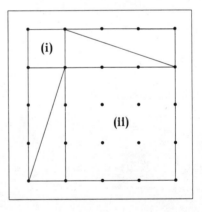

This shows that the area of square (i) plus the area of square (ii) plus four times the area of the triangle equals the area of the large square. Why can you conclude that the area of square (i) plus the area of square (ii) equals the area of the square on the hypotenuse?

(f) Do you see that this proves the Pythagorean theorem for the triangle with legs of lengths 1 and 3?

27. Repeat the above procedure with a triangle having legs of lengths 2 and 3. (You will need to use a 10 × 10-pin geoboard.)

28. Use the diagrams below to explain why the Pythagorean theorem is always true.

(a)

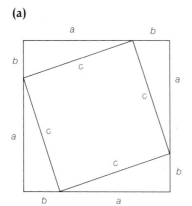

(b)

29. Again construct a right triangle with legs of lengths 1 and 3 on your geoboard. Denote the length of the hypotenuse by c. We are seeking a convenient way to express c. From the Pythagorean theorem we know that $c^2 = 1^2 + 3^2 = 1 + 9 = 10$. That is, c is a number that when multiplied by itself is 10. We say that "c squared equals 10." An equivalent expression is to say that the square root of c equals 10, and write $c = \sqrt{10}$.

Observe that the two expressions $c^2 = 10$ and $c = \sqrt{10}$ mean the same thing, namely $c \times c = 10$.

30. Write an equivalent expression for each of the following.

(a) $c^2 = 5$ _____

(b) $x^2 = 11$ _____

(c) $a = \sqrt{13}$ _____

(d) $y = \sqrt{2}$ _____

31. On the dot paper below and on the following page draw right triangles with legs a and b having the indicated length. Write c as a square root. Draw the square on the hypotenuse and use the method of Exercise 8 to compute its area.

EXAMPLE:

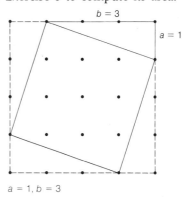

$a = 1, b = 3$
$c = \sqrt{10}$ since $c^2 = 1^2 + 3^2 = 10$
Area = $16 - 1\frac{1}{2} - 1\frac{1}{2} - 1\frac{1}{2} - 1\frac{1}{2} = 10$

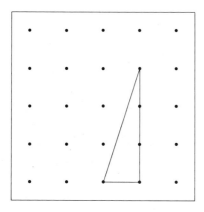

$a = 1, b = 4$

$a = 1, b = 4$

32. Construct line segments of the following length on the geoboard. (*Hint:* Construct a right triangle and use the Pythagorean theorem to determine a right triangle having the appropriate hypotenuse.)

  (a) $\sqrt{2}$ (b) $\sqrt{5}$ (c) $3\sqrt{2}$ (d) $\sqrt{20}$

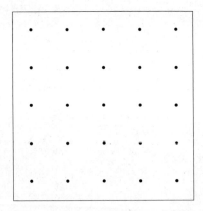

33. Use the tangrams for the following. Assign length 1 to the leg of the smallest triangle and then calculate the dimensions of each of the other tangram pieces.

34. Calculate the area of each of the tangram pieces.

If a square could be constructed from the five smallest pieces, its area would be _____. Hence the side of the square would have length _____. Use this information to help you construct a square from the five smallest pieces. Compare with Exercise 8 of Chapter 12.

35. (a) Remove one of the large triangles from the tangram set. If a square were to be constructed from the remaining six pieces, its area would be _____. Hence, the side of such a square would be _____. Show that it is impossible to construct a line segment of this length from the line segments of the sides of the tangram pieces.

  (b) Repeat (a) above for the other possible ways to remove one tangram piece, trying to prove that it is impossible to construct a square from any size of the tangram pieces. Compare with Exercise 9 of Chapter 12.

■ We have discussed line segments with lengths $\sqrt{2}$, $\sqrt{5}$, and $\sqrt{10}$. What kind of number do these symbols represent? We will show below that these lengths cannot be represented by rational numbers. This will lead us in the next chapter to ask what kind of number is represented by these symbols. We start by obtaining some preliminary information.

36. (a) Square the numbers 1 through 10 and fill in the following table.

Number = k	k^2	Last digit of k^2	Last digit of $2 \times k^2$
1			
2			
3			
4			
5			
6			
7			
8			
9	81	1	2
10			

  (b) Choose a whole number $n > 10$. Verify that the last digit of n^2 appears in the third column and that the last digit of $2n^2$ appears in the fourth column. Repeat for another choice of n.

  (c) Conclude that if n is any whole number, then the last digit of n^2 is one of $\{0, 1, 4, 5, 6, 9\}$.

(d) Conclude that if n is any whole number, then the last digit of $2n^2$ is one of $\{0, 2, 8\}$.

37. Suppose $(m/n)^2 = 2$ and choose m/n in reduced form (i.e., m and n have greatest common factor 1). If $(m/n)^2 = 2$, then $m^2 = 2n^2$, so that the last digit of m^2 equals the last digit of $2n^2$.

(a) From Exercise 36(c), what are the possibilities for the last digit of m^2? _____

(b) From Exercise 36(d), what are the possibilities for the last digit of $2n^2$? _____

(c) Since $m^2 = 2n^2$, the only possibility for the last digit is _____.

(d) Now look at the chart again and conclude that the last digit of n must be 0 or 5 and the last digit of m must be 0.

Hence 5 divides both m and n. But we started out with m/n being in lowest terms. What does this mean? The only assumption we made was that $(m/n)^2 = 2$. Hence this assumption must be incorrect and therefore there is no rational number that squares to 2.

■ We say that $\sqrt{2}$ is an *irrational number*. It represents the length of a line segment but is not the ratio of two whole numbers. In terms of the discussion in Chapter 10, we cannot divide the unit segment into n congruent pieces and put m of them end to end to obtain a segment of length $\sqrt{2}$. Yet surely there is a segment of that length (the hypotenuse of a right triangle with both legs having length 1). Is there some kind of number that represents that length? We will discuss this situation in the next chapter.

38. *Optional.* We conclude these exercises with a neat proof of the Pythagorean theorem discovered by James A. Garfield (1831–1881), the twentieth president of the United States. He discovered the proof when as a member of Congress he was discussing mathematics with some fellow congressmen. (I wonder if any congressmen these days discuss mathematics.) He published his proof in the *New England Journal of Education*. His proof uses the formula for the area of a trapezoid developed in Exercise 14. We will first illustrate Garfield's proof using a right triangle with legs of length $a = 3$ and $b = 1$.

(a) Construct such a triangle on the geoboard and label the hypotenuse c.

(b) Construct a congruent triangle and connect the vertices (dotted line) to form a trapezoid as shown in the next column.

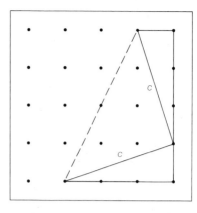

(c) Write the formula for the area of this trapezoid and compute its area. _____

(d) Now compute its area by adding up the area of the three triangles. It is $\tfrac{1}{2}c^2 +$ _____ $+$ _____ $= \tfrac{1}{2}c^2 +$ _____.

(e) If we compare (c) and (d), we see that $\tfrac{1}{2}c^2 =$ _____ or $c^2 =$ _____.

(f) Does $c^2 = a^2 + b^2$ for this triangle?

(g) In your group discuss the following proof for an arbitrary triangle with legs a and b and hypotenuse c. If you get stuck, go over the proof for the triangle with legs of lengths 1 and 3.

Consider the right triangle below. Then

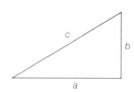

construct a congruent triangle as shown,

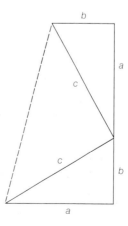

and connect the vertices (dotted line), to form a trapezoid. What is the height of this trapezoid? _____ Write the formula for the area of the trapezoid. _____

Since the height of the trapezoid drawn above is $a + b$, you can use the formula developed in Exercise 14 to conclude that the area of the trapezoid is $\frac{1}{2}(a + b)(a + b) = \frac{1}{2}(a^2 + 2ab + b^2)$. But the area of the trapezoid is also the sum of the areas of the three triangles. This area is $\frac{1}{2}c^2 + \frac{1}{2}ab + \frac{1}{2}ab = \frac{1}{2}c^2 + ab$. Hence $\frac{1}{2}c^2 + ab = \frac{1}{2}(a^2 + 2ab + b^2)$. Therefore, $c^2 + 2ab = a^2 + 2ab + b^2$. We conclude that $c^2 = a^2 + b^2$. Hurray for President Garfield!

Comment: The Pythagorean Theorem

"To this day, the theorem of Pythagoras remains the most important single theorem in the whole of mathematics. That seems a bold and extraordinary thing to say, yet it is not extravagant, because what Pythagoras established is a fundamental characterization of the space in which we move and it is the first time that it was translated into numbers."
—Jacob Bronowski

The importance that Bronowski claims, in the above quotation, for the Pythagorean theorem should not be underestimated. As Jay Hambridge points out, "without it we should not be able to aim our cannon, to sail our ships, to make astronomical calculations, [and] trigonometry would be nonexistent."*

This powerful theorem is also one of the oldest. Pythagoras lived around 540 B.C. But the result was known to the Babylonians 1,200 years earlier although it is unknown whether they knew how to prove it. It is not certain when the Chinese discovered the theorem, but some people think they may have discovered the first proof. It appears in an ancient Chinese manuscript whose date is unknown but which was probably written around 600 B.C. We illustrate it on the geoboard for a right triangle with legs 1 and 3.

Arrange the triangles as shown in Figure 14-24. The resulting figure is a square that has sides equal in length to the length of the hypotenuse. Now rearrange the four triangles as shown in Figure 14-25. (It is helpful to have another geoboard.) Finally, adjoin the center square from the first diagram to the second diagram, as shown in Figure 14-26. The resulting region has the same area as the square on the hypotenuse. But by drawing in the dotted line, as indicated in Figure 14-27, it is clear that the area of this region equals the sum of the areas of the squares of the legs.

39. Prove the Pythagorean theorem on the geoboard using the Chinese method, for a right triangle with legs of lengths 1 and 2.

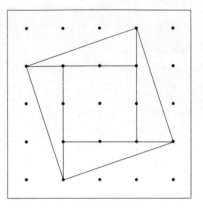

Figure 14-24

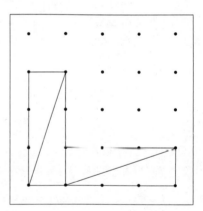

Figure 14-25

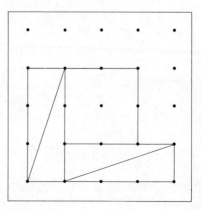

Figure 14-26

Practical Applications of Dynamic Symmetry (Greenwich, CT: Devin-Adair, 1965).

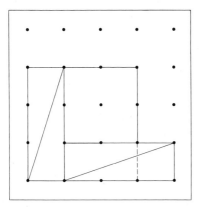

Figure 14-27

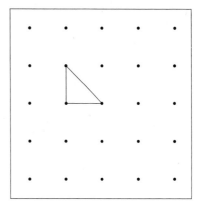

Figure 14-29

■ An algebraic proof of the Pythagorean theorem, using the first diagram (Figure 14-24) of the above proof, was given by the Hindu mathematician Bhaskara in the twelfth century. It is perhaps the shortest proof of the Pythagorean theorem.

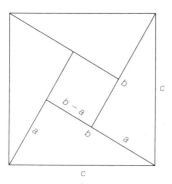

Figure 14-28

Label the diagram as shown in Figure 14-28. Then the large square has area c^2, and the small square $(b - a)^2$. Hence we conclude that

$$4 \times \tfrac{1}{2}ab + (b - a)^2 = c^2$$

Therefore

$$2ab + b^2 - 2ab + a^2 = c^2$$

That is,

$$b^2 + a^2 = c^2$$

As we saw in Exercise 31, the Pythagorean theorem provides a method of constructing on the geoboard line segments of lengths that are not integers. For example, if we construct the triangle in Figure 14-29, the hypotenuse is a number c that when multiplied by itself equals $1^2 + 1^2 = 2$. Thus $c^2 = 2$. We use the symbol $c = \sqrt{2}$ to mean $c^2 = 2$. Now, c cannot be an integer since $1^2 = 1$ and $2^2 = 4$ and there is no integer between 1 and 2. We determined in Exercise 37 that in fact the length of this line is not a rational number; that is, it is impossible to obtain a segment congruent to the hypotenuse by dividing up the unit segment (the leg) into a certain number (say n) of congruent pieces and putting m of them end to end.

If you recall that the rational numbers are dense (i.e., between any two distinct rational numbers there is always another rational number), you might be surprised by this. The followers of Pythagoras certainly were. When they discovered it, they tried to keep it a secret, and members of their cult swore not to reveal it. Nowadays, however, it is public knowledge, and we will give another proof of it. It is rather long, so if you want to understand it, take your time and read it through several times.

Suppose the length of the hypotenuse is m/n, where m/n is in reduced form (i.e., the GCF of m and n is 1). Then the area of the square on the hypotenuse would be $m/n \times m/n = m^2/n^2$. But we know from the Pythagorean theorem that the area of the square is 2. Hence we would have $m^2/n^2 = 2$. This implies $m^2 = 2n^2$, which would mean that m^2 is an even number. Now what about m? If m were odd, then m^2 would be odd, since the product of two odd numbers is odd (see Exercise 44 of Chapter 8). Hence m must be even. Suppose $m = 2k$. Then $m^2 = (2k)^2 = 4k^2$, so we have $2n^2 = 4k^2$. Hence $n^2 = 2k^2$. This says that n^2 is even. Using the same argument as we did above for m, we see that n is even. Hence m and n are both even, which is contrary to choosing m/n in reduced form. This shows it is impossible to find a rational number m/n that squares to 2. Hence the number represented by $\sqrt{2}$ is not a rational number.

Similar arguments show that $\sqrt{3}$, $\sqrt{5}$, $\sqrt{7}$, and many more numbers are not rational. In fact, it can be shown that if n is a whole number and if $\sqrt{n}$ is not a whole number, then $\sqrt{n}$ is not a rational number.

Definition: *Numbers that are not rational are called* irrational numbers.

We will investigate irrational numbers further in the next chapter.

40. *Puzzle Problem—Optional.* Cut out twenty congruent right triangles having legs 3 cm and 6 cm from cardboard or construction paper. Can you arrange them to form a square? (*Hint:* Use the techniques of Exercise 34.) This is a good puzzle to try on a friend.

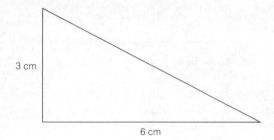

Important Terms and Concepts

Area of rectangle
Area of parallelogram
Area of triangle
Area of trapezoid
Finding the area of regions on the geoboard
Pythagorean theorem
Constructing line segments of length $\sqrt{n}$
Irrational number

Review Exercises

1. (a) Using the tangrams, suppose that the medium-sized triangle is assigned the area 1. Determine the area of each of the other tangram pieces. Also find the area of each figure in Exercise 20.
   (b) Assign the area of the large triangle to be 1 and repeat the above.

2. Find the area of the following shapes.

   (a)
   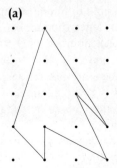

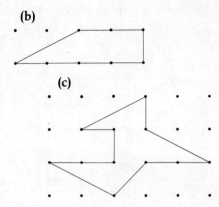

3. (a) Use the dots below to help you construct a parallelogram with one side of length $\sqrt{13}$ and another side of length $\sqrt{5}$. Find the area of the parallelogram.

   (b) How many different parallelograms with sides of these lengths can you find on the geoboard?
   (c) Do any or all of the above parallelograms have the same area?
   (d) If you were not restricted to a geoboard, how many such parallelograms could you find?

4. Suppose you knew that the hypotenuse of a right triangle measured 13 units and one of the sides of the triangle measured 5 units. What would the area of the triangle be? _____

5. (a) What is the length of the longest line segment that would fit into a box 3 cm by 4 cm by 6 cm. (*Hint:* The Pythagorean theorem will be helpful.) _____

   (b) *Optional.* Can you answer the above question for any box of dimensions *a*, *b*, and *c*?

14/Review Exercises

6. Find the lengths of the segments labeled a, b, c, d, and e.

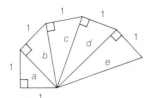

7. What is the length of the second leg of a right triangle if the hypotenuse is $\sqrt{3}$ and one leg is $\sqrt{2}$? _____

8. A ladder 13 feet long is moved from standing vertically against a wall so that its top moves 1 foot lower. How far from the base of the wall will the bottom of the ladder be? _____

9. Make a 3 × 2 rectangle on the geoboard. Investigate the ways this region can be divided into two regions of equal area by a rubber band connecting two nails on the boundary. The band may touch nails in the interior. (Look for a systematic way of doing this investigation. There are more ways than you would think at first.)

10. *Optional.* Show that the median of any triangle divides that triangle into two triangles having equal area.

For Exercises 11–13, write the letter of the correct answer in the blank.

11. If the two parallel sides of a trapezoid have length $2\sqrt{5}$ and $3\sqrt{5}$ and the height is $1/\sqrt{5}$, then the area is _____.
    (a) 5 (b) $2\frac{1}{2}$ (c) 10 (d) $7\frac{1}{2}$ (e) None of these

12. If a parallelogram has an area of 2 and its base is 6, then its height is _____.
    (a) 3 (b) 6 (c) $\frac{2}{3}$ (d) $\frac{3}{2}$

13. If we wanted to use a right triangle to construct a length of $\sqrt{17}$, which pair below could be used as our base and height? _____
    (a) 1, $\sqrt{8}$ (b) 1, 4 (c) 1, $\sqrt{17}$ (d) None of these

For Exercises 14–19, decide whether the statement is true or false. Write T or F in the blank.

_____ 14. If a triangle has sides of lengths 5, $\sqrt{39}$, and 8, then the triangle must be a right triangle.

_____ 15. Any triangle with hypotenuse $\sqrt{2}$ has both legs of length 1.

_____ 16. The diagonal of a square is always a rational number.

_____ 17. The area of any parallelogram constructed on the geoboard is always a whole number.

_____ 18. If you know the lengths of all the sides of a trapezoid, you can find the area.

_____ 19. To find the area of a square, you only need to know the length of one side.

By permission of Johnny Hart and Field Enterprises, Inc.

15

Decimals

"It is perhaps inevitable and certainly probable that you will sometimes feel frustrated while doing mathematics. Whenever you push the limits of your understanding, frustration is a natural feeling. Enjoy the feeling—beat a pillow, jump up and down, yell, growl, have a good cry. Your learning may be easier."—Julian Weissglass

Lab Exercises: Set 1

EQUIPMENT: One compass per student.

We start by trying to approximate $\sqrt{2}$ by rational numbers. Since the calculations with fractions are cumbersome, we develop *decimals*.

In Chapter 14 we saw that there is a line segment whose length is not a rational number; namely, the hypotenuse of a right triangle with legs having length 1 (Figure 15-1). We denoted the length of this line segment by $\sqrt{2}$, and we know that $(\sqrt{2})^2 = 2$.

Figure 15-1

1. (a) Using the unit length 1 of the leg of the triangle above, mark off points 1, 2, 3, 4, 5, and 6 units from point 0 with a compass on the line below.

   0 ───────────────────────────

   (b) Use the compass to indicate a point of distance $\sqrt{2}$ from 0.

   (c) Mark off a point of distance $2\sqrt{2}$ from 0.

   (d) Discuss in your group how to find a line segment of length $\sqrt{3}$ by constructing an appropriate triangle on the right angle shown below. Then indicate the corresponding point on the number line above. (*Hint:* Use the Pythagorean theorem. You want a, b so that $a^2 + b^2 = 3$.)

   (e) Repeat (d) for $\sqrt{5}$, $\sqrt{7}$, and $\sqrt{8}$ on the right angles below.

   (i)   (ii)   (iii)

   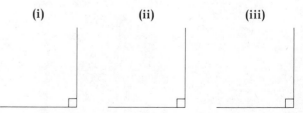

2. (a) Indicate the point 3/2 on the number line in Exercise 1. Is 3/2 larger or smaller than $\sqrt{2}$? _____Verify this algebraically by computing $(3/2)^2$. (*Hint:* If 3/2 were less than $\sqrt{2}$, then $(3/2)^2$ would be less than $(\sqrt{2})^2 = 2$; and if 3/2 were greater than $\sqrt{2}$, then $(3/2)^2$ would be greater than $(\sqrt{2})^2 = 2$.)

(b) We see that $1 < \sqrt{2} < 3/2$. Can you find a rational number that is closer to $\sqrt{2}$ than either 1 or 3/2? _____ What is it? _____ Why?

■ We would like a systematic way of finding rational numbers that get closer and closer to $\sqrt{2}$. If x and y are two numbers, the *average* or *mean* of x and y is $(x + y)/2$. Since the average of two numbers lies between the two numbers, it is a likely candidate for a better approximation.

3. (a) Draw a picture to illustrate why this is true.

(b) Since we know that $1 < \sqrt{2} < 3/2$, we can find the average of 1 and 3/2 to find a closer estimate to $\sqrt{2}$. Calculate the average of 1 and 3/2. _____ Is it less than or greater than $\sqrt{2}$? _____ Why? _____
Complete the following by putting the appropriate whole number in the box.

$$\frac{\Box}{4} < \sqrt{2} < \frac{\Box}{4}$$

(c) Repeat the above to obtain

$$\frac{\Box}{8} < \sqrt{2} < \frac{\Box}{8}$$

(*Note:* In each step only one value of the inequality changes.)

■ The above process could be continued to obtain better and better approximations to $\sqrt{2}$ by rational numbers. However, working with the fractions becomes cumbersome, so it will be useful at this time to develop *decimal notation*.

4. Suppose the unit length is as shown in Figure 15-2. (We choose a longer unit here so that the inaccuracies in our drawings will have minimal effect.) Using the method of Exercise 1(d), construct a line segment of length $\sqrt{2}$ on a separate sheet of paper.

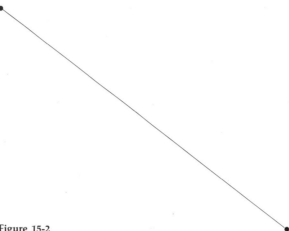

Figure 15-2

5. Now locate the point corresponding to $\sqrt{2}$ on the line in Figure 15-3 on the following page. For convenience, the segment between 1 and 2 has been divided into ten congruent segments. You could have done this using the process of Exercise 13 in Chapter 13, but it would have been time-consuming.

6. You probably found that $\sqrt{2}$ lies between $1\tfrac{4}{10}$ and $1\tfrac{5}{10}$. Can you guess an approximation in hundredths; that is, is $\sqrt{2}$ between $1\tfrac{40}{100}$ and $1\tfrac{41}{100}$ or between $1\tfrac{41}{100}$ and $1\tfrac{42}{100}$, etc.?

If we had precise tools, we could divide the segment between $1\tfrac{4}{10}$ and $1\tfrac{5}{10}$ into ten congruent segments in order to determine this and then continue to repeat the process to locate $\sqrt{2}$ between thousandths, ten-thousandths, etc.

7. Determine an approximation to each of the points on the line in Figure 15-4 on the following page in terms of tenths. Make a guess as to an approximation in terms of hundredths.

8. Use what you know about the ordering of rational numbers from Chapter 10 (i.e., $p/q < r/s$ if and only if $ps < qr$) to approximate each of the following fractions (i) in terms of tenths, and (ii) in terms of hundredths.

EXAMPLE: 3/7

(i) 4/10 < 3/7 since 28 < 30, and 3/7 < 5/10 since 30 < 35. Hence 4/10 < 3/7 < 5/10.

(ii) 42/100 < 3/7 since 294 < 300, and 3/7 < 43/100 since 300 < 301. Hence 42/100 < 3/7 < 43/100.

(a) 5/8

(b) $1\frac{2}{3}$

(c) $6\frac{3}{4}$

■ Since computations with fractions are cumbersome, especially when there are many different denominators, it is convenient to work with fractions that have denominators of 10 or 100 or 1,000, etc. In the seventeenth century, mathematicians developed what is now referred to as *decimal notation*. I think you will agree that it does indeed simplify calculations and that Whitehead's statement at the beginning of Chapter 7 is justified. In decimal notation, we write, for example, 2.1 for $2\frac{1}{10}$, 4.53 for $4\frac{53}{100}$, 0.783 for 783/1,000, etc. We can think of 0.783 as the sum of 7 tenths, 8 hundredths (tenths of tenths), and 3 thousandths (tenths of hundredths):

$$\frac{783}{1,000} = \frac{700}{1,000} + \frac{80}{1,000} + \frac{3}{1,000} = \frac{7}{10} + \frac{8}{100} + \frac{3}{1,000}$$

In general, $a \cdot a_1 a_2 a_3 \ldots a_k$ means

$$a + \frac{a_1}{10} + \frac{a_2}{100} + \frac{a_3}{1,000} + \cdots + \frac{a_k}{10^k}$$

Definition: *The point is called a* decimal point *and the representation is called a* decimal.

9. Write each of the following decimals as fractions.

EXAMPLES:

(i) $1.21 = 1\frac{21}{100} = 121/100$

(ii) $0.011 = 11/1,000$

(a) 5.23 =   (b) 0.007 =

Figure 15-3

Figure 15-4

(c) 0.75 =

(d) 0.5 =

(e) 0.200 =

(f) 0.66 =

10. It should be clear from Exercise 9 that any decimal that stops after a finite number of terms represents a rational number.
    (a) Discuss how you would explain this to a student who was just learning about decimals?

Definition: *A decimal that stops after a finite number of steps is called a* **terminating decimal.**

Every terminating decimal represents a rational number.
    (b) Do you think $\sqrt{2}$ can be represented by a terminating decimal? _____ Discuss in your group and then give a reason for your conclusions.

11. Given a fraction, how do we represent it as a decimal? Represent each of the following as decimals.

    EXAMPLE: 25/10 = 2.5

    (a) 2/10 = _____
    (b) 37/100 = _____
    (c) 928/100 = _____
    (d) 128/1,000 = _____

    If the denominator of the fraction is 10, 100, 1,000, etc. (i.e., a power of 10), then it is easy to find the equivalent decimal.

12. Represent each of the following as a decimal by finding an equivalent fraction having a power of 10 (10, 10^2, ..., 10^k, ...) for the denominator.

    EXAMPLE: 4/5 = 8/10 so 4/5 = 0.8

    (a) 3/20 = _____
    (b) 1/2 = _____
    (c) 7/4 = _____
    (d) 3/8 = _____
    (e) 9/40 = _____
    (f) 1/16 = _____
    (g) 371/400 = _____
    (h) 18/50 = _____

13. Discuss in your group the conditions for a fraction to be equivalent to one having a power of 10 for the denominator. (*Hint:* Factor each one of the denominators in Exercise 12.)

Comment: Computing with Decimals

"For a physicist, mathematics is not just a tool by means of which phenomena can be calculated; it is the main source of concepts and principles by means of which new theories can be created."—Freeman J. Dyson

Our attempt to find rational numbers that approximate $\sqrt{2}$ led us to consider decimals. Decimals represent rational numbers that result from adding "tenths" and "tenths of tenths" and "tenths of tenths of tenths" etc., to a whole number.

For example, 532.347 represents $532 + 3/10 + 4/100 + 7/1{,}000$. Since we write $1{,}000 = 10^3$, $100 = 10^2$ and $10 = 10^1$, we can extend our notation so that $1 = 10^0$, $1/10 = 10^{-1}$, $1/100 = 10^{-2}$, $1/1{,}000 = 10^{-3}$, etc. Then, 532.347 represents $5 \times 10^2 + 3 \times 10^1 + 2 \times 10^0 + 3 \times 10^{-1} + 4 \times 10^{-2} + 7 \times 10^{-3}$. This is a natural extension of the place-value system we developed in Chapter 2.

14. Write each of the following as a sum of multiples of positive and negative powers of 10 as in the example above.

    (a) 217.65 _____
    (b) 11.432 _____
    (c) 2.6008 _____
    (d) 0.00325 _____

■ Addition and multiplication of decimals are easy. All you must do is remember that the numbers represent sums of tenths, hundredths, thousandths, etc. You simply have to add or subtract the corresponding quantities. For example, since $4.325 = 4 + 3/10 + 2/100 + 5/1{,}000$ and $8.162 = 8 + 1/10 + 6/100 + 2/1{,}000$, we have that $4.325 + 8.162 = 4 + 3/10 + 2/100 + 5/1{,}000 + 8 + 1/10 + 6/100 + 2/1{,}000 = 4 + 8 + 3/10 + 1/10 + 2/100 + 6/100 + 5/1{,}000 + 3/1{,}000 = 12 + 4/10 + 8/100 + 7/1{,}000 = 12.487$. It should be clear that the familiar algorithm for adding and subtracting decimals is valid and that we can write

$$\begin{array}{r} 4.325 \\ +\ 8.162 \\ \hline 12.487 \end{array}$$

15. Calculate the following. (*Note:* If you write decimals in two columns for addition or subtraction, be sure to line up the decimal points.)

    (a) 12.73 + 0.827 = _____

(b) $9.68 - 0.0093 =$ _____

(c) $8.957 + 2.364 + 1.003 =$ _____

(d) $1.24 + 0.013 =$ _____

(e) $0.1101 - 0.0003 =$ _____

(f) $4.006 - 1.983 =$ _____

■ The computation of products and quotients with decimals is also straightforward, except that an algorithm to determine the placement of the decimal point is needed. Let us explain the procedure by first developing a few facts about exponential notation.

For any positive number a, recall that $a^2 = a \times a$, $a^3 = a \times a \times a$, and in general

$$a^k = \underbrace{a \times a \times \cdots \times a}_{k \text{ times}}$$

for any k, which is a positive whole number. The number k is called an *exponent*. Now if we define $a^0 = 1$, $a^{-1} = 1/a$, $a^{-2} = 1/(a \times a) = 1/a^2$ and in general $a^{-k} = 1/a^k$ for k, a positive whole number, we can verify the following:

The first law of exponents: $a^m \times a^n = a^{m+n}$ *for any integers m and n.*

For example,

$$a^3 \times a^2 = (a \times a \times a) \times (a \times a)$$
$$= a \times a \times a \times a \times a$$
$$= a^5$$

and

$$a^2 \times a^{-3} = a \times a \times \frac{1}{a \times a \times a}$$
$$= \frac{a \times a}{a \times a \times a}$$
$$= \frac{1}{a}$$
$$= a^{-1}$$

The second law of exponents: $a^m/a^n = a^{m-n}$.

For example,

$$\frac{a^{-5}}{a^3} = \frac{1/a^5}{a^3} = \frac{1}{a^5} \times \frac{1}{a^3} = \frac{1}{a^8} = a^{-8}$$

and

$$\frac{a^2}{a^{-3}} = \frac{a^2}{1/a^3} = \frac{a^2}{1} \times \frac{a^3}{1} = a^5$$

We can use the exponential notation to describe what happens when we multiply a decimal by a power of 10. This leads us directly to the algorithm for multiplying and dividing decimals.

Let us compute the product 324.27×10^3 as follows:

$$324.27 \times 10^3 = (3 \times 10^2 + 2 \times 10^1 + 4 \times 10^0$$
$$+ 2 \times 10^{-1} + 7 \times 10^{-2}) \times 10^3$$
$$= 3 \times 10^2 \times 10^3 + 2 \times 10^1 \times 10^3$$
$$+ 4 \times 10^0 \times 10^3 + 2 \times 10^{-1} \times 10^3$$
$$+ 7 \times 10^{-2} \times 10^3$$
$$= 3 \times 10^5 + 2 \times 10^4 + 4 \times 10^3$$
$$+ 2 \times 10^2 + 7 \times 10^1$$
$$= 324{,}270$$

(Did you notice that we used the distributive property in the above calculation?) We see that the result of multiplying 324.27 by 10^3 is to move the decimal point three places to the right.

16. Compute 42.725×10^2 using the long procedure illustrated above.

■ The product of any decimal and 10^k, where k is positive, is obtained by shifting the decimal point k places to the right.

17. Find the following products.

(a) $1{,}273.457 \times 10^2 =$ _____

(b) $2.1435 \times 10^3 =$ _____

(c) $0.001 \times 10^2 =$ _____

(d) $0.000123 \times 10^7 =$ _____

■ Now let us see what occurs when the exponent of 10 is negative. Compute 732.57×10^{-4}.

$$732.57 \times 10^{-4} = 7 \times 10^2 + 3 \times 10^1 + 2 \times 10^0$$
$$+ 5 \times 10^{-1} + 7 \times 10^{-2}) \times 10^{-4}$$
$$= 7 \times 10^2 \times 10^{-4} + 3 \times 10^1 \times 10^{-4}$$
$$+ 2 \times 10^0 \times 10^{-4} + 5 \times 10^{-1}$$
$$\times 10^{-4} + 7 \times 10^{-2} \times 10^{-4}$$
$$= 7 \times 10^{-2} + 3 \times 10^{-3} + 2 \times 10^{-4}$$
$$+ 5 \times 10^{-5} + 7 \times 10^{-6}$$
$$= 0.073257$$

We see that the decimal point moves four places to the left.

18. Compute 21.3×10^{-2} using the long procedure illustrated above.

■ We can formulate the following general rule: The product of any decimal and 10^k, where k is negative, is obtained by shifting the decimal point k places to the left.

19. Find the following products.

    (a) $7.1 \times 10^{-3} =$ _____

    (b) $134.2 \times 10^{-2} =$ _____

    (c) $0.427 \times 10^{-2} =$ _____

    (d) $1,725.1 \times 10^{-1} =$ _____

■ Suppose we wish to multiply the two decimals 62.79×0.74. We proceed as follows.

$$62.79 \times .74 = (6,279 \times 10^{-2}) \times (74 \times 10^{-2})$$
$$= 6,279 \times 74 \times 10^{-2} \times 10^{-2}$$
$$= 6,279 \times 74 \times 10^{-4}$$

We see that we can do the computation as for whole numbers and then multiply by 10^{-4}, which corresponds to "counting off" four decimal places to the left in the answer. Since $6,279 \times 74 = 464,646$, we have

$$62.79 \times .74 = 464,646 \times 10^{-4} = 46.4646$$

A few examples will convince you that the number of digits after the decimal point in the product equals the sum of the number of digits after the decimal place in the two factors.

20. Find the following.

    (a) $1.9 \times 0.27 =$

    (b) $0.3 \times 0.007 =$

    (c) $7,000 \times 0.002 =$

    (d) $1,000 \times 0.001 =$

■ To find the quotient of two decimals, we first observe that multiplying the dividend and divisor by the same number leaves the quotient unchanged, that is, $a \div b = (a \times c) \div (b \times c)$. The reasoning behind this is the same as proving that $m/n = (m \times p)/(n \times p)$ for m, n, and p whole numbers (see Review Exercise 18 in Chapter 7). Hence, to find the quotient $89.42 \div 0.34$, for example, we reason as follows.

$$89.42 \div 0.34 = (89.42 \times 10^2) \div 0.34 \times 10^2$$
$$= 8,942 \div 34$$

We then can use the division algorithm to proceed. In this case, we obtain $8,924 \div 34 = 263$.

Note that we effect the change easily in the standard division format by shifting the decimal point:

$$.34_\curvearrowright \overline{)89.42_\curvearrowright} \quad \rightarrow \quad 34\overline{)8,942}$$

In this case, the quotient is a whole number. If it were not, we could always leave the answer in the form of a quotient and a remainder. However, since we are working with decimals, it would be desirable to know how to obtain a quotient in decimal form.

Let us consider the quotient $4 \div 2.5$. We reason that $4 \div 2.5 = (4 \times 10) \div (2.5 \times 10) = 40 \div 25$, and use the previously developed multiplication algorithm to proceed as follows.

```
   25)40
      25     | 25 × 1
      ──
      15
      10     | 25 × 0.4
      ──
       5
       5     | 25 × 0.2
              ──────
               1.6
```

Thus $4 \div 2.5 = 1.6$.

Proceeding by the alternative calculation below yields the same result.

```
   25)40
      25     | 25 × 1
      ──
      15
      15     | 25 × .6
      ──    ──────
       0     1.6
```

21. Find the following quotients.

    (a) $3.6 \div 1.2 =$ _____

(b) $16 \div 0.4 =$ _____

(c) $276 \div 0.23 =$ _____

(d) $23 \div 0.8 =$ _____

(e) $177 \div 0.4 =$ _____

(f) $1.97 \div 0.25 =$ _____

■ Although the above procedure provides a workable method for dividing decimals, a simpler algorithm is this: after shifting the decimal point to make the divisor a whole number, perform the division ignoring the decimal point and then enter it directly above the decimal point in the dividend. For example:

Decimal point goes here

$$2.5 \overline{)4.} \quad \rightarrow \quad 25 \overline{)40.0} \atop \underline{25} \atop 150 \atop \text{(1.6)}$$

since $4.0/2.5 = (4.0 \times 10)/(2.5 \times 10) = 40/25$.

Percent

Closely related to the subject of decimals is the terminology and usage of *percent*. The main challenge in dealing with percentage is a semantic one. Once you translate a problem concerning percentage to arithmetical language, the mathematics is straightforward.

Percent notation is a convenient way to express fractions or decimals. Four percent (usually written as 4%) means 4/100 or 0.04. If someone has read 4 pages out of 100 in a book, they have read 4% of the pages. Similarly, if someone misses 1 question out of 25, they miss $1/25 = 4/100 = 4\%$ of the questions.

22. Write each of the following percentages as a fraction and as a decimal.

    EXAMPLE: $10\% = 10/100 = 0.1$

    (a) $8\% =$ _____   (c) $150\% =$ _____

    (b) $3\% =$ _____   (d) $100\% =$ _____

23. Write each of the following fractions as percents.

    EXAMPLE: $3/4 = 75\%$

    (a) $2/5 =$ _____   (c) $3/8 =$ _____

    (b) $66/100 =$ _____   (d) $5/4 =$ _____

■ Very often you will see phrases such as "40% of 70 is 28." This means that 40% times 70 equals 28. To verify this, compute $0.40 \times 70 = 28$. You might be asked a question like "5 is what percent of 40?" To solve this, you must find the solution to $\Box \times 40 = 5$. Multiplying both sides by 1/40, you obtain

$$\Box \times 40 \times \frac{1}{40} = 5 \times \frac{1}{40}$$

So $\Box \times 1 = 5/40 = 0.125 = 12.5\%$.

24. Determine each of the following.

    (a) 20% of 80 _____   (c) 35% of 70 _____

    (b) 38% of 50 _____   (d) 60% of 30 _____

25. (a) 9 is 30% of what? _____

    (b) 15 is 60% of what? _____

26. (a) What percent of 60 is 24? _____

    (b) What percent of 80 is 120? _____

27. A basketball team won 60% of its games last season. If it won 24 games, how many games did it play? _____

28. Suppose you are told that a certain tree is 15 meters high and is expected to grow 20% a year. How tall will it be at the end of the year? _____

Lab Exercises: Set 2

EQUIPMENT: None.

29. Review Exercise 12 and then represent each of the following fractions as decimals.

    (a) $7/40 =$ _____   (c) $11/400 =$ _____

    (b) $9/16 =$ _____   (d) $7/8 =$ _____

■ It is easy to determine the decimal representation of certain fractions; namely, those whose denominators have only 2 and 5 as factors. Now we wish to find a way to represent other fractions.

30. Find the decimal representation to five decimal places of each of the following fractions.

    EXAMPLE: 36/11

$$11 \overline{)36}$$

33	11×3
3 0	
2.2	$11 \times .2$
.80	
.77	$11 \times .07$
.030	
.022	$11 \times .002$
.0080	
.0077	$11 \times .0007$
.00030	
.00022	$11 \times .00002$
.00008	3.27272

15/Lab Exercises: Set 2

(a) 4/9

(b) 95/22

31. (a) What do you notice about the above? Discuss your observations.

(b) Can you predict what the digit in the sixth decimal place would be in each of the above? _____ Why? _____

Definition: *A decimal such as the one obtained for 36/11 is called a* repeating decimal *since it never stops. We write it as* $3.\overline{27}$, *where the* ¯ *over the 27 indicates that the digits 27 are to be repeated endlessly:*

$$3.\overline{27} = 3.27272727 \ldots 27 \ldots$$

32. (a) Write 5.131313 . . . 13 . . . using the above notation. _____

(b) Write 0.167167 . . . 167 . . . using the above notation. _____

33. Find repeating decimals that represent each of the following rational numbers.

(a) 2/3 = _____   (d) 4/11 = _____
(b) 5/7 = _____   (e) 16/111 = _____
(c) 5/12 = _____   (f) 6/13 = _____

■ If the "repeating" part of a repeating decimal is zero, we call the decimal a *terminating decimal* as before. For example, we say that $0.2\overline{0} = 0.2$ is a terminating decimal.

34. Discuss in your group how to show that every rational number is represented by either a repeating infinite decimal or a terminating decimal.

35. We now consider whether every repeating infinite decimal represents a rational number. Discuss the procedure outlined below, which finds the rational number represented by the repeating infinite decimal $63.\overline{471}$.

Let $\quad x = 63.\overline{471}$
Then $10x = 634.\overline{714}$
$\quad\quad 100x = 6,347.\overline{147}$
$\quad\quad 1,000x = 6,471.\overline{471}$

Notice that $1,000x$ and x have the same repeating part to the right of the decimal point. Subtracting x from $1,000x$, we get

$$999x = 63,471.\overline{471} - 63.\overline{471}$$
$$= 63,408$$

Hence

$$x = \frac{63,408}{999} = \frac{21,136}{333}$$

(We assume here that, when subtracting repeating decimals with identical "tails," the "tails" will cancel out.)

36. Use the above procedure to find the rational number represented by each of the following.

(a) $0.\overline{17}$ = _____   (b) $5.\overline{23}$ = _____

(c) $0.\overline{3} =$ _____ (e) $16.\overline{32} =$ _____

(d) $0.\overline{2} =$ _____ (f) $0.\overline{7} =$ _____

37. (a) Find a way to specify an infinite decimal that is not repeating.

(b) Do you think it is a rational number or not? _____

Comment: Terminating, Repeating, and Nonrepeating Decimals

"Mathematics and natural philosophy are so useful in the most familiar occurrences of life, and are so peculiarly engaging and delightful, as would induce everyone to wish an acquaintance with them. Besides this, the faculties of the mind, like the members of the body, are strengthened and improved by exercise. Mathematical reasonings and deductions are, therefore, a fine preparation for investigating the abstruse speculations of the law."—Thomas Jefferson

Let us summarize some information about decimals from the above exercises.

Every rational number has a decimal representation either as a terminating decimal or as a repeating decimal.

The argument that proves this is easy once you believe that the division algorithm can be extended to decimals. For then, you can find the decimal representation of m/n, by dividing n into m, using this algorithm. For example, to compute the decimal representation of $29/13$, we do the following division.

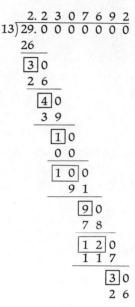

The remainders, which are enclosed in boxes above, are always less than the divisor (13 in the above example). Hence there are only a finite number of possibilities for these numbers. So, at some stage, either a remainder is zero or a remainder is repeated. If a remainder is zero, then the rational number is represented by a terminating decimal. If a remainder repeats, then from that stage on the decimal will repeat. In the above example, we have $29/13 = 2.\overline{230769}$.

We can also determine when the above division process results in a remainder of zero. If the denominator b of a fraction a/b is divisible only by 2 or 5 (or both), eventually the division process will yield a remainder of zero. The easiest way to confirm this is to observe that if b is divisible only by 2 or 5, then $b = 2^k \times 5^l$ (where $k \geq 0$ and $l \geq 0$). Thus, for some positive integer x, $bx = 10^m$. Therefore, $a/b = ax/bx = ax/10^m$, which is clearly a terminating decimal. We can therefore say:

If the denominator of a fraction is divisible only by 2 or 5 (or both), then it is represented by a terminating decimal. Otherwise it is represented by a repeating decimal.

38. For each of the following, find x such that $bx = 10^m$ for some m.

(a) $b = 40$ _____   (d) $1{,}000$ _____

(b) $b = 8$ _____   (e) 640 _____

(c) $b = 400$ _____ (f) $3{,}200$ _____

The representation of rational numbers as terminating decimals is reversible. You verified in the lab exercises that:

Every terminating decimal and every repeating decimal represents a rational number.

The relationship between rational numbers and decimals is now complete. However, we saw earlier that there are line segments that have lengths that are not rational numbers. We called these numbers *irrational numbers*. We saw that we could indicate these numbers on the number line and then approximate these points by rational points through successive division into tenths.

Thus we found that $1.4 < \sqrt{2} < 1.5$ and indicated that we could continue to obtain $1.41 < \sqrt{2} < 1.42$, $1.414 < \sqrt{2} < 1.415$, and so on, to $1.41421356 < \sqrt{2} < 1.51521357$. Of course, we would not do this geometrically, but rather would use the algebraic technique developed in Exercise 2. We say that 1.4142135 approximates $\sqrt{2}$ to seven decimal places. Although we cannot write down an infinite decimal for $\sqrt{2}$, we know that no terminating or repeating decimal represents it, and we can think of it being represented by an infinite decimal that we can approximate by a finite decimal to any desired accuracy.

39. *Optional.* Use an electronic calculator and the process in Exercise 2 to approximate $\sqrt{5}$ to three decimal places.

■ In conclusion, we can think of points on the number line that represent irrational numbers as having infinite nonrepeating decimals associated with them. Conversely, given an infinite nonrepeating decimal, we can locate a point corresponding to it. For example, if the number is 6.43152 . . . to five decimal places, we can divide the segment between 6 and 7 into tenths and say the point lies between 6.4 and 6.5. Then we divide the tenth into tenths and locate the point between 6.43 and 6.44. Continuing in this manner, we obtain a better and better approximation. It can be shown that in fact an infinite decimal does determine exactly one point.

Observe that it is easy to specify decimals that are neither terminating nor repeating. For example, we can start by writing 0.10, then 0.10100, then 0.101001000, etc. At each stage, add a 1 and some zeros, the number of zeros increasing by one at each stage. This is clearly nonrepeating and nonterminating, so it cannot represent a rational number.

40. Find another procedure that specifies a nonrepeating infinite decimal.

Important Terms and Concepts

Average
Decimal notation
Decimal point
Representing fractions as decimals
Addition, subtraction, multiplication, and division of decimals
Terminating decimal
Repeating decimal
Infinite nonrepeating decimal
Laws of exponents

Review Exercises

1. Suppose you are given $3/2 < \sqrt{3} < 2$.

   (a) Use the technique from the lab exercises to fill in the boxes below.
   $$\frac{\Box}{4} < \sqrt{3} < \frac{\Box}{4}$$

   (b) Use the above information to help you make a better approximation below.
   $$\frac{\Box}{8} < \sqrt{3} < \frac{\Box}{8}$$

2. Given that $2 < \sqrt{5} < 2\frac{1}{4}$, make two more approximations, as you did in (a) and (b) above.

3. Find a fraction less than $\sqrt{3}$ but greater than $\sqrt{2}$.

4. Fill in the boxes below with $<$ or $>$ so that (a), (b), and (c) form true number statements. Then use this information to help you approximate each of these fractions in terms of tenths and hundredths.

   (a) $\frac{5}{6} \Box \frac{8}{10}$ $\quad \frac{\Box}{10} < \frac{5}{6} < \frac{\Box}{10}$

   $\frac{\Box}{100} < \frac{5}{6} < \frac{\Box}{100}$

   (b) $\frac{2}{3} \Box \frac{6}{10}$ $\quad \frac{\Box}{10} < \frac{2}{3} < \frac{\Box}{10}$

   $\frac{\Box}{100} < \frac{2}{3} < \frac{\Box}{100}$

(c) $3\frac{3}{8} \square \frac{34}{10}$ $\frac{\square}{10} < 3\frac{3}{8} < \frac{\square}{10}$

$\frac{\square}{100} < 3\frac{3}{8} < \frac{\square}{100}$

5. Represent the following fractions in an equivalent form so that they have a power of 10 for the denominator. Then write each one as a decimal.

   (a) $\frac{7}{25}$ (c) $\frac{3}{160}$

   (b) $\frac{4}{80}$ (d) $\frac{8}{200}$

6. Find the following products.

   (a) $0.23 \times 10^4 = $ _____
   (b) $5 \times 10^3 = $ _____
   (c) $0.1 \times 10^{-4} = $ _____
   (d) $7.6235 \times 10^{-2} = $ _____
   (e) $0.00003126 \times 10^6 = $ _____
   (f) $123.1 \times 10^2 = $ _____

7. Fill in the boxes below.

   (a) $5^6 \times 5^8 = 5^{\square}$ (c) $4^3 \times 4^{-5} = 4^{\square}$
   (b) $5^6 \div 5^8 = 5^{\square}$ (d) $4^3 \div 4^{-5} = 4^{\square}$

8. Find the answers to the following problems. (Use powers of 10 to express the answers.)

   (a) $0.000002 \times .00006 = $ _____
   (b) $12.1 \times 0.00001 = $ _____
   (c) $16 \times 1.5 = $ _____
   (d) $0.1 \times 213.476 = $ _____

9. Change the following fractions into repeating decimals.

   (a) $\frac{2}{9} = $ ___   (b) $\frac{4}{7} = $ ___   (c) $\frac{4}{15} = $ ___

10. Change the following repeating decimals into fractions.

    (a) $.\overline{12} = $ ___   (b) $0.3\overline{16} = $ ___

    (c) $0.\overline{131} = $ ___

11. Explain why no terminating or repeating decimal represents $\sqrt{2}$.

12. Susan has planted a vegetable garden. She finds that gophers have eaten 25% of the carrots. If she has harvested 300 carrots, how many did the gophers eat?

13. In a math class 26% of the students are also studying geography. If 13 are studying geography, how many students are there in the math class?

14. Make up an infinite decimal that does not terminate and does not repeat. _____

In Exercises 15–22, decide whether the statement is true or false. Write T or F in the blank.

___ 15. If you subtract a rational number from an irrational number, your answer will always be an irrational number.

___ 16. If you add two irrational numbers, you will always get an irrational number.

___ 17. A terminating decimal can never represent an irrational number.

___ 18. Integers are rational numbers.

___ 19. If you divide any number by an irrational number, your answer will always be an irrational number.

___ 20. A fraction is represented by a terminating decimal if and only if the denominator is divisible by 10.

___ 21. There is no best decimal approximation of $\sqrt{3}$.

___ 22. The infinite decimals that do not repeat and do not terminate represent a special group of rational numbers.

23. *Puzzle Problem—Optional.* Express each of the numbers 1 through 10 using four 2's and the operations $+$, $-$, $\times$, $\div$ and exponentiation (raising a number to a power). For example, $5 = (2 \div 2) + 2 + 2$. (*Hint:* If you have difficulty expressing the number 7 in this way, think about why the problem is in this particular chapter.)

16

Measurement: The Metric System

"The Universe has its endless gamut of great and small, of near and far, of many and few."—D'Arcy Thompson

Lab Exercises: Set 1

EQUIPMENT: One meter stick and two containers of C-rods for each group; one or two trundle wheels for the class.

You start by discussing some basic ideas about measurement. Then you will measure some lengths using different units and study the system of names for the metric units.

1. In each of the following, describe what attribute is being intuitively measured.

   (a) "I'm not going out there. It's too cold." _____

   (b) "Sure you can lift it. It's pretty light." _____

   (c) "I'll never make it. It's just too late." _____

   (d) "My dog's a klutz. He trips over his own shadow." _____

   (e) "That hill's too steep to climb." _____

   (f) "The soup won't fit in that container." _____

   (g) "They say she's the lightest person on the track team." _____

   (h) "Wow! Did you see how fast that bicycle was going?" _____

2. Each person relate a memory of measuring something. Explain carefully what attribute you were measuring, how you measured it, and why you did it that way.

■ The concept of measurement is an integral part of everyone's experience. However, most people don't actually carry around measuring devices but rather use words to express general comparisons. For example, when a person says, "It's too far to walk," she or he is speaking in a general way about a measure of distance, using previous experience with distances to make an intuitive judgment that this distance is too far to walk.

3. List four other attributes that are measurable.

   (a) _____

   (b) _____

   (c) _____

   (d) _____

4. Suppose you lived in ancient times before our present system of measurement was developed. How might you express to another person

   (a) An amount of time _____

   (b) The distance to a lake _____

   (c) The weight of a sack of grain _____

   (d) The amount of water to use when making a loaf of bread _____

■ Although in casual conversation it suffices to say that the distance to a lake is "a stone's throw," or "1,000 steps," these specifications would not satisfy a highway engineer. As society becomes more complex, it becomes necessary to be more precise in our communication about quantities.

In order to express any amount fairly accurately, we need something with which to compare it; that is, we need a unit of measure. Historically, people would simply agree on a unit of measure. In order to measure length, for example, it was easiest for people to choose units that were available on their bodies. The ancient Egyptians used such measures as

A digit (the width at the middle of a middle finger)
A palm (the width of four fingers)
A span (the distance between the tip of the thumb and the tip of the little finger)

5. *Each* person in the group estimate the length of a rectangular table in digits, palms, and spans. Then check your estimates by measuring the length with each unit. Compare your results with the rest of your group's. Discuss the advantages and disadvantages of these units of measurements.

Unit	Estimate	Measurement
Digits		
Palms		
Spans		

6. A *cubit* is an ancient unit of length originally defined as the distance from the elbow to the end of the middle finger. Using your body, answer the following.

   (a) How many spans make a cubit? _____

   (b) How many palms make a span? _____

   (c) How many digits make a palm? _____

   (d) How many digits make a cubit? _____

7. If a piece of rope measures 27 cubits in length, how long is it in digits? _____

8. The people of Camelot were fortunate in that the units derived from King Arthur's body were such that each unit was an exact multiple of the next smaller unit. It worked out so that 6 digits equaled 1 palm, 2 palms equaled 1 span, and 3 spans equaled 1 cubit.

   (a) Complete the following.

   (i) 1 span = _____ digits

   (ii) 1 cubit = _____ digits

   (iii) 4 cubits = _____ digits

   (iv) 1.5 spans = _____ digits

   (v) 3.2 cubits = _____ palms

   (vi) 0.8 spans = _____ digits

   (vii) 1.6 cubits = _____ digits

   (b) Discuss why the people of Camelot were fortunate. What is the advantage of having each unit an exact multiple of the next smaller?

9. Why would it be even more convenient if each unit were *the same* multiple of the next smaller unit? For example, suppose that 5 digits = 1 palm, 5 palms = 1 span, and 5 spans = 1 cubit. Discuss.

10. The land of Camelot is considering whether to establish a new system of measurement. One proposal is to use the magician Merlin's magic wand as a unit, and to define new units as: 1 wand = 10 pencils, 1 pencil = 10 fingers, and 1 longwand = 10 wands.

    (a) Complete the following.

    (i) 1 wand = _____ fingers

    (ii) 1 longwand = _____ fingers

    (iii) 4 longwands = _____ fingers

    (iv) 1.5 wands = _____ fingers

    (v) 3.2 longwands = _____ pencils

    (vi) 0.8 wands = _____ fingers

    (vii) 1.6 longwands = _____ fingers

(b) Your group is appointed as mathematical consultant to King Arthur's court. What are the arguments in favor of this new system?

(King Arthur, of course, likes the convenience of having the units readily available on his body, and needs to be convinced of the virtues of this new system.) Compare the computations in Exercise 10(a) with those in Exercise 8(a).

■ In the metric system the basic unit of measure is the *meter*. A meter is a little bigger than the width of a classroom doorway—or for most adults, it is about the distance from the left shoulder to the tip of the right hand when the right arm is outstretched. The abbreviation for meter is m.

11. Guess how many meters long your room is. _____ Use a trundle wheel (Figure 16-1), a meter stick, or a metric tape measure to measure your room's length. _____

12. Have someone in your group use chalk to mark off an arbitrary distance on the floor. Each person estimate this distance in meters. _____ What is the actual measure of the distance? _____

13. Many people use their pace as a means of estimating distances. Mark off 10 meters in your room (or the hall outside) and walk that distance, counting your paces. Do it five times and record how many paces make 10 meters: _____, _____, _____, _____, _____. What is the average? _____

Figure 16-1

14. Pace off the length of your hallway, the width of the room, or some other fairly long distance. How many paces is it? _____ Use this to estimate the length in meters. _____ Use a measuring tool to measure the distance. _____ How far off were you in your estimate? _____

15. Use the meter stick to find a 1-meter measure on your body. Describe this measure. _____

_____

(This may be useful to you someday.)

■ Of course, the meter isn't always the handiest of units. For example, it would be unwise to try to measure the growth of your pet caterpillar with a 1-meter trundle wheel (especially if you want it to continue

growing). What you need is a smaller unit. In the metric system the smaller units are

1/10 meter
1/100 meter
1/1,000 meter

To name these smaller units, the following Latin prefixes are used.

1/10 deci-
1/100 centi-
1/1,000 milli-

Hence

1/10 meter = 1 decimeter
1/100 meter = 1 centimeter
1/1,000 meter = 1 millimeter

The centimeter is the most practical unit for measuring small lengths, so we will concentrate on developing familiarity with that unit. You can look up the others when you need to. The abbreviation for centimeter is cm. Remember that 1 cm = 1/100 m.

16. How many centimeters are there in 1 meter? _____

■ Figure 16-2 depicts a 15-centimeter ruler. Use it to measure the length of the white C-rod. _____

17. (a) What are the measures of all the C-rods?

W _____ P _____ Bl _____ O _____
R _____ Y _____ Bk _____
LG _____ DG _____ Br _____

(b) From your experience with C-rods, you probably already have a good intuitive sense of small metric lengths. For example, without looking at the rods, try to draw a line below, which measures 1 centimeter. Use a white rod to see how close you are.

(c) Without looking at the C-rods, try to draw line segments having lengths of 3 centimeters, 5 centimeters, and 10 centimeters. Again see how close you are by comparing your segments with the appropriate C-rod.

18. Use the meter stick to measure the height of each person in the group. (For some reason the general practice is to express people's height in centimeters rather than meters.) What is your height in centimeters? _____ .

19. Now measure one another's armspan in centimeters. What is your armspan? _____ How close is it to your height?

■ Meters and centimeters are good for measuring caterpillars, people's heights, or dimensions of rooms. But if you want to measure the distance between cities, or the distance to the moon, it would be convenient to have larger units. The larger metric units are named with Greek prefixes. They are

10 deka-
100 hecto-
1,000 kilo-

Thus,

10 meters = 1 dekameter
100 meters = 1 hectometer
1,000 meters = 1 kilometer

The kilometer turns out to be the most practical unit for measuring large distances. We will concentrate on developing some familiarity with it. The abbreviation for kilometer is km.

20. The distance from Los Angeles to New York is approximately three times the distance from New York to St. Louis. The distance from Los

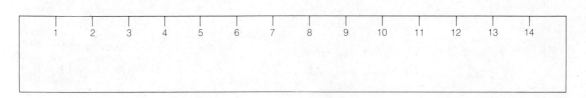

Figure 16-2

Angeles to New York is approximately 4,800 km. What is the distance between New York and St. Louis?

21. It is proposed that the new speed limit on U.S. highways be set at 90 kilometers per hour. The distance between Los Angeles and San Francisco is 650 kilometers. If you drive at the speed limit, about how long would it take you to drive from Los Angeles to San Francisco?

22. The current world record for the 1,000-meter run is 2 minutes, 13.9 seconds, and is held by Rick Wolhuter of the United States. How long would it take him to run a kilometer?

23. Mount Everest is 8,547 meters tall, and Mt. McKinley is 6,193 meters tall. What is the difference of these heights in kilometers?

Comment: The Metric System

"And there went out a champion out of the camp of the Philistines named Goth of Goliath, whose height was six cubits and a span."—I Samuel 17:14

From earliest times, humans have been interested in measuring quantities—length, area, volume, weight, time, loudness, velocity, intensity of earthquakes—just to name a few. In the above set of lab exercises we focused on the measurement of length.

In order to measure any quantity, it is necessary to choose a unit. How should we do this? Many units are possible. You saw in the lab above that if everyone chose the most convenient unit for this—some part of their body—then everyone's measurements would differ, which turns out not to be very convenient at all.

The most commonly used units of measurement in the world today comprise a system called the *metric system*. Although the United States still uses the English system of measurement, it seems inevitable that we will "go metric."

A federal commission has been instituted to coordinate the transition to the metric system, and the California State Board of Education, for one, has instructed that all materials used to teach measurement employ the metric system and avoid computational conversion to the English system. As teachers, you will undoubtedly be teaching the metric system.

The metric system was originally proposed by the French Academy of Science in 1791. The people who proposed it were searching for the most convenient system possible. They decided that

1. Each unit should be ten times the next smaller unit (making the arithmetic easier).
2. The names for the different units should have some relation to one another (so users wouldn't have to memorize unrelated names like inch, foot, yard, rod, mile).
3. The basic units of measure for such things as distance, weight, volume, temperature should all be based on things from nature and interrelated with one another.

In order to build their system, they decided to choose as a unit of length one ten-millionth (1/10,000,000) of the distance from the equator to a pole. They called this unit the *meter,* derived from a Greek word meaning "to measure." The measurements were made along a meridian of longitude from Dunkirk, France, to a town near Barcelona, Spain.

Scientists all over the world rapidly adopted the metric system, but it was not until 1878 that twenty nations sent representatives to a conference. This conference agreed to define the meter to be the distance between two finely engraved lines on a platinum-iridium rod (which expands and contracts very little with changes in temperature). The rod was kept in France, with copies made available to each of the nations participating in the "Metric Convention"—as it came to be known. Since 1878, almost

By permission of Selby Kelly.

every nation in the world has adopted the metric convention. Only the United States and a few African countries still use the English system of measurement. England adopted the metric system in 1975.

In 1960, a new standard was adopted for the meter. A meter is now defined as 1,650,763.73 wavelengths of the light emitted by heated krypton gas. The lengths of these waves are always the same and can be used to determine a meter in any well-equipped scientific laboratory. Of course, the length of the new unit is the same as the length of the old.

After choosing the standard unit of length, the proposers of the metric system proceeded to build a system of measurement. They knew that their larger units were going to be 10 meters, 100 meters, and 1,000 meters, and that the smaller units would be 1/10 meter, 1/100 meter, and 1/1,000 meter. In order to make the names for these units simple, they decided to use Greek prefixes for 10, 100, 1,000, and Latin prefixes for 1/10, 1/100, 1/1,000. The names of these units are

kilometer	1,000 meters
hectometer	100 meters
dekameter	10 meters
meter	1 meter
decimeter	0.1 meter
centimeter	0.01 meter
millimeter	0.001 meter

When you were mathematical consultants in Camelot in Exercise 20 above, you probably noticed the ease of converting from one unit to another when using a Base 10 measurement system, i.e., a system in which each larger unit equals 10 of its predecessor. Let us briefly explore how this works with length in the metric system.

Suppose a room is 9 meters long and you want to know how long it is in centimeters. Since 1 meter = 100 centimeters, 9 meters = 9 × 100 centimeters = 900 centimeters. This employs

A basic principle of measurement: *If any equality of measure is multiplied by a positive number, the equality is preserved.*

As another example, suppose a length of rope measures 326 centimeters. How long is it in meters? Since we know that 1 centimeter = 0.01 meter, we can multiply by 326 to obtain 326 centimeters: 326 × 0.01 meters = 3.26 meters.

You may have observed that these results can be accomplished simply by moving the decimal point. Thus, to convert centimeters to meters, move the decimal point two places to the left. To convert meters to centimeters, move the decimal point two places to the right. Similar rules are true of the other conversions between units of metric length. However, you do not need to memorize these rules. If you know the relationship between the units and remember the basic principle of measurement stated above, you can figure out the rule. For example, suppose you wish to convert centimeters to kilometers. You know that (1) 1 meter = 100 centimeters, and (2) 1,000 meters = 1 kilometer. Multiplying (1) by 1,000, you obtain 1,000 meters = 100,000 centimeters. Now, using (2) you see that 1 kilometer = 100,000 centimeters. Written with the decimal point, this is 1.0 kilometer = 100,000.0 centimeters. Thus, to convert from kilometers to centimeters, move the decimal point five places to the right.

24. Complete each of the following.
    (a) 24 millimeters = _____ meters
    (b) 0.13 meter = _____ centimeters
    (c) 25 centimeters = _____ millimeters
    (d) 186 millimeters = _____ centimeters
    (e) 24 kilometers = _____ millimeters
    (f) 24 decimeters = _____ kilometers
    (g) 321 dekameters = _____ kilometers
    (h) 15 hectometers = _____ decimeters

25. Complete each of the following.
    (a) To convert decimeters to millimeters, move the decimal point _____ places to the _____.
    (b) To convert dekameters to kilometers, move the decimal point _____ places to the _____.
    (c) To convert kilometers to centimeters, move the decimal point _____ places to the _____.
    (d) To convert hectometers to decimeters, move the decimal point _____ places to the _____.

■ The basic principle of measurement is also used when converting from the metric system into the English system or vice versa. For example, 1 inch = 2.54 centimeters (approximately). To find out the metric measure of a foot, we multiply this equality by 12, to obtain 12 inches = 12 × 2.54 centimeters = 30.48 centimeters.

26. A rough rule for converting feet to meters is to multiply the number of feet by .3. Justify this rule.

27. How many inches are there in a centimeter?

28. How many feet in a meter?

29. How many miles in a kilometer?

Area

When we measure the area of a region using the metric system, the unit of area we use depends on the unit of length we have chosen. Suppose we measure length in centimeters. Then the unit of area is a square with sides of length 1 centimeter. We call this unit of area a *square centimeter*, which we abbreviate as cm^2. For example, the rectangle in Figure 16-3 has an area of 15 cm^2 (15 square centimeters) since it has length 5 cm and width 3 cm.

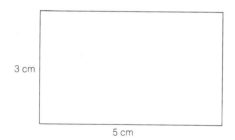

Figure 16-3

Similarly, if length is measured by meters, the unit of area is a square meter (abbreviated m^2) and if by kilometers the unit of area is a square kilometer (abbreviated km^2). Thus, a rectangle measuring 6 m by 8 m has an area of 48 m^2 (48 square meters).

30. A rectangular region is being considered by the U.S. Forest Service as an addition to a wilderness region. It measures 50 km by 26 km. What is its area? _____

31. A triangle has sides 5 cm and height 4 cm. What is its area? _____

■ Suppose we want to find the area of a square with sides of length 100 cm. We can compute the area of the square in two ways. First, the area is 100 cm × 100 cm = 10,000 cm^2. On the other hand, since 100 cm = 1 m, the square is 1 m × 1 m. Hence it has area 1 m^2. This shows that 1 m^2 = 10,000 cm^2.

32. How many square meters are there in a square kilometer? _____

Lab Exercises: Set 2

EQUIPMENT: Two meter sticks for each group, two metric tape measures for each group (or a roll of adding-machine tape).

The first exercise will give you some appreciation of the merits of the metric system. You will then discover the sources of inaccuracy inherent in any measurement.

33. (a) If carpet costs $6 a square meter, what will it cost to carpet a floor measuring 251 cm by 350 cm? (*Hint:* 251 cm = 2.51 m)

(b) If carpet costs $6 per square yard, how much will it cost to carpet a rectangular floor measuring 8 ft, 3 in. by 11 ft, 6 in.? (*Hint:* 1 sq. yd = 9 sq. ft = 1,296 sq. in.)

(c) Which computation is easier?

34. Each person in your group measure the length and width of your room with a meter stick. It is important that each person do this independently.

(a) length = _____ cm

(b) width = _____ cm

Use the chart on the following page to summarize

the results for each person (A, B, C, D, E) in your group.

Meter stick

	A	B	C	D	E
Length					
Width					

35. Are the results exactly the same? _____ (In general, they will not be.) Discuss in your group why they are not. List below all the sources of error that you think might influence the measurement. Be sure to include the errors from the process of measuring, as well as from the instrument.

36. Two people in your group now measure the width and length of your room, using a metric tape measure. Record the results below. (If you do not have a metric tape measure, you can make one from a roll of adding-machine tape.)

Tape measure

	A	B
Length		
Width		

Do these results agree with the previous ones? _____ Why might they differ? _____
Where are the sources of error in using a tape measure?

Do you think the meter stick or tape gives a more accurate measure? _____ How would using the trundle wheel compare in accuracy? _____ Why? _____

37. (a) Using Exercise 36, what is your best estimate of the length of your room, to the nearest centimeter? _____ On the basis of your measurement, what is the shortest possible length of the room? _____ What is the longest? _____ Express your answer as a number ± possible error. (For example, if the shortest possible length is 548 cm, and the longest is 550 cm, then you would write 549 ± 1 cm. The length is 549 cm with a possible error of 1 cm.)

$$\text{length} = \_\_\_\_ \pm \_\_\_\_ \text{ cm}$$

(b) What is your best estimate of the width of your room, to the nearest centimeter? _____ On the basis of your measurements, what is the shortest possible width? _____ What is the longest? _____ Express your answers in the form used above:

$$\text{width} = \_\_\_\_ \pm \_\_\_\_ \text{ cm}$$

38. The perimeter of a rectangle is the sum of the lengths of the sides.
   (a) On the basis of your measurements, what is the longest possible perimeter of your room? _____
   (b) What is the shortest? _____
   (c) Express the perimeter as _____ ± _____ cm.
   (d) How does the error in the perimeter compare with the error in measuring the sides? _____
   (e) Complete the following: The possible error of the sum of two or more measurements is equal to the _____ of their individual errors.
   (f) What would be the possible error for the difference of two measurements?

39. On the basis of your answers to Exercise 37:
   (a) What is your best estimate of the area of your room? _____
   (b) What is the greatest possible area? _____
   (c) What is the least possible area? _____
   (d) Express the area in square centimeters and square meters in the form _____ ± _____ cm^2 and _____ ± _____ m^2.

■ In order to see what the error will be when two approximate numbers are multiplied, it is necessary to introduce the concept of *percentage* (or *relative*) *error*. Suppose the length of a line is measured to be 25 cm with a possible error of 1 cm. We write that the length is 25 cm ± 1 cm. The percentage error is 1/25 = 4/100 = 0.04 or 4 percent.

In general, to find the percentage error, divide the possible error by the measurement and express the resulting decimal as a percent.

40. (a) When astronomers say that the distance from the earth to the sun is 93,000,000 miles, they mean that it is between 92,500,000 miles and 93,500,000 miles. That is, the distance is 93,000,000 ± 500,000. What is the percentage error? ____

  (b) A wooden beam is measured as 125 cm ± 0.5 cm. What is the percentage error? ____

■ What happens to the percentage error when we multiply two approximate numbers? Suppose a rectangle is measured and one side has length $a \pm e_1$ and the other side has length $b \pm e_2$. The minimum area of the rectangle is $(a - e_1)(b - e_2) = ab - e_1 b - e_2 a + e_1 e_2$.

41. (a) What is the maximum area of this rectangle?

Since e_1 and e_2 are very small compared to a and b, the product $e_1 e_2$ can be omitted without significantly affecting the result. Hence we can express the area as

$$ab \pm (e_2 a + e_1 b)$$

  (b) Show that the relative error is

$$\frac{e_2}{b} + \frac{e_1}{a}$$

We can conclude that the **percentage error** *of the product of two measurements is approximately equal to the sum of the percentage error of each factor.*

42. Suppose the sides of a rectangle are measured to be 100 ± 0.5 cm and 200 ± 0.5 cm.

  (a) What is the percentage error of each side? ____ ____

  (b) What is the percentage error of the area? ____

Comment: Errors in Measurement

"Indulge your passion for science . . . but let your science be human."—David Hume

The lab work above should have convinced you that the process of measuring anything involves inaccuracy. Although, theoretically we can achieve any desired degree of precision, absolute accuracy is unobtainable. What are the possible sources of error? An article in the National Council of Teachers of Mathematics Yearbook suggests that the sources of error can be grouped into four categories:*

1. Assumptions about the object to be measured.
2. Choice of the measuring instrument to be used.
3. Use of the measuring instrument.
4. Calculations, if any, that are made.

Let us analyze your lab work with reference to these sources of error.

1. In measuring the length of your room, you probably measured along some line that you assumed to be straight. A likely choice would be the intersection of a wall and the floor. But the wall or the floor or both may not be perfectly flat, so the line may not be straight.

  Also, in measuring the area, we tacitly assumed that the room was a rectangle. But it may be a little distorted. Perhaps it is a parallelogram or some other quadrilateral.

2. The scale on the measuring instrument obviously limits the accuracy of the measurement. Think about the differences that would have been introduced if your meter stick did not have any indications of centimeters or millimeters, or if you had only a trundle wheel to work with. It might also be that there is an error in the calibration of the meter stick (or other instrument).

3. In measuring a length, you probably found that different people obtained different results. This is due to the human error in using the instrument. Within this category are errors due to mistakes on the part of the person doing the measurement as well as the inevitable variability in reading instruments. An example of a mistake in using the meter

---

*Donald Kerr and Frank Lester, "An Error Analysis Model for Measurement" (NCTM Yearbook, 1976), pp. 105–122.

stick would be to not accurately mark the floor when moving the meter stick.

4. Computational errors include mistakes in arithmetic as well as errors due to rounding off in your calculations. For example, an example of the latter might occur if a rectangle is measured to be 3.74 m by 8.24 m and you wish to find its area to the nearest 0.1 m². One person might round off the measurements and compute $3.7 \times 8.2 = 30.34$ and give the area as approximately 30.3 m². However, if she had an electronic calculator, she would probably compute $3.74 \times 8.24 = 30.8176$ and give the area as approximately 30.8 m². A full analysis of the rounding-off process is interesting, but beyond the scope of the present chapter.

Before rounding off the chapter, let us discuss how errors accumulate. Then we will briefly discuss the metric units for measuring volume, weight, and temperature.

Recall that we can express a measurement as a number plus or minus the possible error. For example, if we know a length to be between 224 and 228 centimeters, we can write it as 226 ± 2 cm (or in meters as 2.26 ± 0.02 m).

Now suppose we measured the lengths of two beams and were going to put them together end to end. What would be the error in the total length if we added the two lengths? If one beam is 328 ± 0.5 cm long and the other beam is 241 ± 0.5 cm long, we know the new length is 569 ± 1 cm. This is so because the first beam is between 327.5 cm and 328.5 cm and the second beam is between 240.5 cm and 241.5 cm. Adding the two extremes, we see that the new length is between 568 cm and 570 cm, which is expressed as 569 ± 1 cm. We see that the possible error of the sum of two measurements is equal to the sum of the individual errors. (Note that if we were to measure the new length again, we would undoubtedly reduce the possible error, but we are not considering that possibility.)

Now let us consider the difference of two measurements. Suppose we have a beam that is 438 ± 1 cm long and we cut a new beam off this, which we measure to be 125 ± 1 cm. What can we say about the length of the remaining beam? Since the original beam is between 437 cm and 439 cm, and the piece removed is between 124 cm and 126 cm, we see that the shortest possible length of the remaining piece is 437 cm − 126 cm = 311 cm, and the longest possible length is 439 cm − 124 cm = 315 cm. Thus the length of the new beam can be expressed as 313 ± 2 cm. We see that the possible error of the difference of two measurements is equal to the sum of the individual errors. If we have more than two measurements, the above reasoning extends to that case. We can thus state:

The possible error in the sum or difference of two or more measurements is equal to the sum of the individual errors.

43. (a) Two students measure the length and width of a rectangular court. One student is given a ruler marked in meters. The second student is given a ruler marked in centimeters. The first says the length is 22 ± 0.5 m and the width is 10 ± 0.5 m. The second says the length is $2,238 \pm 10$ cm and the width is $1,035 \pm 10$ cm. Explain how both could be right.

(b) The same students are then asked to estimate the perimeter of the court. The first says it is 64 m, and the second that it is 6,546 cm. A third student lays a rope around the perimeter of the court, stretches it out straight, and measures it. This student says the perimeter is 6,575 cm. How do you explain the different answers to the students?

(c) If the third student had obtained an answer of 6,612 cm, would your explanation be different?

■ When we multiply numbers representing measurements, the analysis of the possible error becomes more complex. If, for example, a room is measured to be 600 ± 1 cm by 900 ± 1 cm, then its largest possible dimensions are 601 cm by 901 cm and its smallest possible dimensions are 599 cm by 899 cm. Hence the largest possible area is 601 cm by 901 cm = 541,501 cm² and the smallest possible area is 599 cm × 899 cm = 538,501 cm². We see that the area is close to 540,000 cm², but the possible error in the product is certainly not the product of the possible

16/Comment

errors. In order to analyze this situation, we introduce the concept of percentage error.

Definition: *The percentage error is the possible error divided by the measurement expressed as a percent.*

Thus the percentage error for the 600 cm measurement is $1/600 = 0.167\%$, and the percentage error for the 900 cm measurement is $1/900 = 0.111\%$. The sum of these two is 0.278%, and 0.278% of $540,000$ cm^2 is $1,501.2$ cm^2, which is *approximately* the difference between $540,000$ cm^2 and each of the two extreme measurements above.

The percentage error in the product of two measurements is approximately equal to the sum of the percentage errors of each factor.

Since you verified this in Exercise 41, we need not repeat it here.

44. Because of a fuel and water shortage, you have decided to fill your swimming pool with sand. The measurements of the pool, with the percentage errors, are:

Length:	12 meters	0.001 percentage error
Width:	6 meters	0.002 percentage error
Depth:	2 meters	0.005 percentage error

    What volume of sand should you order from the sand company so as to be sure that the pool will be filled?

45. The sides of a rectangle measure 80 ± 0.5 cm. Find the percentage error of each side, and the percentage error in the area.

■ Let us now make a brief survey of the metric units for volume, weight, and temperature.

The unit of volume is called a *liter*. It is the amount of space in a cube that measures 10 centimeters on each side. Such a cube is made up of 1,000 cubes measuring 1 centimeter on a side, so we see that 1 liter = 1,000 cubic centimeters. Since a milliliter is 1/1,000 of a liter, 1 milliliter = 1 cubic centimeter (Figure 16-4). If you can obtain a 1-liter bottle and make a cardbox cube 10 cm × 10 cm × 10 cm, you can fill one with sand and pour it into the other to see that they hold equal volumes.

When choosing a unit of measure for weight, the creators of the metric system decided to take 1 milliliter of water and let its weight be called 1 *gram*. This became the unit of weight. A gram is a very small amount—almost exactly the weight of a dollar bill.

To measure weight, one of the most commonly

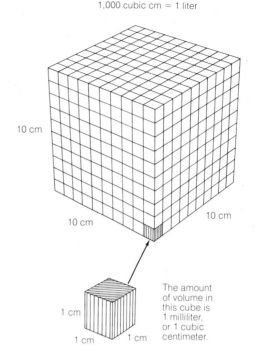

Figure 16-4

used units is the *kilogram*. It is equivalent to about 2.2 pounds. What is your weight in kilograms? _____

To measure temperature, the metric system uses a system devised by the Swedish astronomer Anders Celsius, who assigned the freezing point of water to be zero degrees (0°), and the boiling point of water to be 100°), as shown in Figure 16-5. Gradations between are measured in terms of degrees *Celsius* in his honor. (He called it the *centigrade* scale.) In order to distinguish this scale from the Fahrenheit scale, we usually write a C after a temperature. Thus water boils at 100°C, or it is 20°C outside today.

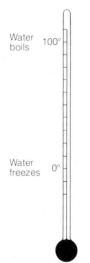

Figure 16-5

In conclusion, it is important to remember that the metric system is not something that you learn from a chapter in a book, or from a single class, or from a special metric workshop. Rather, you learn the metric system by using it. The goal in this chapter was to give you a start. Hopefully you now have enough background information so that when you encounter metrics again, you will pick it up a little easier. Eventually, when road signs are changed to kilometers, when fruit and meat are purchased in grams and bottled beverages are marked in liters, when measuring cups, tape measures, and bathroom scales are all converted to metrics, then all of us will use metrics as a part of our everyday lives, and it will be easier to learn the finer points of the metric system.

Important Terms and Concepts

Units of measure

Metric units for length, volume, and weight

System of names for the metric units

Converting from one unit to another

Sources of error

Percentage error

The error of the sum or product of two measurements

"If any equality of measure is multiplied by a positive number, the equality is preserved."

Review Exercises

1. In order to measure volume in Camelot, it is proposed that a *jar* be used as the unit. It is found that Merlin's *hat* contains exactly 10 jars of water and 10 of Merlin's hats will fill a *basket*. Furthermore, 10 baskets fill up a *barrel*. (In Camelot, everything is perfect!) So Camelot has the following system of volume measurement.

    1 hat = 10 jars
    1 basket = 10 hats
    1 barrel = 10 baskets

    Complete the following.

    (a) 5 baskets = _____ jars

    (b) 5 barrels = _____ jars

    (c) 1/2 of a hat = _____ jars

    (d) 1/2 of a hat = _____ baskets (Express as a decimal.)

2. Suppose Camelot decides to adopt Merlin's system of measurement (Exercise 10). It is found that 1 wand = 5 palms. Using the system of Exercise 8, complete the following conversion table.

    (a) 1 longwand = _____ palms

    (b) 1 wand = _____ spans

    (c) 1 pencil = _____ digits

    (d) 6 pencils = _____ palms

3. A road sign says that you are 300 miles from San Francisco. What will it say after it is changed to the metric system? _____

4. A playground has the shape of a regular pentagon, as shown below. One side is measured and found to be 82.3 m ± 0.05 m. Express the perimeter, indicating the possible error. If the playground is to be fenced, what is the minimum amount of fencing necessary to be sure that there will be enough to enclose the playground? _____

5. A rectangle measures 48 cm by 3 m. What is its area? _____

6. A cement firm measures a rectangular sidewalk and finds its length 40 m ± 0.1 m and its width 1 m ± 0.01 m. In order to make sure they have enough cement, they need to know the possible percentage error in the area. What is it? _____

7. How did the originators of the metric system define the following?

    (a) Meter _____

    (b) Liter _____

    (c) Gram _____

    (d) Degrees Celsius _____

8. Goliath of Goth measured 6 cubits and a span in height. If a cubit is 45 cm and a span is 22 cm, how tall was Goliath in centimeters? _____ How tall in feet and inches? _____

9. What is the volume in liters of each of the following?

    (a) _____

1 cm × 1 cm × 1 cm

16/Review Exercises

(b) _____

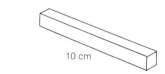

(c) _____

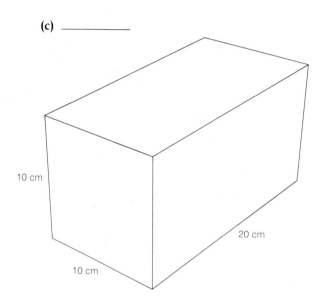

10. Indicate the equal volumes and tell how much each volume of water would weigh. (The drawings are not drawn to scale.)

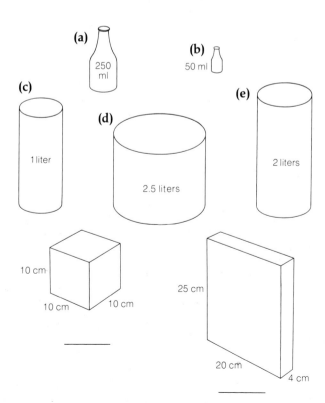

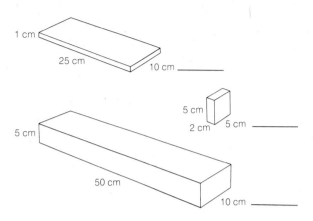

11. **(a)** Which weighs more—a kilogram of feathers or a kilogram of lead?

(b) Which weighs more—a liter of feathers or a liter of lead?

12. Using a Celsius thermometer, water freezes at 0°C and boils at 100°C (Figure 16-6a). Using a Fahrenheit thermometer, water freezes at 32°F and boils at 212°F (Figure 16-6b). How many degrees are there from freezing to boiling in each system?

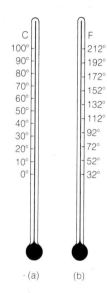

Figure 16-6

(a) Celsius _____

(b) Fahrenheit _____

Can you make any observation about the difference in the degree between the two systems?

Look at the two thermometers in Figure 16-6. If normal body temperature is 98.6°F, what would you estimate it to be in Celsius? _____

13. A kilogram is approximately 2.2 pounds. If tomatoes are selling for 89 cents per pound, what would be their price if they were selling by the kilogram? _____

14. (a) A turkey weighs 14 pounds, 12 ounces. What will it cost if it sells at 82¢ per pound? _____
    (b) A turkey weighs 6.7 kilograms. What will it cost if it sells at $1.81 per kilogram? _____

15. A paper clip weighs about 1 gram. A bicycle weighs about 15 kilograms. How many paper clips would you need to equal the weight of the bicycle? _____

16. Suppose that in a certain gear, one rotation of the pedals of your bicycle will carry you 4 meters. How many rotations must you pedal in order to go 1 kilometer? _____ 10 kilometers? _____ About how many times per second would you have to pedal in order to go 330 meters per second (about the speed of sound)? _____

By permission of Johnny Hart and Field Enterprises, Inc.

17 Tessellations: Mathematics in Art and Floor Tilings

"Let no one who is not a mathematician read my work."—Leonardo da Vinci

Lab Exercises: Set 1

EQUIPMENT: One pair of scissors and one straightedge per group. One set of pattern blocks (see page xviii) for every two groups.

In these exercises you will investigate some properties of polygons that intrigued the Dutch graphic artist M. C. Escher. We begin by obtaining information about the angles of polygons and then study how they fit together.

1. What polygons do you see in this reproduction of the Escher lithograph *Reptiles* (Figure 17-1)?

Figure 17-1

■ We begin our inquiry by studying the angles of a polygon. In order to simplify our discussion let us assume in the chapter that all polygons under consideration are *convex*.

In Exercise 24 of Chapter 13, we learned that the sum of the angles of any triangle is 180°. In the following exercises, we will investigate the sum of the angles of other convex polygons.

2. Draw a convex pentagon on a piece of scrap paper and extend the sides as indicated in Figure 17-2. The angles P, Q, R, S, and T are called the *interior angles* of the pentagon (or simply the *angles* of the pentagon). The angles marked P', Q', R', S', and T' are called *exterior angles* of the pentagon. Label the angles as shown. Cut out the angles P', Q', R', S', and T', and piece them together. What is the sum of the exterior angles in degrees? _____ (Note: At each vertex we choose to extend a particular side. If we had extended the other side, another exterior angle would have been formed. But both these angles have the same measure. Do you see why?)

Figure 17-2

3. Figure 17-3 pictures a hexagon. Indicate whether each of the designated angles is an interior angle of the hexagon or an exterior angle of the hexagon.

A _____   D _____
B _____   E _____
C _____   F _____

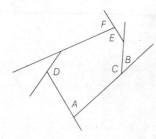

Figure 17-3

4. Refer to the pentagon PQRST in Figure 17-4. Denote by $(\angle P)°$ the measure of angle P in

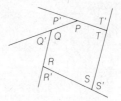

Figure 17-4

degrees; $(\angle P')°$, the measure of angle P', etc. Fill in the blanks below.

(a) $(\angle P)° + (\angle P')° = $ _____
(b) $(\angle Q)° + (\angle Q')° = $ _____
(c) $(\angle R)° + (\angle R')° = $ _____
(d) $(\angle S)° + (\angle S')° = $ _____
(e) $(\angle T)° + (\angle T')° = $ _____

Now add the left-hand sides of these equations together and then add together the right-hand sides and set the sums equal to get

$[(\angle P)° + (\angle Q)° + (\angle R)° + (\angle S)° + (\angle T)°]$
$+ [(\angle P')° + (\angle Q')° + (\angle R')° + (\angle S')° + (\angle T')°]$
$= $ ___ + ___ + ___ + ___ + ___ = ___

Now from Exercise 2 above, the sum of the exterior angles equals 360°.

$(\angle P')° + (\angle Q')° + (\angle R')° + (\angle S')°$
$\qquad\qquad + (\angle T')° = 360°$

Therefore
$(\angle P)° + (\angle Q)° + (\angle R)° + (\angle S)°$
$\qquad\qquad + (\angle T)° + 360° = 900°$

Hence we obtain
$(\angle P)° + (\angle Q)° + (\angle R)° + (\angle S)° + (\angle T)°$
$\qquad\qquad = $ _____

You have found the sum of the angles of a pentagon.

Now write the above equation as

$(\angle P)° + \cdots + (\angle T)° = \square \times 180°$

5. (a) Draw a hexagon on a piece of scrap paper and repeat Exercise 2. The sum of the exterior angles of the hexagon is _____. Compare your result with the result in Exercise 2. Discuss in your group an explanation for why the answers are the same.

(b) Repeat the procedure of Exercise 4 to find the sum of the angles of the hexagon.

Polygon	No. of sides	No. of diagonals from one vertex	No. of triangles	Measure of the sum of the angles
Triangle	3	0	1	180°
Quadrilateral	4			
Pentagon	5	2	3	
Hexagon				
Octagon				
Decagon				
n-gon	n			

(c) Make a guess as to the sum of the angles of a polygon with n sides. (We shall call this an n-gon.) The sum of the angles of an n-gon is

$\square \times 180° - 360°$
$= \square \times 180° - (2 \times 180°)$
$= (\quad) \times 180°$

■ In the following exercises we use a different method to determine the sum of the angles of a polygon. It is often true in mathematics that there is more than one way to demonstrate a particular fact or do a particular problem. (Try to remember that when you teach.) We will use the fact that the sum of the angles of a triangle is 180°.

6. Consider the polygon in Figure 17-5. The diagonals of the polygon at the vertex P are drawn as dotted lines.

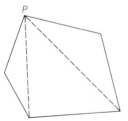

Figure 17-5

(a) How many diagonals are there? _____

(b) How many triangles are formed? _____

(c) Discuss why the sum of the interior angles of the pentagon equals the sum of the angles of all the triangles.

(d) The measure of this sum is _____°.

Definition: *A polygon whose sides are all the same length is an* **equilateral** *polygon. A polygon whose angles all have the same measure is an* **equiangular** *polygon. A* **regular** *polygon is one that is equilateral and equiangular.*

7. Use the procedure in Exercise 6 to complete the following table.

8. (a) Is every equilateral polygon an equiangular polygon? _____ Justify your answer by either showing why it is true or by providing a counterexample.

(b) Is every equiangular polygon an equilateral polygon? _____ Justify your answer.

(c) Answer (a) and (b) above for triangles. Is every equilateral triangle equiangular? _____ Justify your answer. (*Hint:* An equilateral triangle is isosceles. What do you remember about the base angles of an isosceles triangle?) Do you think that every equiangular triangle is equilateral? _____ (No justification is necessary.)

(d) What is a regular quadrilateral called? _____

9. (a) In Exercises 5 and 7 you showed that the sum of the angles of an n-gon is $(n-2) \times 180°$. We can use the fact that each angle of a regular polygon has the same measure to calculate the measure of each angle. For example, if $n = 7$, the sum of the angles of the regular septagon is _____°. Since each angle has the same measure, we divide this number by 7 and see that the measure of each angle is _____°.

(b) Use this procedure to calculate the measure of an angle of a *regular*

(i) Triangle _____ (A regular triangle is usually called an *equilateral* triangle.)

(ii) Quadrilateral (square) ____

(iii) Pentagon ____

(iv) Hexagon ____

(v) Septagon ____

(vi) Octagon ____

10. Take out your pattern blocks and play with them for a few minutes. Make some designs with them.

11. Pretend that you wish to tile a floor with the pattern blocks. Can you tile a floor using just one shape? More than one? Have you seen any of these designs in other places? Do any occur in nature? (*Hint*: There is a honey of an answer to this one.)

■ A *tiling* of a plane is any arrangement of polygons that will cover the entire plane without overlapping. Tilings that exhibit regularities are the most interesting. We will say that two vertices of a tiling are congruent if the arrangement of line segments emanating from one vertex would fit exactly on the arrangement of line segments emanating from the other vertex. If every vertex of the tiling is congruent to every other vertex, we call the tiling a *tessellation* of the plane. In other words, a tessellation of a plane is a covering of the plane with polygons so that each vertex looks the same as any other vertex; that is, the lines and the angles of any vertex match. For example, Figure 17-6 is a tessellation of the plane with parallelograms, and Figure 17-7 is a tessellation of the plane with squares.

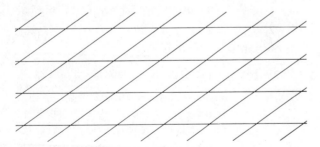

Figure 17-6

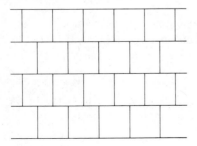

Figure 17-7

However, Figure 17-8 is a tiling but *is not* a tessellation since the vertices are not congruent. Circle two non-congruent vertices in Figure 17-8.

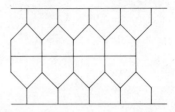

Figure 17-8

12. (a) Show that each pattern block can be used to tessellate the plane.
    (b) Find two different tessellations using the square.
    (c) Find two different tessellations using the triangle.
    (d) Sketch your results to (b) and (c) below.

■ In Exercise 12 you showed that the regular (equilateral) triangle, regular quadrilateral (square), and regular hexagon will each tessellate the plane. A tessellation by regular polygons all of which are the same shape is called a *regular tessellation*, if the corners of any polygon touch only corners of the other polygons. Thus the tessellation of squares shown in Figure 17-7 is not regular, but the tessellation in Figure 17-9 is regular.

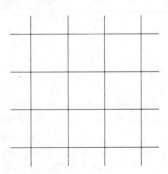

Figure 17-9

13. (a) Find a tessellation with triangles that is not regular.
    (b) Do you think that there can be a tessellation with hexagons that is nonregular? _____

Definition: *A tessellation by regular polygons of two or more shapes, such that corners of polygons touch only corners, is called a* semiregular tessellation.

14. (a) Use the pattern blocks to find a semiregular tessellation of hexagons and triangles.
    (b) Can you find a different tessellation using hexagons and triangles?
    (c) How many other semiregular tessellations can you find? (Try for about five to ten minutes.)

15. Use the pattern blocks to find semiregular tessellations using
    (a) Triangles and squares (there are two).
    (b) Hexagons, triangles, and squares. (This tessellation, made of small clay tiles, appears on the floor of a castle in Windsor, England. The builder had done a lot of decorative work during the reign of Queen Victoria.)
    (c) The design in Figure 17-10 is derived from one of the semiregular tessellations. Which one is it?

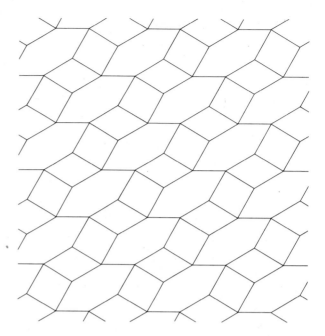

Figure 17-10

■ For Exercise 17 and for Lab Exercises: Set 2, you will have to do Exercise 16 at home. Bring the polygons you construct to class.

16. Use Figure 17-11 to construct the indicated polygons from cardboard or construction paper. Construct three dodecagons (12-sided polygons),

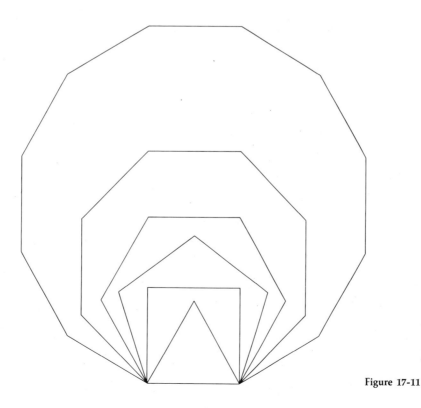

Figure 17-11

three octagons, six hexagons, four pentagons, six squares, and six triangles. (One good way is to use a compass point or pin to punch holes through the vertices onto the construction paper. Then connect the vertices with a straight edge.)

17. You have found three semiregular tessellations. Use the polygons you constructed above to find three more.

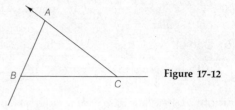

Figure 17-12

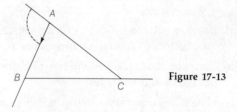

Figure 17-13

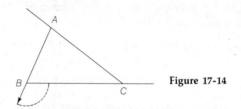

Figure 17-14

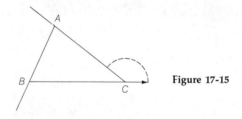

Figure 17-15

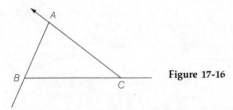

Figure 17-16

Comment: Regular Polygons

"I often seem to have more in common with mathematicians than with my fellow artists."—M. C. Escher

The device used in Exercises 1, 2, and 3 to find the sum of the angles of a polygon is an interesting one. Let us look at it more closely for the case of a triangle.

Consider the triangle ABC and extend three of the sides to form three exterior angles, as shown in Figure 17-12. At angle A, place an arrow on the extended side. If we rotate the arrow through the exterior angle at A, our diagram becomes as shown in Figure 17-13. If we now slide the arrow along side AB so that it starts at vertex B, we have Figure 17-14. Now rotate the arrow through the exterior angle at B and slide it along $\overline{BC}$ to obtain Figure 17-15. If we now rotate the arrow through the exterior angle at C and translate along $\overline{CA}$, the arrow has returned to its original position (Figure 17-16). The arrow has been rotated through 360°. Since there is no rotation at all involved in sliding along the sides, we can conclude that the sum of the measures of the exterior angles of a triangle is 360°.

Since each exterior angle and its adjacent interior angle form a straight line, the sum of these two angles is 180°. Therefore the sum of the measures of all the interior and exterior angles is $3 \times 180°$. Hence the sum of the measures of the angles of a triangle is $3 \times 180° - 360° = 180°$.

In general, the same procedure can be used to show that the sum of the measures of the angles of a polygon with n sides (an n-gon) is $(n - 2) \times 180°$.

18. If the sum of the measures of the angles of a polygon is 1,800°, how many sides does it have?

■ Let us now apply the above formula to regular polygons. A regular polygon has congruent sides and congruent angles. In particular, the measure of each angle is equal. Therefore, we can compute the measure of each angle by dividing the sum of the measures by the number of sides. For example, since the sum of the measures of a triangle is 180°, each angle of a regular (equilateral) triangle must measure $180°/3 = 60°$. In general, the measure of each angle of a regular n-gon is

$$\frac{(n - 2) \times 180°}{n} = 180° - \frac{360°}{n}$$

19. (a) If the measure of each angle of a regular polygon is 174°, the polygon has ____ sides.

(b) If the measure of each angle of a regular polygon is 144°, the polygon has _____ sides.

(c) If the measure of each angle of a regular polygon is $x°$, how many sides does the polygon have?

20. What is the measure of an exterior angle of a regular polygon with 6 sides? _____ With 20 sides? _____

Lab Exercises: Set 2

EQUIPMENT: Polygons from Set 1, scrap paper or file cards (four 3 × 5 file cards per group), one pair of scissors, and one straightedge for each group.

In these exercises we will conclude our study of regular tessellations and then proceed to consider tessellations with parallelograms, triangles, and quadrilaterals.

21. (a) In Exercise 12 you found that the equilateral triangle, the square, and the regular hexagon will tessellate a plane. Will the regular pentagon tessellate a plane? Use the pentagons you constructed in Exercise 16 to find the answer.

    (b) Will the regular octagon tessellate a plane? _____ The regular decagon? _____

    (c) Can you make a guess as to what conditions are necessary for a regular polygon to tessellate the plane?

22. Complete the following.

    (a) For any tessellation, the sum of the measures of the angles around a vertex will add up to _____ degrees. Since each of the angles of a regular polygon has the same measure in a regular tessellation, this number must _____ 360°. From Exercise 9, we see that the only regular polygons that satisfy this criterion are the (i) _____, (ii) _____, and (iii) _____.

    (b) Why do we not have to consider tessellations by regular polygons with more than six sides? (*Hint:* What measurements greater than 120° would divide 360°?)

23. Argue that any parallelogram will tessellate a plane. Make a sketch below.

24. Will any of the shapes in Figure 17-17 tessellate a plane? Why or why not?

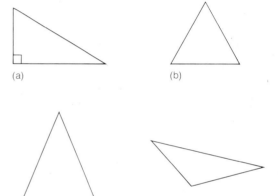

Figure 17-17

25. Can you find a triangle that will not tessellate a plane? Explain. (*Hint:* Use Exercise 23.)

26. Let us explore tessellations with quadrilaterals.

    (a) Cut out eight congruent convex quadrilaterals from scrap paper or index cards. (The index cards work better. Cut carefully so that the quadrilaterals are actually congruent.) Try to form a tessellation with these quadrilaterals.

    (b) Cut out eight nonconvex quadrilaterals. Do you think they will tessellate? If you proceed the same way you did in (a), you will find a tessellation. (Most people do not think a nonconvex quadrilateral will tessellate a plane—ask your friends.)

27. *Optional.* The shape in Figure 17-18 is called a *Greek cross.* Will it tessellate a plane? Use a portion of the square grid paper below.

Figure 17-18

28. *Optional.* A figure formed from five squares, each square sharing at least one side with another square, is called a *pentomino.* The Greek cross is a pentomino. How many pentominos do you think there are? _____ Find as many as you can, using the square grid paper on the following page to record your answers.

29. *Optional.* Use the grid paper on the following page to investigate which pentominos will tile a plane. Can you find any that will tile in more than one way? You can make some pretty designs by coloring your tiles different colors.

Comment: Tessellations in Mathematics and Art

"This [the regular division of the plane] is the richest source of inspiration that I have ever struck; nor has it yet dried up. The symmetry-drawings . . . show how a surface can be regularly divided into, or filled up with, similar-shaped figures which are contiguous to one another, without leaving any open spaces. The Moors were past masters of this. They decorated walls and floors, particularly in the Alhambra of Spain, by placing congruent, multi-coloured pieces of majolica together without leaving any spaces between. What a pity it is that Islam did not permit them to make 'graven images.' They always restricted themselves, in their massed tiles, to designs of an abstract geometrical type. Not one single Moor-

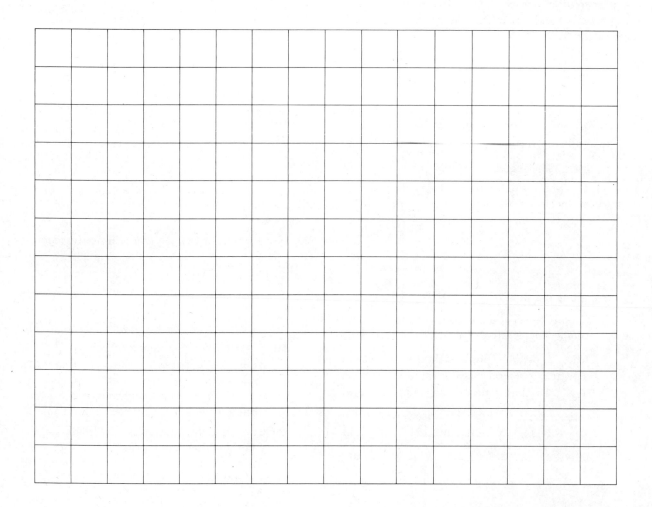

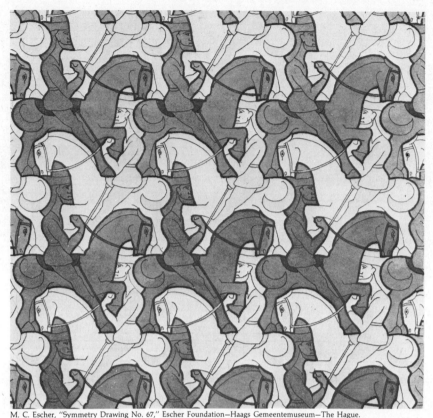

M. C. Escher, "Symmetry Drawing No. 67," Escher Foundation—Haags Gemeentemuseum—The Hague.

Figure 17-19

ish artist, to the best of my knowledge, ever made so bold (or maybe the idea never occurred to him) as to use concrete, recognisable, naturalistically conceived figures of fish, birds, reptiles or human beings as elements in their surface coverage."—M. C. Escher

Tessellations and tilings provide an imaginative way to teach many geometrical concepts. In addition, they exhibit an exciting connection between mathematics and art. Figure 17-19 (and others in this chapter) indicate how tessellations and tilings inspired the graphic artist M. C. Escher. These and other topics are explored in the following books:

Bezuzska, S., M. Kenney, and L. Silvey, *Tessellations: The Geometry of Pattern* (Palo Alto, CA: Creative Publications, 1977).

Bourgoin, J., *Arabic Geometrical Pattern and Design* (New York: Dover Publications, 1973).

Bruno, E., *The Magic Mirror of M. C. Escher*, trans. J. E. Brigham (New York: Ballantine Books, 1976).

Escher, M. C., *The Graphic Work of M. C. Escher* (New York: Ballantine Books, 1967).

Locher, J. L., ed., *The World of M. C. Escher* (New York: Harry N. Abrams, Inc., 1971).

MacGillavry, C. H., *Fantasy and Symmetry: The Periodic Drawings of M. C. Escher* (New York: Harry N. Abrams, Inc., 1976).

Mold, J., *Tessellations* (Cambridge: Cambridge University Press, 1969).

Ranucci, E., and J. Teeters, *Creating Escher-Type Drawings* (Palo Alto, CA: Creative Publications, 1977).

A method for producing an Escher-type tiling based on the regular tessellation by triangles appears at the end of this chapter. You might enjoy creating one.

Let us summarize what we know about tessellations: (1) In any tessellation, the sum of the measures of the angles at any vertex will add up to 360°. (2) In a regular tessellation, all of the angles will be congruent and the measure of the angle of the polygon must divide 360°. There are only three regular polygons that satisfy this condition—the triangle, the square, and the hexagon. Each yields a tessellation.

If we consider semiregular tessellations, we must consider all the ways of arranging two or more regular polygons around a point. It can be shown that there are exactly seventeen ways of doing this. This is done in two articles: (1) "On the Occasional Incompatibility of Algebra and Geometry" by M. Farrell and E. Ranucci (*Mathematics Teacher,* October 1973, p. 491), and (2) "A Simple Sorting Sequence" by D. Duncan and B. Littwiller (*Mathematics Teacher,* April 1974, p. 311). The first article shows that only eleven of the seventeen actually yield tessellations. These are the three regular tessellations and the eight semi-

17/Comment

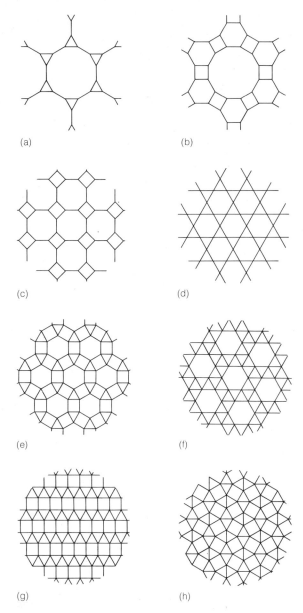

Figure 17-20

regular tessellations, the latter of which are shown in Figure 17-20.

The question of regular and semiregular tessellations is therefore solved. However, there are many other interesting tessellations to explore, some of which have been indicated in the laboratory exercises.

The more general question of which shapes *tile* the plane (remember that, for tiling, the vertices need not be congruent) is still open. That strange things can happen is evidenced in Figure 17-21, of a tiling by enneagons (nine-sided polygons) arranged in a double spiral.

30. *Optional.* If you want to experiment with tessellations on your own, a good way to begin is to

Figure 17-21 A spiral tiling by Heinz Voderberg.

read the article "Designs with Tessellations" by Evan M. Maletsky in the April 1974 issue of the *Mathematics Teacher*, or the book *Creating Escher-Type Drawings* by E. Rannuci and J. Teeters. The following is taken from the Maletsky article.*

Step 1. Start with equilateral triangle *ABC*. Mark off the same curve on both sides, *AB* and *AC*, as shown in Figure 17-22a. Mark off another curve on side *BC* that is symmetric about the midpoint *P* (Figure 17-22b). If you choose the curves carefully, as Escher did, an interesting figure suitable for tiling will be formed (Figure 17-22c).

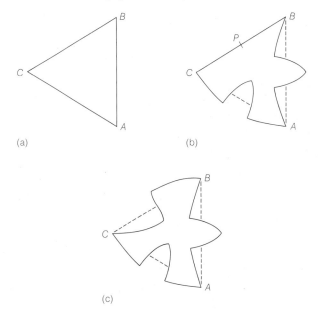

Figure 17-22

---

*Reprinted from the *Mathematics Teacher*, April 1974 (vol. 67, p. 338), copyright © 1974 by the National Council of Teachers of Mathematics. Used by permission.

Step 2: Six of these figures accurately fit together about a point, forming a hexagonal array. Trace and cut out one of the basic figures and show how it can be used to continue the tiling over the entire plane (Figure 17-23).

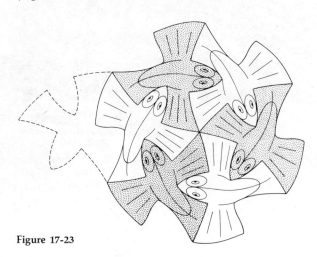

Figure 17-23

Important Terms and Concepts

Interior angles of a polygon
Exterior angles of a polygon
Equilateral polygon
Equiangular polygon
Regular polygon

Tiling
Tessellation
Regular tessellation
Semiregular tessellation
Tessellations with other shapes

Review Exercises

For Exercises 1–5, write the letter of the correct answer in the blank.

1. The exterior angles of a convex polygon always add up to _____.
   (a) 180°   (b) 360°   (c) 270°
   (d) A number that depends on the number of sides of the polygon

2. The number of triangles made by drawing all the possible diagonals from one vertex in a (convex) polygon with n sides is _____.
   (a) n (b) $n - 1$ (c) $\frac{n}{2} - 2$
   (d) $n - 2$ (e) None of these

3. To find the sum of the interior angles of a 22-sided figure, we multiply _____.
   (a) 24×180 (b) 22×180 (c) 20×180
   (d) Don't multiply: it is always 360.

4. In a *regular* polygon with 22 sides, each *interior* angle would measure _____ degrees.
   (a) 180   (b) $16\frac{4}{11}$ (c) $163\frac{7}{11}$ (d) $196\frac{4}{11}$

5. In a regular polygon with 22 sides, each *exterior* angle would measure _____ degrees.
   (a) $16\frac{4}{11}$ (b) $163\frac{7}{11}$ (c) 22   (d) 20

6. If a regular polygon has interior angles that each measure 170°, how many sides does it have? _____

7. If a regular polygon has exterior angles that each measure 6°, how many sides does it have? _____

8. Suppose you have a box of tiles that are all regular octagons, and you want to use them to tile a floor. You recall from your math background that octagons alone cannot tile a floor, so you have to order another tile to fill in the spaces. Naturally you want to spend as little money as possible, so you have to decide the shape and size of the best kind of tile to do the job. What tile would it be? _____

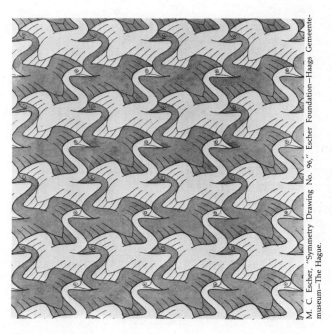

Figure 17-24

17/Review Exercises

9. Draw a semiregular tessellation using squares and triangles.

10. If someone believed that every parallelogram tiles a plane, how could you convince him that every triangle tiles a plane?

11. How would you explain to someone that there are only three regular tessellations? Include an explanation of each term or concept.

12. Decide whether each of the following statements is true or false. Write T or F in the blank.

    _____ (a) Every three-sided figure will tessellate a plane.

    _____ (b) Every four-sided figure will tessellate a plane.

    _____ (c) There is no regular polygon with more than six sides that will tessellate a plane.

Mary Fairfax Somerville (1780–1872), denied access to education by her parents, taught herself mathematics with a little assistance from her brother's tutor. She translated and expanded Laplace's work on the motions of the solar system. She wrote three other treatises, including *Physical Geography,* the first important book on this subject.

18

Transformations

"[Geometry] that held acquaintance with the stars,/And wedded soul to soul in purest bond/Of reason, undisturbed by space or time."—William Wordsworth

Lab Exercises: Set 1

EQUIPMENT: One Mira per student and one ruler per group.

In these exercises you will investigate the process of performing an operation on the points in a plane. Those operations that preserve the distance between points will be given special attention.

1. Place the Mira on line $\overleftrightarrow{m}$ below and find the image of the points *A, B, C,* and *D*. Recall that we call these points the *images* of *A, B, C,* and *D* *through the line* $\overleftrightarrow{m}$ (see Chapter 13). Does every point have an image through the line $\overleftrightarrow{m}$? _____ Find the image of *E* and *F*.

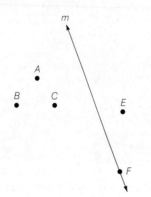

■ We call the above operation *reflection through the line* $\overleftrightarrow{m}$. The operation associates with each point in the plane in the Mira.

There are other operations that permit us to associate with each point in the plane another point in the plane. We will consider two such operations in Exercises 2 and 3.

2. Let *P* be chosen as below. Associate with a point *A* the point *A'* on the ray $\overrightarrow{PA}$ so that the distance from *P* to *A'* is twice the distance from *P* to *A*. We will call *A'* the *image* of *A*, by analogy with the operation of reflection.*

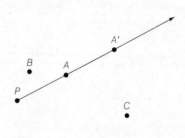

---

*Observe that this expands our use of the word *image*. It no longer necessarily refers to the result of reflecting a point through the Mira.

(a) Find the image of B and C and label them B' and C' respectively. (*Hint:* Draw $\overrightarrow{PB}$ and $\overrightarrow{PC}$.)

(b) Find the point D whose image is D'. This association defines a transformation of the plane.

3. Let P be a point on the plane. Associate with each point A the image A' on the ray $\overrightarrow{PA}$ so that the distance from P to A' is 2/3 the distance from P to A.

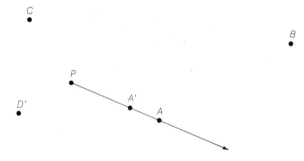

(a) Find the image of B and C.
(b) Find the point whose image is D'.

■ It would be convenient to have a word to describe those operations on a plane that associate points with points. Mathematicians use the word *transformation* for this phenomenon.

Definition: *A* transformation *of the plane is an operation that associates each point with an image such that two different points always have different images.*

If a transformation associates A' with A, we say the transformation sends A to A' and we write

$$A \to A'$$

The operation of reflection through a line in a plane is a transformation of the plane, as are the operations in Exercises 2 and 3 above.

4. (a) Can you think of a difference between a reflection and the operation in Exercise 2 or Exercise 3?
   Discuss in your group. (*Hint:* Consider in Exercises 1, 2, and 3 the triangles △ABC and their images △A'B'C'.)

(b) One difference has to do with the concept of *distance-preserving*. In Exercise 5 of Chapter 13, you saw that reflections are distance-preserving transformations. Discuss whether the transformation of Exercise 2 is a distance-preserving transformation. State why it is or is not.

(c) Discuss whether the transformation of Exercise 3 is a distance-preserving transformation. State why it is or why it is not.

■ To show that a transformation is distance-preserving, you must verify that if A and B are any points such that $A \to A'$ and $B \to B'$, then the length of $\overline{AB}$ equals the length of $\overline{A'B'}$. To show that it is not distance-preserving, you <u>must</u> find two points A and B such that the length of $\overline{AB}$ is not equal to the length of $\overline{A'B'}$. Review your answers to Exercise 4(b) and (c) with this in mind.

5. Suppose we have a transformation that takes each point to a point three units to the right and two units down.

   (a) Perform this transformation on each of the shapes in Figure 18-1 on the following page and indicate their images.
   (b) Discuss whether this is a distance-preserving transformation.

■ The transformation in Exercise 5 is an example of a translation.

Definition: *A* translation *is a transformation of the plane such that the image of every point is a fixed distance in a given direction from the original point.*

Schematically, a translation looks like Figure 18-2. A translation is determined by specifying the horizontal and vertical distances of the image point from the original point. We can use an ordered pair of integers—written $\overrightarrow{(a, b)}$—to indicate this. Thus, the translation in Exercise 5 is determined by $\overrightarrow{(3, -2)}$, which means that each image point is three units to the right and two units down from the original point.

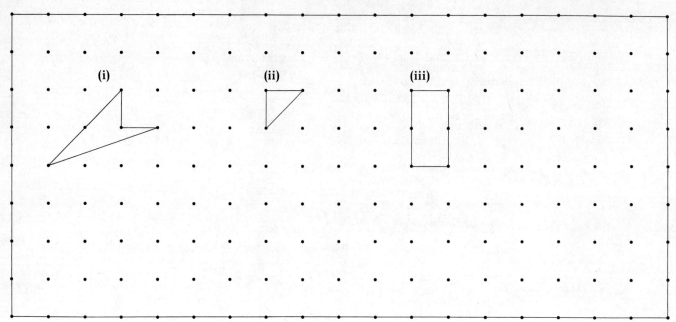

Figure 18-1

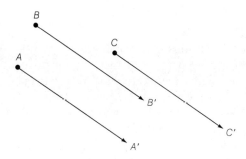

Figure 18-2

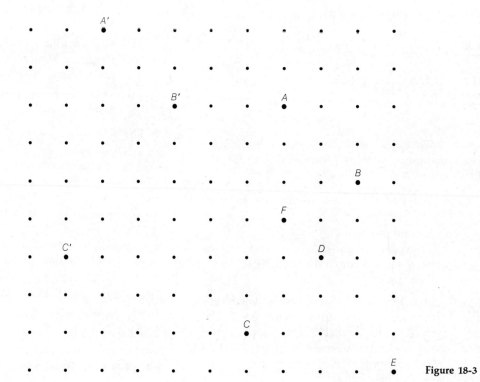

Figure 18-3

For the translation determined by $\overrightarrow{(-5, 2)}$, each image point would be obtained by moving five units to the left and two units up from the original point.

For example, in Figure 18-3, A', B', and C' denote the image of A, B, and C under the translation determined by $\overrightarrow{(-5, 2)}$.* For simplicity we shall refer to this transformation as the translation $\overrightarrow{(-5, 2)}$.

6. Find the images of D, E, and F in Figure 18-3 under the translation $\overrightarrow{(-5, 2)}$.

7. Indicate the image of P in Figure 18-4 under each of the following translations.

   (a) $\overrightarrow{(1, 3)}$
   (b) $\overrightarrow{(-2, -1)}$
   (c) $\overrightarrow{(-3, -3)}$
   (d) $\overrightarrow{(0, 4)}$
   (e) $\overrightarrow{(0, 0)}$
   (f) $\overrightarrow{(3, -4)}$

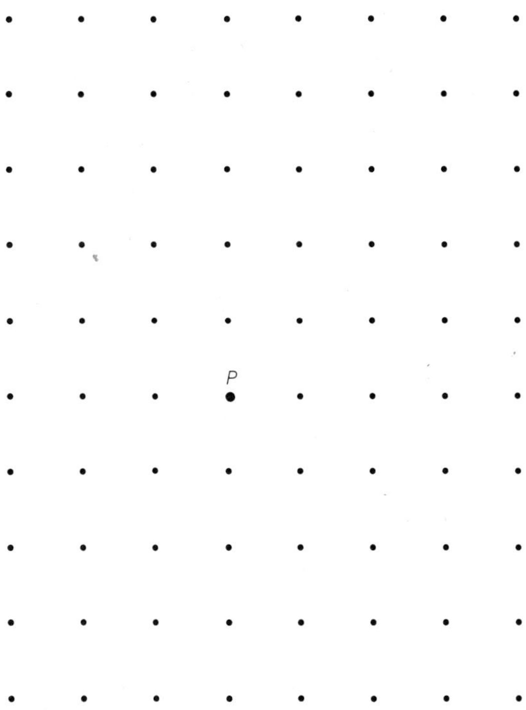

8. Find the point in Figure 18-5 that has P as its image under each of the following translations.

   (a) $\overrightarrow{(3, 0)}$
   (b) $\overrightarrow{(-4, -3)}$
   (c) $\overrightarrow{(0, 4)}$
   (d) $\overrightarrow{(-2, 1)}$
   (e) $\overrightarrow{(3, -4)}$
   (f) $\overrightarrow{(2, 1)}$

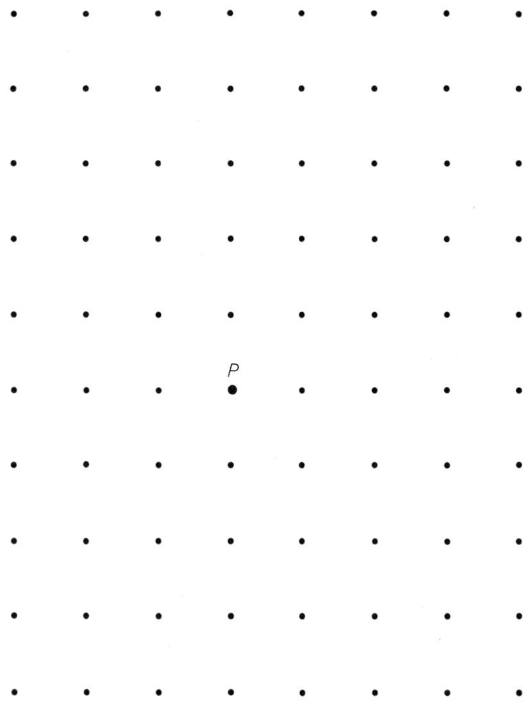

Figure 18-4

9. Consider the translation $\overrightarrow{(4, -2)}$. Let A, B, C, and D be as indicated in Figure 18-6, and let A', B', C', and D' denote their respective images. Make a 10×10-pin geoboard by putting four geoboards together.

   Find A', B', C', and D' and verify by inspection that:

   (a) $\triangle ABC$ is congruent to $\triangle A'B'C'$.
   (b) $\overline{AB}$ is parallel to $\overline{A'B'}$.
   (c) The segments $\overline{AA'}$, $\overline{BB'}$, $\overline{CC'}$, $\overline{DD'}$ are all congruent.
   (d) $\overline{AA'}$ is parallel to $\overline{BB'}$.

---

*The word *under* in the phrase "under the translation" is used in the sense of "receiving the action or application of," as in the phrases "the topic under discussion," "go under the surgeon's knife," "machinery under repair," and so on.

Figure 18-5

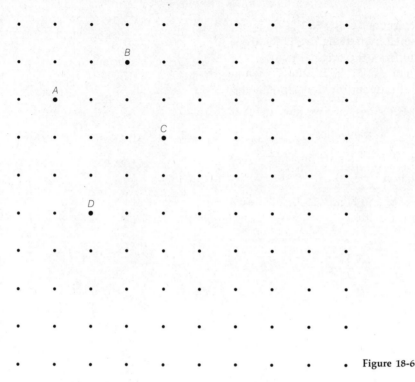

Figure 18-6

■ Statements (a) through (d) in Exercise 9 are true in general. That is, under a translation:

(1) A figure and its image are congruent.
(2) A segment and its image are parallel (or possibly collinear).
(3) Segments formed by points and their images are congruent.
(4) Segments formed by points and their images (such as $\overline{AA'}$ and $\overline{BB'}$) are either parallel or collinear).

10. Each member of your group choose one of the above statements and convince the rest of the group of its truth.

11. For the translation in Exercise 9, find two points R and S so that the segments $\overline{RS}$ and $\overline{R'S'}$ are collinear.

12. Let l and m be the lines indicated in Figure 18-7. Let R_l be that reflection in l and R_m the reflection in m. Let A' denote the image of A under the reflection R_l and A'' the image of A' under the reflection R_m; i.e., A'' is the result of doing R_l to A and then R_m to the image of A. Do the same for B', B'' and C', C''. Use the Mira to find A'', B'', and C''. Draw $\triangle ABC$ and $\triangle A''B''C''$. Make sure everyone agrees.

Definition: *This defines a new transformation that we denote $R_l \cdot R_m$, called the* **product** *of R_l and R_m.*

To find the image of a point P under $R_l \cdot R_m$, find the image of P under R_l and then find the image of this new point under R_m. Observe that we are using the phrase "under the reflection" in the same sense as we used "under the translation," that is, "receiving the action of."

13. In Figure 18-8, A, B, and C are placed as in Exercise 12. Indicate the points A'', B'', C'' that you found. Can you describe a single transformation that sends A to A'', B to B'', and C to C''? _____

_____

(*Hint:* Draw the triangles $\triangle ABC$ and $\triangle A''B''C''$.) You have found a single transformation that does the same thing as the product $R_l \cdot R_m$.

■ The *product* of two reflections is an abstract concept that describes a new operation on the points in the plane. Given any point in the plane, we can find its image under $R_l \cdot R_m$. (We do R_l and then R_m.) In Exercise 13 you found a single transformation that equals the *product* of the two reflections of Exercise 12.

14. In Figure 18-9, choose a point P between l and m, a point Q to the right of m, and a point R on m. Find the images of P, Q, R under $R_l \cdot R_m$. Check that your results are consistent with the results of performing the single transformation you found in Exercise 13.

15. Now consider the transformation $R_m \cdot R_l$. To find

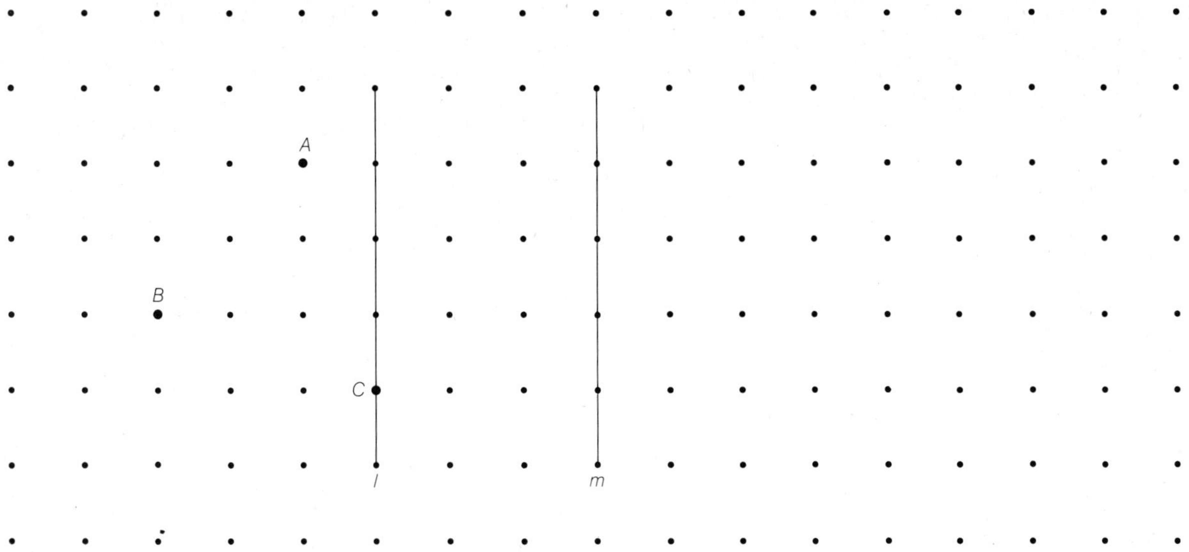

Figure 18-7

Figure 18-8

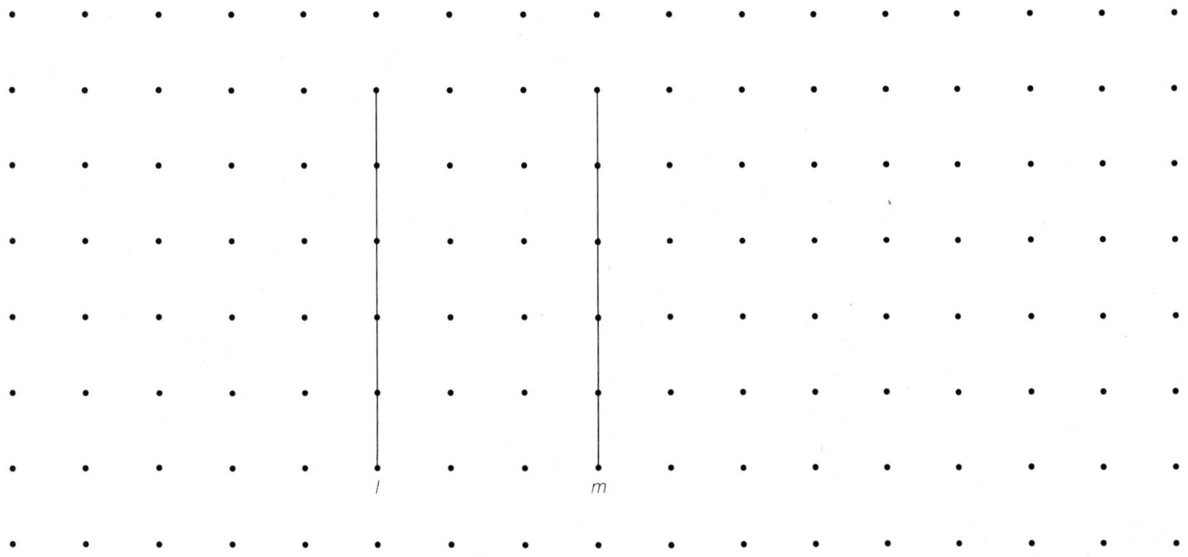

Figure 18-9

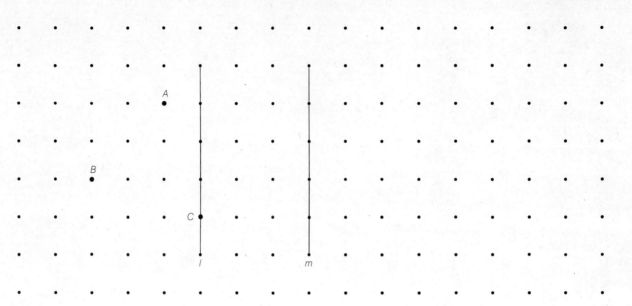

Figure 18-10

the image of a point under this transformation, we first reflect in m and then reflect in l. (See Figure 18-10.) Find the images of A, B, C, P, Q, R under $R_m \cdot R_l$. Discuss whether $R_m \cdot R_l$ is the same transformation as $R_l \cdot R_m$.

16. Describe a single transformation that is equal to $R_m \cdot R_l$.

17. $\overleftrightarrow{l}$ and $\overleftrightarrow{m}$ are shown below. Use the Mira to find the images of A, B, C, D, E under the transformation $R_l \cdot R_m$. Can you describe in words a single transformation equal to $R_l \cdot R_m$? Discuss this in your group. If you succeed, write the description below. If not, proceed to Exercise 18.

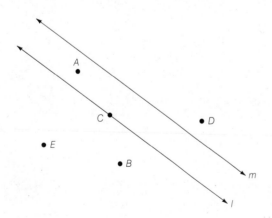

18. Let A'', B'', C'', D'', E'' denote the images of A, B, C, D, E under the transformation $R_l \cdot R_m$ in Exercise 17. Measure the lengths of $\overline{AA''}, \overline{BB''},$ $\overline{CC''}, \overline{DD''}, \overline{EE''}$. They are _____. How does this compare with the distance between $\overleftrightarrow{l}$ and $\overleftrightarrow{m}$? _____ Now, can you describe in words a single transformation equal to $R_l \cdot R_m$?

19. Complete the following: If $\overleftrightarrow{l}$ and $\overleftrightarrow{m}$ are parallel lines, then $R_l \cdot R_m$ is the same transformation as a _____ in the direction from l to m a distance _____ times the distance between $\overleftrightarrow{l}$ and $\overleftrightarrow{m}$.

20. Draw two lines $\overleftrightarrow{l}$ and $\overleftrightarrow{m}$ in Figure 18-11 so that the product $R_l \cdot R_m$ equals the translation $\overrightarrow{(0, 2)}$.

21. This exercise is an application of reflections. A pumping station is to be constructed at a river bank to serve two towns, A and B. The two towns wish to construct the station at the point X such that $m(\overline{AX}) + m(\overline{XB})$ is a minimum. How can we find X? (Discuss this before reading on.)

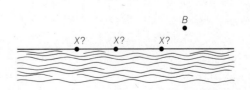

Figure 18-11

Figure 18-12

Suppose we represent the river on the 10 × 10-pin geoboard with a rubber band between P and Q and place A and B as shown in Figure 18-12. Show that if B' is the image of B under a reflection in the line $\overleftrightarrow{PQ}$, and if X is any point on this line, then $m(\overline{AX}) + m(\overline{XB}) = m(\overline{AX}) + m(\overline{XB'})$. Now find X such that $m(\overline{AX}) + m(\overline{XB'})$ is a minimum. Describe how you found the point X. Change the positions of A, B, and the river, and repeat.

Comment: Translations and Reflections

"Plato said that God geometrizes continually."—Plutarch

We have been studying transformations of the plane.

Definition: *A transformation is an operation on the plane that associates each point with an image such that two different points always have different images.*

Some transformations have the additional property that they preserve distances between points; that is, the distance between A and B is equal to the distance between the images of A and B. We call such a transformation a *distance-preserving* transformation.

Although not all transformations preserve distance (for example, the transformations in Exercises 2 and 3), some very important types of transformations are distance-preserving. In Chapter 13 you studied reflections using the Mira. If you did Review Exercise 10 in that chapter, however, you discovered that reflections can be described without reference to the Mira. To find the image of a point A through the line $\overleftrightarrow{m}$, first draw a line $\overleftrightarrow{l}$ perpendicular to $\overleftrightarrow{m}$. Let P be the intersection of $\overleftrightarrow{l}$ and $\overleftrightarrow{m}$. Let A' be the point on $\overleftrightarrow{m}$ such that the segments $\overline{A'P}$ and $\overline{AP}$ have the same length (Figure 18-13). In other words, the image of A is the point on the perpendicular from A to $\overleftrightarrow{m}$, which is the same distance as A from $\overleftrightarrow{m}$.

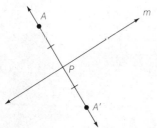

Figure 18-13

It can be proven using the side-angle-side congruence axiom that reflections, as so defined, are distance-preserving. We need not present the full proof here, but if you are so inclined, you might try to find it on your own. It is not too difficult.

Another type of distance-preserving transformation is a *translation*.

In a translation, the image of every point is a fixed distance in a given direction from the original point.

We saw in the lab exercises that a translation can be specified by an ordered pair of integers $(\overrightarrow{a, b})$ where a is the (directed) horizontal distance and b is the (directed) vertical distance associated with the translation. For example, the translation indicated in Figure 18-14 is specified by $(\overleftarrow{-2, 1})$, as shown in Figure 18-15.

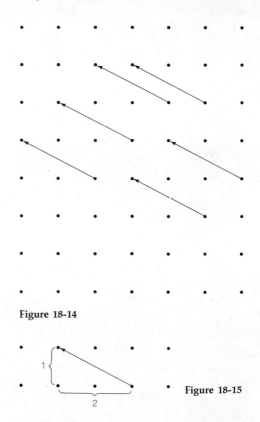

Figure 18-14

Figure 18-15

22. Find the ordered pair that specifies the transformation indicated by each of the arrows below.

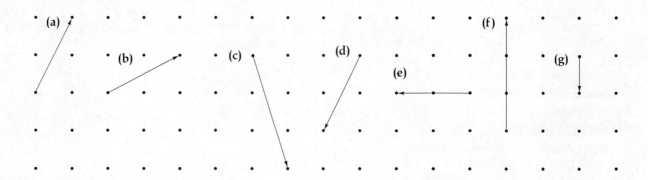

Figure 18-16

23. Indicate the image of P in Figure 18-16 under the translation determined by each of the following.
    (a) $\overrightarrow{(4, ^-5)}$
    (b) $\overrightarrow{(^-3, ^-6)}$
    (c) $\overrightarrow{(1, 1)}$
    (d) $\overrightarrow{(0, ^-3)}$
    (e) $\overrightarrow{(0, 0)}$
    (f) $\overrightarrow{(^-2, 5)}$
    (g) $\overrightarrow{(0, 5)}$

24. Find the point in Figure 18-17 on the following page that has P as its image under each of the following translations.
    (a) $\overrightarrow{(^-2, 4)}$
    (b) $\overrightarrow{(^-4, ^-5)}$
    (c) $\overrightarrow{(0, 0)}$
    (d) $\overrightarrow{(2, 2)}$
    (e) $\overrightarrow{(3, ^-2)}$
    (f) $\overrightarrow{(0, 5)}$
    (g) $\overrightarrow{(^-4, 0)}$

■ One of the most intriguing things about transformations is that they can be combined to form new transformations. Thus if R_l and R_m denote reflections in lines $\overleftrightarrow{l}$ and $\overleftrightarrow{m}$ respectively, $R_l \cdot R_m$ is the transformation obtained by first reflecting in $\overleftrightarrow{l}$ and then reflecting the images obtained in $\overleftrightarrow{m}$. We saw in the lab exercises that if $\overleftrightarrow{l}$ and $\overleftrightarrow{m}$ are parallel lines, then $R_l \cdot R_m$ is the same transformation as a translation of twice the distance between $\overleftrightarrow{l}$ and $\overleftrightarrow{m}$ in the direction from $\overleftrightarrow{l}$ to $\overleftrightarrow{m}$.

Combining transformations in this manner is not restricted to reflections. For example, we can form $\overrightarrow{(2, 3)} \cdot \overrightarrow{(1, 2)}$. To find the image of a point A under this transformation, we find the image under $\overrightarrow{(2, 3)}$ and then find the image of this point under $\overrightarrow{(1, 2)}$. In Figure 18-18 on the following page, A' is the image of A under $\overrightarrow{(2, 3)}$ and A'' is the image of A' under $\overrightarrow{(1, 2)}$.

25. Find the image of B and C under $\overrightarrow{(2, 3)} \cdot \overrightarrow{(1, 2)}$.

26. Find a single translation equal to $\overrightarrow{(2, 3)} \cdot \overrightarrow{(1, 2)}$.

27. Find a single translation equal to each of the following.
    (a) $\overrightarrow{(1, ^-3)} \cdot \overrightarrow{(2, 4)}$ _____
    (b) $\overrightarrow{(^-2, 5)} \cdot \overrightarrow{(1, 1)}$ _____
    (c) $\overrightarrow{(^-2, 4)} \cdot \overrightarrow{(3, ^-4)}$ _____

28. Complete the following: If $\overrightarrow{(a, b)}$ and $\overrightarrow{(c, d)}$ are translations, then $\overrightarrow{(a, b)} \cdot \overrightarrow{(c, d)} =$ _____.

Figure 18-17

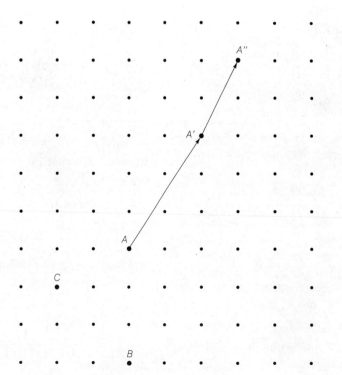

Figure 18-18

Lab Exercises: Set 2

EQUIPMENT: Two protractors and two compasses for each group, one Mira for each person, tracing paper or other semitransparent paper.

In these exercises we will explore rotations of the plane and show that every rotation is a product of two reflections. We will then examine how distance-preserving transformations can be written as the product of three reflections.

29. Consider Figure 18-19.

    (a) If the plane were rotated 90° clockwise around the point O, where would A end up? Use tracing paper or any other method to find the image of A and indicate it by A'.

    (b) Choose points B and C and find their images under the same rotation.

    (c) Draw △ABC and △A'B'C'.

■ Clearly, a *rotation* is a transformation since it sends points to points. In the following exercise we will investigate a 30° rotation.

30. We associate with the point X, below, the image X' such that $\overline{OX}$ and $\overline{OX'}$ are congruent and ∠XOX' measures 30° (counterclockwise orientation). Use a protractor and compass to find X'. Find Y' and Z' in the same manner.

■ This transformation is called a rotation of 30° counterclockwise. O is called the *center of the rotation*.

31. Verify for the above rotation that $\overline{XY}$ is congruent to $\overline{X'Y'}$, that $\overline{YZ}$ is congruent to $\overline{Y'Z'}$, and that XZ is congruent to $\overline{X'Z'}$. Do you think that every rotation is a distance-preserving transformation?

Definition: *A rotation of x° about O is the transformation obtained by associating with each point P the point P' such that* $(\angle POP')° = x°$ *and* $\overline{OP'}$ *is congruent to* $\overline{OP}$.

32. Discuss why each of the following statements is true.

    (a) A figure and its image under a rotation are congruent.

    (b) A point and its image under a rotation are equidistant from the center.

    (c) Parallel lines are still parallel after a rotation.

33. Let $\overleftrightarrow{l}$ and $\overleftrightarrow{m}$ be as indicated in Figure 18-20 on the following page. Use the Mira to find the images of A, B, C, and D under the product $R_l \cdot R_m$. That is, first reflect A in $\overleftrightarrow{l}$ to obtain A' and then reflect A' in $\overleftrightarrow{m}$ to obtain A'', etc. Describe a single transformation equal to $R_l \cdot R_m$. If you don't see it at first, find the images of a few more points.

34. (a) Let $\overleftrightarrow{l}$ and $\overleftrightarrow{m}$ be as indicated in Figure 18-21 on the following page and indicate the images of A, B, C, D under the product $R_l \cdot R_m$.

Figure 18-19

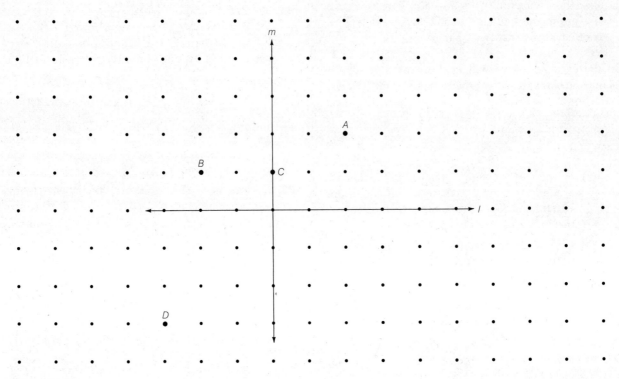

Figure 18-20

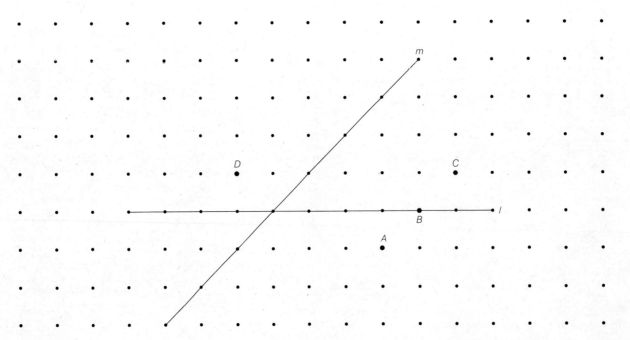

Figure 18-21

(b) Describe a single transformation equal to $R_l \cdot R_m$. (A piece of tracing paper might be useful to explore this.)

35. (a) Guess the answer to the following: If lines $\overleftrightarrow{l}$ and $\overleftrightarrow{m}$ intersect at an angle of $x°$, then $R_l \cdot R_m$ equals the transformation that is a rotation of _____° around the point of intersection.

(b) Use a protractor to test your guess where $\overleftrightarrow{l}$ and $\overleftrightarrow{m}$ intersect at an angle of $30°$.

36. Draw two lines $\overleftrightarrow{l}$ and $\overleftrightarrow{m}$ below so that $R_l \cdot R_m$ equals a rotation of $70°$ around the point P.

37. Below are (i) three points A, B, C, and (ii) their images A', B', C' under a distance-preserving transformation. Use the following procedure to find three reflections R_k, R_l, R_m whose product $R_k \cdot R_l \cdot R_m$ sends A, B, C to A', B', C' respectively.

(a) Find a line $\overleftrightarrow{k}$ such that R_k sends A to A'. Let $\overline{B}$ be the image of B and $\overline{C}$ the image of C.

(b) Find a line $\overleftrightarrow{l}$ such that R_l sends $\overline{B}$ to B'. What happens to A'? _____ Let $\overline{\overline{C}}$ be the image of $\overline{C}$.

(c) Find a line $\overleftrightarrow{m}$ such that R_m sends $\overline{\overline{C}}$ to C'. What happens to A' and B'?

(i)

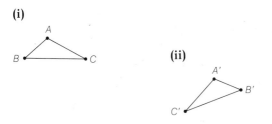

(ii)

■ If you did the above carefully, you should have observed that certain important points stayed fixed when R_l and R_m were applied. This leads us to the following.

Definition: *A point A is a fixed point of a transformation if it is identical with its image under the transformation.*

38. Describe the set of fixed points for each of the following transformations.
(a) A reflection in a line $\overleftrightarrow{l}$
(b) A $45°$ rotation about a point O
(c) A translation of one unit to the right

Comment: Products of Reflections

"Indeed the modern developments of mathematics constitute not only one of the most impressive, but one of the most characteristic, phenomena of our age. It is a phenomenon, however, of which the boasted intelligence of a 'universalized' daily press seems strangely unaware; and there is no other great human interest, whether of science or of art, regarding which the mind of the educated public is permitted to hold so many fallacious opinions and inferior estimates."—C. J. Keyser (1908)

We have studied three types of distance-preserving transformations: reflections, translations, and rotations. We have seen that each of these has the property that a figure and its image are congruent. The same is true for every distance-preserving transformation. That is why these transformations are often called "congruences." The term *isometry* is also used since "iso" is the Greek prefix for equal and "metry" means measure.

From the lab work you should be convinced of the importance of reflections. We can summarize the results of our investigations as follows.

A translation is the product of two reflections in parallel lines. The lines are perpendicular to the direction of the translation. The distance between them is one-half the distance of the translation.

A rotation is the product of two reflections in intersecting lines. The point of intersection is the center of rotation. The angle between the lines is one-half the angle of rotation.

It can be shown that a distance-preserving transformation is completely determined once you know the images of any three noncollinear points. In Exercise 37 you found that a product of three reflections is sufficient to send A, B, C to their images A', B', C'. This is true in general.

Any distance-preserving transformation is equal to the product of at most three reflections.

Of course, the transformation may be a reflection or the product of two reflections. The above rule states simply that you will never need more than three.

39. Find three lines $\overleftrightarrow{k}, \overleftrightarrow{l}, \overleftrightarrow{m}$, such that $R_k \cdot R_l \cdot R_m$ sends A to A', B to B', and C to C'.

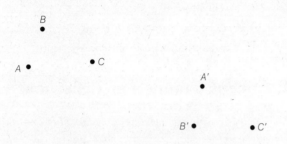

■ The next time you look in a mirror, think of what you have learned about reflections.

Drawing by Chas. Addams; © 1957 The New Yorker Magazine, Inc.

Important Terms and Concepts

Transformation
Image of a point
Reflection
Translation
Rotation
Distance-preserving transformation
Product of two transformations

Review Exercises

1. Find the image of the parallelogram $ABCD$, under the translation $(-2, -1)$.

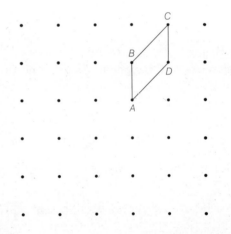

2. Find the image of $\triangle PQR$ under the product $R_t \cdot R_q$.

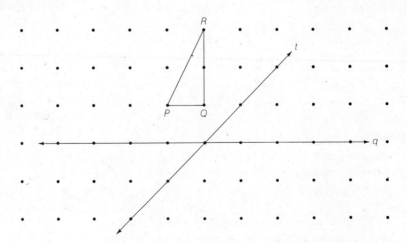

18/Review Exercises

3. Find the image of △LMN under a clockwise rotation of 90° about point P by finding lines l and m so that $R_l \cdot R_m$ is the desired rotation.

4. Given segment AB, find a translation such that $\overline{A'B'}$ is collinear with $\overline{AB}$.

5. (a) Suppose you left your geoboard sitting with △ABC on it, and when you return you notice that someone has moved your triangle to the position A'B'C' on the board. Your first reaction is to investigate what transformation

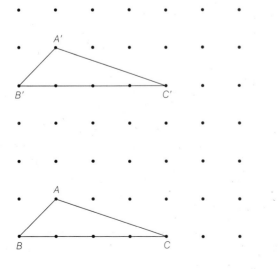

has been performed. If you wish to express this transformation as a simple translation, what translation would it be? _____

(b) Suppose you wish to express it in a second way, as the product of two reflections. What transformation would it then be? _____
(*Hint:* You will have to draw and label your lines of reflection.)

6. Consider a transformation that sends △ABC to △A'B'C'.

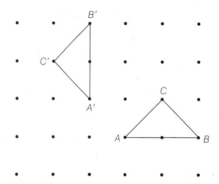

(a) Express this transformation as the product of a translation and a rotation. _____
(b) Express it as the product of two reflections. _____
(c) Express it as a single rotation about some point. _____

7. Draw two lines $\overleftrightarrow{l}$ and $\overleftrightarrow{m}$ such that $R_l \cdot R_m$ equals the translation $\overrightarrow{(1, 2)}$.

8. Suppose $\overleftrightarrow{l}$ and $\overleftrightarrow{m}$ are perpendicular lines, as in the diagram on the following page. Describe the image of a point P under reflection in $\overleftrightarrow{l}$ if

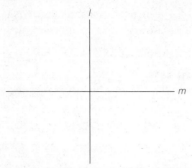

(a) P is 2 cm to the left of $\overleftrightarrow{l}$ and 3 cm above $\overleftrightarrow{m}$.

(b) P is 2 kilometers to the right of $\overleftrightarrow{l}$ and 10 meters below $\overleftrightarrow{m}$.

(c) P is on $\overleftrightarrow{l}$, 8 meters above $\overleftrightarrow{m}$.

9. Decide whether each of the following statements is true or false. Write T or F in the blank.

_____ (a) A figure and its image have the same size and shape under any translation.

_____ (b) The only points that are fixed under the transformation R_l are the points on line $\overleftrightarrow{l}$.

_____ (c) If lines $\overleftrightarrow{l}$ and $\overleftrightarrow{m}$ are parallel, the fixed points for $R_l \cdot R_m$ are all the points on $\overleftrightarrow{l}$ and all the points on $\overleftrightarrow{m}$.

_____ (d) If lines $\overleftrightarrow{l}$ and $\overleftrightarrow{m}$ intersect at point Q and form a 30° angle, the transformation $R_l \cdot R_m$ will be equivalent to a clockwise rotation of 60° about point Q.

10. Given three parallel lines $\overleftrightarrow{l}, \overleftrightarrow{m}, \overleftrightarrow{n}$ below, find a line t such that the product of the reflections R_l, R_m, R_n equals the reflection R_t.

11. Which of the figures below are congruent to each other? _____ Why?

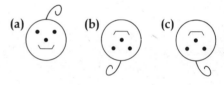

12. Find the image of the triangle ABC under the transformation $T \cdot R_l$ where T is a translation three units to the right and R_l is a reflection in the line l; that is, first translate $\triangle ABC$ three units to the right and then reflect it in l.

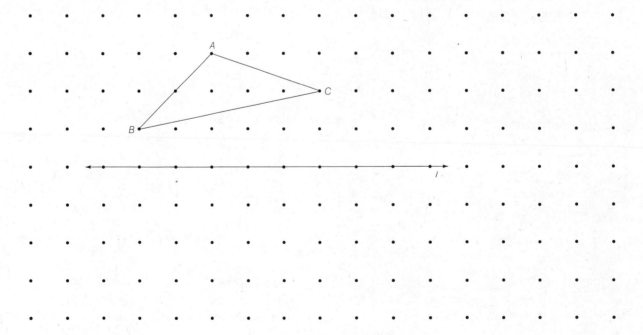

19
Symmetry

"Symmetry, as wide or as narrow as you may define its meaning, is one idea by which humanity through the ages has tried to comprehend and create order, beauty, and perfection."—Hermann Weyl

Lab Exercises: Set 1

EQUIPMENT: One Mira for each student and one set of pattern blocks for every two groups.

These exercises explore symmetry. We will start by doing some visual activities using the Mira and pattern blocks and then explore some more abstract concepts.

1. Draw two designs below that have symmetry and one design that does not. Discuss your concept of symmetry. Try to explain what symmetry is.

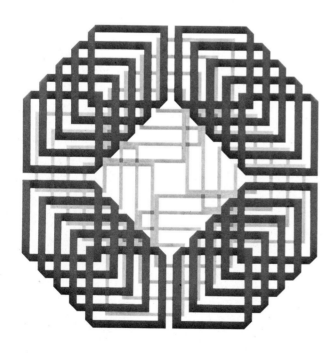

Figure 19-1 "Four into Four into Eight" by Rene Parola. Used by permission of the artist.

2. Place your Mira on the rectangle below so that the reflected image of the rectangle matches the original rectangle. One way you can do this is by placing the Mira on the dotted line. Such a line is called a *line of symmetry* for the figure. Can you find any other lines of symmetry for the rectangle? If so, draw them. Record the number of lines of symmetry in the space provided.

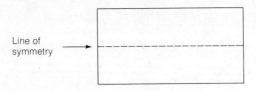

Number of lines of symmetry in this rectangle: _____

3. Indicate all of the lines of symmetry for each of the following figures and record the number in the space provided.

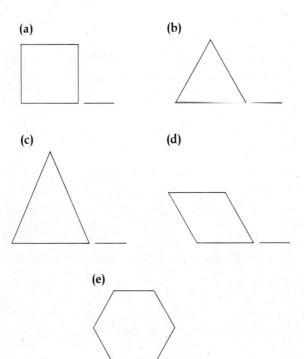

4. How many lines of symmetry does a circle have? _____

5. Draw a triangle below with no line of symmetry.

6. Make a design with the pattern blocks that has exactly one line of symmetry; exactly two; exactly three. Make one with as many lines of symmetry as you can.

7. Reproduce with pattern blocks the design shown in Figure 19-2. Discuss the symmetry of this figure. Does it look symmetrical? _____ How many lines of symmetry does it have? _____ Check your answer with a Mira or a mirror.

8. Although the figure you constructed in Exercise 7 has no lines of symmetry, it still looks symmetrical. Try to describe in words why this figure is symmetrical.

9. The reason the figure in Exercise 7 looks symmetrical even though it has no lines of symmetry is that if it is rotated 120° or 240° or 360°, it is indistinguishable from its image. We say the figure has three *rotational symmetries*. Can you draw a figure that has more than three rotational symmetries but no lines of symmetry? Sketch it below.

10. *Optional.* Fold a sheet of paper (or construction paper) in half lengthwise and then in half again along with the width (Figure 19-3a). Use a scissors to cut shapes from the paper (Figure 19-3b). Unfold and investigate the symmetry of the design (Figure 19-3c). Try folding your paper in other ways to see how that affects the symmetry. Your students will enjoy doing this with colored paper.

■ Symmetry is also important in mathematics, physics, and chemistry. The use of this concept in these fields requires a more precise treatment than the intuitive one given above. We will investigate below how to make more precise (in a mathematical sense) exactly what we mean by symmetry.

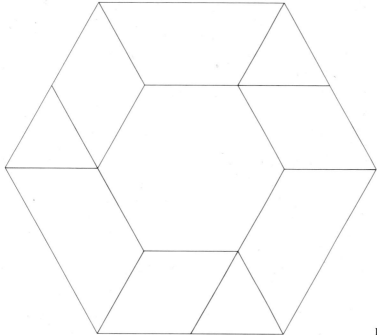

Figure 19-2

Figure 19-3a

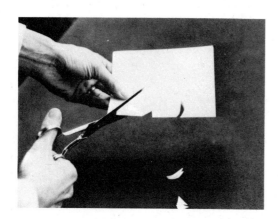

Figure 19-3b

Figure 19-3c

The method of doing this is to use the concept of distance-preserving transformation studied in Chapter 18.

Roughly speaking, a figure is "symmetrical" or "has symmetry" if there are distance-preserving transformations of the plane that leave the whole figure unchanged; i.e., the image of the figure under the transformation is the figure itself. For example, in Exercise 2 the image of the rectangle under reflection through the indicated line is the rectangle again.

11. What is a distance-preserving transformation? Try to recall the definition before looking it up in Chapter 18.

Definition: *A distance-preserving transformation that leaves a figure as a whole unchanged is called a* symmetry transformation of the figure *or a* symmetry of the figure.

For example, a clockwise rotation of 120° about *P* is a symmetry transformation of the shape in Figure 19-4.

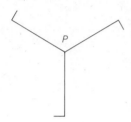

Figure 19-4

12. (a) Describe two other symmetry transformations of the above figure.

    (b) Are there any lines of symmetry for the figure? _____

■ If the symmetry transformations of a figure are reflections, we say it has *reflectional* (or *bilateral*) symmetry. If a figure has a symmetry transformation that is a rotation (except 360° or multiples of 360°), we say the figure has *rotational* symmetry. (The reason for the 360° exception is that *every* figure is left unchanged by a rotation through 360° or multiples of 360°, since such a transformation is equivalent to not moving the figure at all; we do not consider a figure symmetrical if it possesses only this trivial symmetry transformation.) For example, look at the letters in Figure 19-5.

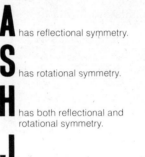

Figure 19-5

13. On the geoboard, construct a figure that has both reflectional and (nontrivial) rotational symmetry. Construct a figure that has reflectional symmetry but no (nontrivial) rotational symmetry. Record your results in Figure 19-6.

14. Discuss in your group and then describe the symmetry transformations of the rectangle in Exercise 2. (There are four of them.)

Figure 19-6

15. Discuss in your group and then describe the eight symmetry transformations of the square.

16. *Optional.* What are the symmetry transformations of the designs you made in Exercise 10? Investigate how the way you fold the paper affects the number of symmetries of the design.

17. Can you draw a figure that has a translational symmetry (i.e., has a translation as a symmetry transformation)?

   The image of a figure under a symmetry must coincide with itself. What does this imply about the extent of a figure that has a translation as a symmetry?

18. (a) The figure below has translational symmetry. (Assume that the pattern goes on indefinitely in both directions.) Describe a translation that is a symmetry.

    (b) What other symmetries does the figure have?

Comment: The Concept of Symmetry

"What immortal hand or eye/Dare frame thy fearful symmetry?"—William Blake

There are many interesting activities that explore the intuitive concepts of symmetry. Some of these are indicated in the optional exercises below.

19. *Optional.* Explore the symmetry of the letters of the alphabet.

20. *Optional.* Collect wallpaper designs and investigate their symmetry.

21. *Optional.* Collect leaves, flowers, insects (or pictures of them), and investigate their symmetry.

22. *Optional.* Obtain a Spirograph (available in most toy stores) and draw symmetrical designs.

23. *Optional.* Look for symmetry in art and architecture. Collect pictures or make sketches.

Figure 19-7

Figure 19-8

■ We have seen that the intuitive idea of symmetry can be made precise by using the concept of a distance-preserving transformation. We said that a distance-preserving transformation is a symmetry of a figure if it leaves the figure as a whole unchanged. Observe that this does not imply that the individual points in the figure necessarily coincide with their images, only that the set of image points coincides with the original figure. For example, reflection in the line in Figure 19-9 is a symmetry of the indicated equilateral triangle.

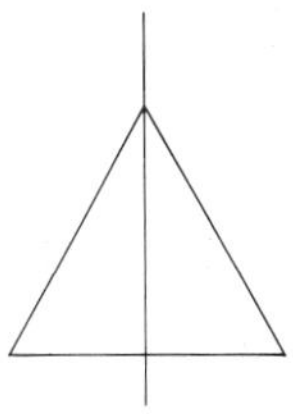

Figure 19-9

24. Find five other symmetries of the equilateral triangle in Figure 19-9. Which one is the trivial symmetry?

25. Describe the symmetries of one of the snowflakes pictured in Figure 19-10. Do all the snowflakes have the same symmetries?

Figure 19-10 Courtesy National Oceanic and Atmospheric Administration; W. A. Bentley, photographer.

By permission of Johnny Hart and Field Enterprises, Inc.

19/Comment

26. Describe the symmetries of the carved lacquer tray pictured in Figure 19-11. It was made in Japan during the Yüan-Ming Dynasty (fourteenth century).

Figure 19-11 Courtesy of the Zauho Press.

27. Describe the symmetries of the carved lacquer tray in Figure 19-12. It too was made during the Yüan-Ming Dynasty.

Figure 19-12 Courtesy of the Zauho Press.

28. Describe the symmetries of the tray in Figure 19-13. (Look carefully!) It was made in Japan during the Southern Sung-Yüan Dynasty (thirteenth to fourteenth century).

Figure 19-13 Courtesy of the Zauho Press.

■ Figures that possess translational symmetry must be considerably different from figures with rotational and reflectional symmetries. Suppose that the image of a figure under a translation of a distance d coincides with that figure. Consider a point P in the figure. Then the image P' of P is also in the figure. Now P' is a distance d from P. But since P' was already in the figure, its image, which is a distance $2 \times d$ away from P, is also in the figure. Continuing in this way, we see that if a figure has translational symmetry, it must be infinite in extent.

Although we cannot draw such a figure (our paper is finite), we can imagine what it looks like and draw a section of it. For example, the infinite strip in Figure 19-14 has the translation 1 cm to the right as a symmetry. Observe that it has also as symmetries the translations 2 cm, 3 cm, etc. to the right, and similarly, translations 1 cm, 2 cm, 3 cm, etc. to the left.

Figure 19-14

This figure has only translations as symmetries, but a figure may possess rotational and reflectional symmetries as well as translational symmetries. For example, the figure in Exercise 18 has as symmetries the following.

(1) Reflection through the horizontal line midway between the border.
(2) Reflection through each vertical line that bisects a diamond or that bisects one of the horizontal segments (an infinite number of reflections).
(3) 180° rotation around the centers of each diamond or around the midpoint of each segment (an infinite number of rotations).

29. Describe the symmetries of each of the designs in Figure 19-15. (Assume they extend indefinitely.)

■ Some books to look at in order to investigate symmetry further are as follows.

Gardner, M., *The Ambidextrous Universe* (New York: Mentor, 1969).
Holden, A., *Shapes, Space and Symmetry* (New York: Columbia University Press, 1971).
Parola, R., *Optical Art: Theory and Practice* (New York: Reinhold Book Corp., 1969).
Steinhaus, H., *Mathematical Snapshots* (New York: Oxford University Press, 1968).

"Mirror Cards" (manufactured by McGraw-Hill Book Company) are a series of activities to help young people explore symmetry.

Lab Exercises: Set 2

EQUIPMENT: For each person, an equilateral triangle (congruent to the one in Figure 19-16) cut from construction paper or cardboard.

In these exercises we will explore what happens when we form the product of two symmetries. It will be helpful if you work through this set of exercises on your own before doing them in the laboratory.

We can illustrate the three rotational symmetry operations of the equilateral triangle by marking an x in one corner and then showing the result of performing the rotation. Thus if $\mathscr{A}$ denotes a 120° clockwise rotation of the plane about the center of the triangle, we have Figure 19-16. Similarly, if $\mathscr{B}$ denotes a 240° clockwise rotation around the center, we have Figure 19-17. And if $\mathscr{I}$ denotes a 360° rotation (the transformation that takes each point to itself), then we have Figure 19-18.

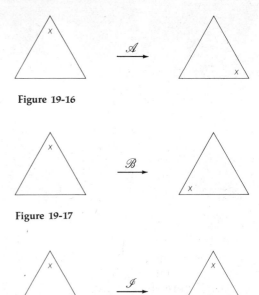

Figure 19-16

Figure 19-17

Figure 19-18

The product of two symmetry transformations is defined to be the result of performing them in succession. Thus if $\mathscr{A}$ represents a rotation of 120° clock-

Figure 19-15

19/Lab Exercises: Set 2

wise and $\mathscr{B}$ a rotation of 240° clockwise, then $\mathscr{A} \cdot \mathscr{B}$ represents the transformation resulting from first doing $\mathscr{A}$ and then $\mathscr{B}$. Similarly $\mathscr{A} \cdot \mathscr{A}$ represents the transformation of performing $\mathscr{A}$ twice. Since performing $\mathscr{A}$ twice amounts to rotating through 240°, we have $\mathscr{A} \cdot \mathscr{A} = \mathscr{B}$. The motion $\mathscr{A} \cdot \mathscr{B}$ results in rotating the triangle through 360°, which is the same as not moving at all. Thus $\mathscr{A} \cdot \mathscr{B} = \mathscr{I}$. The algebra of symmetry transformations is founded on the principle that the product of any two symmetry transformations is again a symmetry transformation.

30. Complete the "multiplication" table below for the symmetry transformations $\mathscr{I}, \mathscr{A}, \mathscr{B}$. (The operation on the left side is done first.)

31. Is the product of two elements of $\{\mathscr{I}, \mathscr{A}, \mathscr{B}\}$ in the set $\{\mathscr{I}, \mathscr{A}, \mathscr{B}\}$? _____ Since it is, we say that $\{\mathscr{I}, \mathscr{A}, \mathscr{B}\}$ is *closed* under this operation. Is the operation of multiplication commutative for this set of symmetry transformations? _____

■ We saw earlier that there are three symmetry transformations of the equilateral triangle that are rotations. Let us now investigate the *reflections* that are symmetry transformations of the triangle. A convenient way to keep track of the reflections physically (since we cannot actually reflect the points) is to flip the triangle around the line of reflection. To see how this works, mark the other side of your triangle in some way. For simplicity, we will put a circle at the vertex on the other side of the vertex marked x. In order to specify the reflections, we must specify three lines in the plane; these will be the *reflection lines* (Figure 19-19). Place the triangle on the paper so its center is at the intersection point of the lines F, G, H.

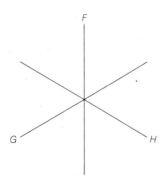

Figure 19-19

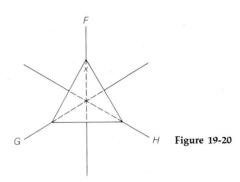

Figure 19-20

The x is visible at the top (Figure 19-20). A reflection in the line F can be thought of as flipping the triangle around the line F (as if F were an axle). We can then picture the reflectional symmetry transformation $\mathscr{F}$ as in Figure 19-21.

(1) $\mathscr{F}$ (reflection in the line F):

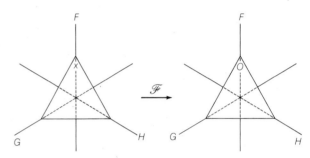

Figure 19-21

Note: You can now see the circle at the top. Note that the lines F, G, and H do not move.

The other reflectional symmetry transformations are depicted in Figures 19-22 and 19-23. (We have omitted some of the lines for convenience.)

(2) $\mathscr{G}$ (reflection in the line G):

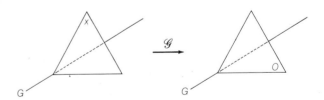

Figure 19-22

(3) $\mathscr{H}$ (reflection in the line H):

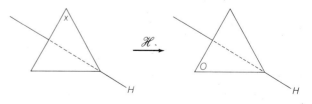

Figure 19-23

Remember the triangle you cut out is a physical model and that an actual triangle is not meant to have any thickness. These flips, therefore, are not actual reflections but rather a way of physically keeping track of them.

Recall that the product of two symmetry transformations is the result of doing them in succession, reading from left to right. What is $\mathscr{F}\cdot\mathscr{F}$? Clearly, if you do the same reflection twice, you have the identity. So $\mathscr{F}\cdot\mathscr{F}=\mathscr{I}$. Similarly, $\mathscr{G}\cdot\mathscr{G}=\mathscr{I}$ and $\mathscr{H}\cdot\mathscr{H}=\mathscr{I}$. Let us try a more complicated one. What is $\mathscr{A}\cdot\mathscr{F}$? (Remember, this means do $\mathscr{A}$ first and then $\mathscr{F}$.) So we do what is shown in Figure 19-24 and see that performing the symmetry transformation $\mathscr{A}$ and the symmetry transformation $\mathscr{F}$ is equivalent to performing $\mathscr{H}$. Thus $\mathscr{A}\cdot\mathscr{F}=\mathscr{H}$.

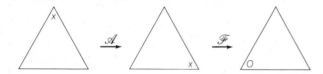

Figure 19-24

32. What is $\mathscr{B}\cdot\mathscr{G}$? ____

33. What is $\mathscr{G}\cdot\mathscr{B}$? ____

34. Is $\mathscr{B}\cdot\mathscr{G}=\mathscr{G}\cdot\mathscr{B}$? ____

35. Complete the "multiplication" table for the six symmetry transformations of an equilateral triangle, $\mathscr{I},\mathscr{A},\mathscr{B},\mathscr{F},\mathscr{G},\mathscr{H}$, by entering in the row marked $\mathscr{A}$ and the column $\mathscr{F}$ the product $\mathscr{A}\cdot\mathscr{F}$.

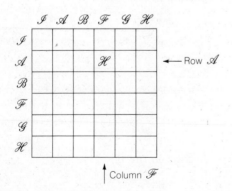

36. Are the symmetry transformations of the triangle closed under this operation of multiplication; that is, is the product of any two of them a symmetry transformation of the triangle? ____

37. Is the operation of multiplying symmetry transformations of a triangle commutative? ____

The symmetry $\mathscr{I}$ has the property that for every symmetry $\mathscr{X}$, $\mathscr{I}\cdot\mathscr{X}=\mathscr{X}\cdot\mathscr{I}=\mathscr{X}$. It is called the **identity**.

38. Consider the rectangle below. Let $\mathscr{L}$ denote reflection in the line l, $\mathscr{M}$ denote reflection in the line m, $\mathscr{R}$ a rotation of 180° around the center P of the rectangle, and $\mathscr{I}$ a rotation of 360° (equivalent to a rotation of 0°) around P:

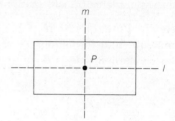

Cut out a rectangle from scrap paper and use it to help you complete the "multiplication" table below.

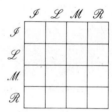

Is there an identity? ____ Is the system commutative? ____

39. Show that each symmetry transformation of an equilateral triangle has an *inverse*; i.e., for each symmetry $\mathscr{X}$ there is a symmetry transformation $\mathscr{Y}$ such that $\mathscr{X}\mathscr{Y}=\mathscr{Y}\mathscr{X}=\mathscr{I}$. What is the inverse of $\mathscr{A}$? ____ Of $\mathscr{B}$? ____ Of $\mathscr{F}$? Of $\mathscr{G}$? ____ Of $\mathscr{H}$? ____

40. Consider the "multiplication" table in Exercise 38. Find the inverse of each symmetry transformation.

   (a) The inverse of $\mathscr{L}$ is ____.

   (b) The inverse of $\mathscr{M}$ is ____.

   (c) The inverse of $\mathscr{R}$ is ____.

   (d) The inverse of $\mathscr{I}$ is ____.

41. Develop the multiplication table for the symmetry transformation of the symbol ⊢⊐ and write it below. (*Note:* This symbol was used by both the Native American and Oriental civilizations long before its mirror image ⊏⊣ was adopted by the National Socialist Party in Germany.)

42. *Optional.* As we have seen, different figures in the plane will have different sets of symmetry transformations associated with them. If we consider designs that cover the entire plane, we obtain infinite sets of symmetry transformations. For example, the symmetry transformations of the tessellation with squares below includes the three classes of symmetries.

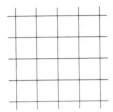

(1) Rotations of 90°, 180°, 270°, and 360° around the center and corner of every square.

(2) Reflections in the diagonals of every square.

(3) Translations determined by $(\vec{m}, \vec{n})$ where m and n are positive or negative integers (where we assume that the sides of the squares have length 1).

Can you find any others?

43. *Optional.* List some of the symmetry transformations of the following tessellation.

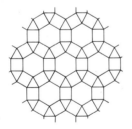

44. *Optional.* In his book *Symmetry* (Princeton University Press, 1952), Hermann Weyl states, "Beauty is bound up with symmetry." Discuss what he meant by this. (The book is excellent and deserves to be looked at if only for its illustrations.)

45. *Optional.* Read the poem below. Why does the author refer to God as a symmetry? What word in the passage most contrasts with symmetry? Do you have any other observations about the poem?

> God, Thou great symmetry
> Who put a biting lust in me
> From whence my sorrows spring,
> For all the frittered days
> That I have spent in shapeless ways
> Give me one perfect thing.
> Anna Wickham, "Envoi"

Comment: The Algebra of Symmetry

"Symmetry is not merely a descriptive nicety . . . it penetrates to the harmony in nature."—Jacob Bronowski

In the exercises above you investigated the symmetries of figures. Any figure in a plane has associated with it a set of symmetries. They are the distance-preserving transformations that map the figure onto itself. A symmetry, $\mathscr{S}$, of a figure is characterized by two properties: (1) $\mathscr{S}$ is a distance-preserving transformation, and (2) the figure coincides with its image under $\mathscr{S}$.

Given any two symmetries, $\mathscr{S}_1$ and $\mathscr{S}_2$, of a figure, we can form the product $\mathscr{S}_1 \cdot \mathscr{S}_2$. Recall from Chapter 18 that this means we do $\mathscr{S}_1$ first and then $\mathscr{S}_2$. It is easy to see that if $\mathscr{S}_1$ and $\mathscr{S}_2$ are symmetries, then so is $\mathscr{S}_1 \cdot \mathscr{S}_2$, since if $\mathscr{S}_1$ and $\mathscr{S}_2$ preserve distances, then performing one after the other will also preserve distances. Finally, if $\mathscr{S}_1$ and $\mathscr{S}_2$ each leave the figure unchanged (as a whole), then performing each in succession will also leave the figure unchanged.

These facts enable us to consider what mathematicians call the *group of symmetries* of a figure. This consists of the set of symmetries, together with the rule for forming products.

If the set of symmetries is small, we can write down the multiplication table for the group of symmetries. For example, the table for the group of symmetries of the equilateral triangle is given in Figure 19-25.

The group of symmetries of a figure has some interesting properties. First, the identity transformation (the image of every point is that point) is always in the group. This symmetry has the property that $\mathscr{I} \cdot \mathscr{X} = \mathscr{X} \cdot \mathscr{I} = \mathscr{X}$ for every $\mathscr{X}$ in the group. Also, every symmetry $\mathscr{X}$ in the group has an inverse $\mathscr{Y}$

	𝓘	𝓐	𝓑	𝓕	𝓖	𝓗
𝓘	𝓘	𝓐	𝓑	𝓕	𝓖	𝓗
𝓐	𝓐	𝓑	𝓘	𝓗	𝓕	𝓖
𝓑	𝓑	𝓘	𝓐	𝓖	𝓗	𝓕
𝓕	𝓕	𝓖	𝓗	𝓘	𝓐	𝓑
𝓖	𝓖	𝓗	𝓕	𝓑	𝓘	𝓐
𝓗	𝓗	𝓕	𝓖	𝓐	𝓑	𝓘

Figure 19-25

such that $\mathcal{X} \cdot \mathcal{Y} = \mathcal{Y} \cdot \mathcal{X} = \mathcal{I}$. It can also be shown that multiplication of symmetries is *associative;* that is, if $\mathcal{W}, \mathcal{X}, \mathcal{Y}$ are symmetries, then $\mathcal{W} \cdot (\mathcal{X} \cdot \mathcal{Y}) = (\mathcal{W} \cdot \mathcal{X}) \cdot \mathcal{Y}$. Observe, however, that multiplication of symmetries is not always commutative. For example, $\mathcal{B} \cdot \mathcal{F} \neq \mathcal{F} \cdot \mathcal{B}$ in the multiplication table for the symmetries of the equilateral triangle.

46. Using the table of the symmetries of the equilateral triangle, verify that $\mathcal{A} \cdot (\mathcal{B} \cdot \mathcal{F}) = (\mathcal{A} \cdot \mathcal{B}) \cdot \mathcal{F}$.

47. Verify that $\mathcal{R} \cdot (\mathcal{M} \cdot \mathcal{L}) = \mathcal{R} \cdot (\mathcal{M} \cdot \mathcal{L})$ using the table you computed in Exercise 38.

48. Is the multiplication in the group of symmetries of the rectangle commutative? _____

■ If we consider different figures, we can obtain a wide variety of symmetry groups. They provide interesting examples of mathematical systems since they are simply defined, possess many nice properties, and are important to the study of the structure of the physical world.

49. *Optional.* On a piece of paper trace the regular pentagon shown below. Cut it out and color each side a different color.

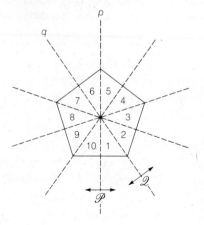

(a) Show how to move Region 1 into Region 5 by a suitable combination of the reflection in the lines p and q.

(b) Show how to move Region 1 into Region 7 by a suitable combination of the reflection in the lines p and q.

(c) Describe the symmetry transformation $\mathcal{P} \cdot \mathcal{Q}$ where $\mathcal{P}$ is the reflection in p and $\mathcal{Q}$ is the reflection in q.

(d) Can every symmetry transformation be written as a combination of p's and q's? _____ (No justification is necessary.)

Important Terms and Concepts

Symmetry
Line of symmetry
Reflectional symmetry
Rotational symmetry
Translational symmetry

Multiplication of symmetries
Identity element
Inverse
Group of symmetries

Review Exercises

1. Describe all the symmetry transformations for each of the following. (This includes reflections and rotations. For rotations, be sure to state the point of rotation.)

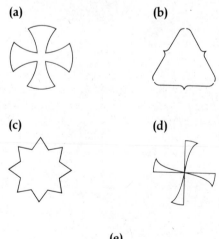

(a)   (b)

(c)   (d)

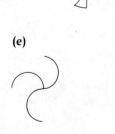

(e)

2. Consider the rhombus below. Let $\mathscr{L}$ denote the reflection in line *l*. Let $\mathscr{M}$ denote the reflection in line *m*. Let $\mathscr{A}$ denote a rotation of 180° around point P, and let $\mathscr{B}$ denote a rotation of 360° around point P. (You may want to cut out a rhombus like the one in the picture to help you work this out).

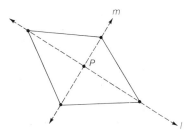

   (a) $\mathscr{L} \cdot \mathscr{A} = $ _____

   (b) $\mathscr{A} \cdot \mathscr{L} = $ _____

   (c) $\mathscr{L} \cdot \mathscr{M} = $ _____

   (d) Is $(\mathscr{L} \cdot \mathscr{M}) \cdot \mathscr{A} = \mathscr{L} \cdot (\mathscr{M} \cdot \mathscr{A})$? _____

   (e) What is the inverse of $\mathscr{A}$? _____

3. Show that if a figure has reflectional symmetry about two perpendicular lines, then it also has 180° rotational symmetry about the point of intersection of the two lines.

4. Draw a figure consisting of two circles that has exactly two reflectional symmetries and one nontrivial rotational symmetry.

5. Investigate the symmetries of a square. Write down the multiplication table for the symmetries of the square.

6. Draw a figure consisting of exactly four rotational symmetries (including the identity).

7. Consider the tessellation below made up of equilateral triangles. Which of the following are symmetry transformations of this tessellation?

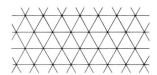

   (a) A rotation of 120° around any point of intersection.
   (b) A horizontal translation through the distance of length equal to the length of any side of the triangles.
   (c) A rotation of 60° around the center of any triangle.
   (d) A reflection in the altitudes of any of the triangles.
   (e) A reflection in any line in the figure.

8. What are the symmetry transformations of the design below?

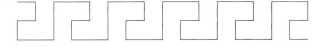

9. Decide whether each of the following statements is true or false. Write T or F in the blank.

   _____ (a) A rectangle has two rotational symmetries.

   _____ (b) A triangle has three lines of reflection.

   _____ (c) A figure that has rotational symmetries must also have reflectional symmetry.

   _____ (d) A rotation can always be expressed as the product of two reflections.

   _____ (e) A translation can always be expressed as the product of two reflections.

_____ **(f)** A reflection can always be expressed as a product of two translations.

_____ **(g)** Most of the transformations in this chapter are distance-preserving, but there are transformations that are not.

10. Let $\mathscr{I}, \mathscr{A}, \mathscr{B}, \mathscr{F}, \mathscr{G}, \mathscr{H}$ be the symmetries of the equilateral triangle, as in Exercise 35. Compute the following products in two ways:

    (i) Use a triangle and actually perform the transformations physically.

    (ii) Use the multiplication table.

    (a) $(\mathscr{A} \cdot \mathscr{B}) \cdot \mathscr{F} = $ _____

    (b) $\mathscr{F} \cdot (\mathscr{F} \cdot \mathscr{A}) = $ _____

    (c) $\mathscr{G} \cdot \mathscr{H} \cdot \mathscr{F} = $ _____

11. *Optional.* Construct a tetrahedron (a four-sided solid with each face an equilateral triangle) and investigate its symmetries.

12. *Optional.* Construct a cube and investigate its symmetries.

Sonyá Corvin-Krukovsky Kovalévsky (1850–1891) left Russia because women were not allowed to attend universities there. She studied in Germany and later became a professor at the University of Stockholm. She wrote many original papers in analysis and won the famous Prix Bordin of the French Academy of Sciences.

20
Probability

"The true logic of this world is in the calculus of probabilities.—James C. Maxwell

Lab Exercises: Set 1

EQUIPMENT: One die and two coins for each person.

We start with a discussion of the different uses of the word "probability," and then proceed to explore the mathematical concept of probability using dice and coin flipping.

1. (a) Each person think of a sentence that uses the word "probable," "probably," or "probability." Explain your use of the word.
   (b) Below are three statements. Discuss in your group the use of the word "probability" in each sentence.
      (i) The *probability* of rolling two 6's with a pair of dice is 1/36.
      (ii) The *probability* that a child born in the United States will be a boy is slightly better than half.
      (iii) The *probability* is high that the president will be reelected.

■ The use of the word "probability" differs in each of the above statements. In (i) we are concerned with calculating the ratio of certain outcomes to the total possible outcomes. The statement (ii) is based on statistical information about previous births. The statement (iii) is a statement of belief or judgment.

   (c) Make up another sentence that illustrates each of the above uses of the word "probability."

■ Although each of these statements uses the word "probability," the mathematical theory of probability is concerned primarily with statements similar to (i). Therefore, in this chapter we will concern ourselves with analyzing certain processes with a view to predicting the likelihood of a certain event.

2. Suppose you were to roll a die 30 times. What would you guess as to the number of times a 3 would appear? _____ A 6? _____ An even number? _____ Answer the same questions for 60 rolls, for 120 rolls, and for 480 rolls.

3. What *percentage* of the rolls would be a 3? _____

4. (a) Each person perform 30 rolls of a die. Record the frequency of occurrence of each number

in the table below and the approximate percentage of the total number of rolls.

30 Rolls of a die

Outcome	1	2	3	4	5	6
Frequency						
Percentage						

(b) Which number occurred most often? _____

(c) Would you expect this number to appear most often if you repeated the experiment? _____ Compare the table with the predictions you made in Exercise 2. Are there differences? _____ Why might these be? _____

5. (a) Obtain the data from the other people in your group and record the information below.

30 Rolls of a die

Outcome	1	2	3	4	5	6
Frequency						
Percentage						

30 Rolls of a die

Outcome	1	2	3	4	5	6
Frequency						
Percentage						

30 Rolls of a die

Outcome	1	2	3	4	5	6
Frequency						
Percentage						

30 Rolls of a die

Outcome	1	2	3	4	5	6
Frequency						
Percentage						

(b) What do you notice about these data? _____

(c) Do any of the experiments agree with your prediction? _____ Summarize the results for all rolls in the table below, entering the number of rolls in the space provided. Compute the percentage that each number appears.

_____ Rolls of the die

Outcome	1	2	3	4	5	6
Frequency						
Percentage						

(d) Compare the observed percentages with the predicted percentages. What do you observe?

6. (a) Obtain the data from the other groups in your class and summarize the results in the table below, entering the total number of rolls in the space provided.

_____ Rolls of the die

Outcome	1	2	3	4	5	6
Frequency						
Percentage						

(b) Compare these percentages with your predictions. What happens as the number of rolls increases?

7. Suppose you are going to roll a *pair* of dice. How many ways are there for the dice to come up?

_____ (*Hint:* Suppose one die is red and the other is white. How many ways are there for the red die to come up? _____ The white die? _____) A complete analysis of this experiment will be carried out in the following exercise.

8. Since the possibilities for the red and white die are {1, 2, 3, 4, 5, 6}, we can fill out the table on the following page by entering the sum in the appropriate square. We will then have displayed all the possible ways the dice can turn up, and the resulting sum.

20/Lab Exercises: Set 1

White die

	1	2	3	4	5	6
1						
2						
3						
4						
5						11
6						

■ Exercises 9–15 involve tossing two dice.

9. Use the chart to determine the number of ways there are of rolling a 7. _____ An 11. _____ The probability of rolling a 7 is equal to

$$\frac{\text{Number of ways of rolling a 7}}{\text{Total number of ways the dice can occur}}$$

10. What is the probability of rolling a 7? _____ An 11? _____

11. Express each of the above probabilities as a percent. _____ _____

12. If you rolled the dice 100 times, about how many times would a 7 occur? _____ An 11? _____

13. Fill in the table at the bottom of the page, using the information in Exercise 8.

14. If you and a friend were to each roll a pair of dice simultaneously, would you be willing to bet that you would roll a 5 before your friend would roll a 4? _____

15. What is the probability of rolling a 7 or an 11? _____

■ In the experiment of rolling two dice, it was easy to determine the number of outcomes of each event and the total number of outcomes. Other situations may be less routine. For example, consider the problem of determining the probability of being dealt a straight flush in a poker game. It would take a long time to list all the possible 5-card hands for a deck of 52 cards. In fact, there are 2,598,960 5-card hands and of these 40 are straight flushes. So the probability of being dealt a straight flush is

$$\frac{40}{2{,}598{,}960} = \frac{1}{64{,}974}$$

Fortunately, we can determine this without listing all the possibilities. It is not very difficult to develop the mathematics to do this, but we will limit ourselves to simpler problems.

In the following exercises we will develop the tools to answer questions of the following type.

(1) If a coin is tossed 10 times, what is the probability that it will come up heads every time?

(2) In a family with 3 children, what is the probability that at least one is a boy?

16. Let us begin by tossing two coins. Guess the probability of getting two heads. _____ Based on your guess, what percentage of a large number of tosses of two coins would be two heads? _____ Each person perform the experiment 10 times and record the data below by entering H or T in the appropriate space.

For example, if two heads were flipped on the first toss, you would write H/H next to (a). If a head and a tail were flipped, you would write H/T or T/H. If two tails were flipped, you would write T/T.

(a) _____ _____ (f) _____ _____
(b) _____ _____ (g) _____ _____
(c) _____ _____ (h) _____ _____
(d) _____ _____ (i) _____ _____
(e) _____ _____ (j) _____ _____

Sum of two dice	2	3	4	5	6	7	8	9	10	11	12
Number of occurrences											
Probability											
Probability expressed as a percent											

Record the information below, together with the totals from your group and lab.

	Your tosses	Total of group	Total for lab
Total outcomes with both coins heads			
Total outcomes with both coins tails			
Total outcomes with both coins one head and tail			

In what percentage of the outcomes for the entire lab do two heads appear? _____ Two tails? _____ One head and one tail? _____ How does this compare with your guess? _____

Comment: Basic Concepts of Probability

"It is remarkable that a science which began with the consideration of games of chance should have become the most important object of human knowledge."—The Marquis de Laplace

In the exercises above you were studying the outcome of certain experiments—tossing a die, tossing a pair of dice, tossing two coins. The *theory of probability* provides a mathematical model for certain experiments or processes like these. The model is an idealization: we disregard the possibility of a die disintegrating as it hits the table or being lost; we do not consider the likelihood of a coin landing on its edge. Although these events might happen, they are quite unlikely. In order to simplify our model, we regard them as impossible. Our goal is to be able to predict the outcome of a certain experiment or process if that experiment is repeated a large number of times.

Consider the experiment of rolling one die (Exercise 2). One way of predicting the outcomes is to assume that it is equally likely that one of the six numbers 1, 2, 3, 4, 5, 6 would occur and hence, for example, a 3 would appear "on the average" about 1/6 (or approximately 16.7 percent) of the time. If one were to guess at how many times a 3 would appear in 30 rolls, the best guess would be 1/6 of 30, or 5 times. As you no doubt observed in your experiments, the actual number of times a 3 appears in 30 rolls may vary considerably from 5. However, as the number of rolls increases, the variation from the predicted percentage of 16.7 percent will become smaller. (Of course, if the die is "loaded" or biased in some way, then our model will not accurately predict the outcome, even if the number of rolls becomes very large, since we are assuming that each outcome is equally likely.)

The rational number (fraction) that expresses the likelihood of an event occurring is called the *probability* of an event. It is defined as:

$$\text{Probability of an event} = \frac{\text{Number of successful outcomes}}{\text{Total number of outcomes}}$$

Using this definition, we can determine the probability of possible outcomes of rolling two dice by examining the table in Exercise 8. For example, since there are six ways of rolling a 7 (six successful outcomes) and 36 possible outcomes, the probability of rolling a 7 is 6/36 = 1/6. Since there are two ways of rolling an 11, the probability of rolling an 11 is 2/36 = 1/18. To determine the probability of rolling a 7 or an 11, we observe that this means we must consider a successful event to be a 7 or an 11. Hence there are 6 + 2 = 8 successful events. Therefore, the probability of rolling a 7 or an 11 is 8/36 = 2/9.

17. What is the probability of rolling an even number with a pair of dice? _____ An odd number? _____ A prime number? _____

18. What is the probability of rolling 13 with a pair of dice? _____

19. *Optional.* Suppose three dice are tossed. What is the probability of rolling a 3? _____ A 4? _____ A 2? _____

20. A jar contains 20 red marbles, 30 white marbles, and 50 blue marbles. If one marble is selected at random, what is the probability that it is red? _____ Blue? _____ White or blue? _____

■ Consider the experiment (Exercise 16) of flipping two coins. Let us try to calculate the theoretical probability for the outcome of two heads in this experiment. At first thought, you might not realize that there are *four* possible outcomes, namely: (1) head,

head; (2) head, tail; (3) tail, head; and (4) tail, tail. Thus the probability of getting two heads is 1/4.

21. Compare this with your experimental results.

Is there a discrepancy? _____ If so, explain why.

■ In experiments of this kind, a tree diagram is often useful to visualize the outcomes. Thus, in the flipping of two coins, the possibilities are shown in Figure 20-1. Note that the terms "first" and "second" do not mean that we are necessarily flipping them in order. It is used only to distinguish between the two of them. We could use a red coin and white coin as well.

Figure 20-1

22. The tree diagram for tossing three coins is shown in Figure 20-2. What is the probability of getting three heads? _____ Two heads? _____ One head? _____ Zero heads? _____

Figure 20-2

23. Draw a tree diagram showing the outcomes for tossing four coins. What is the probability of getting four heads? _____ Three heads? _____ Two heads? _____ One head? _____ Zero heads? _____

■ Now consider the problem of determining the probability of flipping ten heads in succession. One way of doing this problem would be to draw a tree diagram for the ten flips.

There is a mathematical approach to this problem that eliminates the necessity of drawing the tree diagram. Let us look at two flips. The probability of flipping a head on the first toss is 1/2, on the second toss is 1/2, and thus the probability of flipping two heads is $1/2 \times 1/2 = 1/4$. For three flips the probability is $1/8 = 1/2 \times 1/2 \times 1/2 = 1/2^3$. For ten flips the probability would be $1/2 \times 1/2 \times \cdots \times 1/2$ (ten times) or $1/2^{10} = 1/1{,}024$.

24. If 50 coins are flipped, what is the probability they will all come up heads? _____ Do you notice anything peculiar about Figure 20-3?

■ In general, if the probability of one event is p and the probability of a second event is q and the success of the second event does not depend on the success of the first event, then the probability of the first and the second event occurring is $p \times q$.

Figure 20-3

Multiplication principle for probabilities: *To find the probability of two events happening in succession, multiply the probabilities for each event.*

EXAMPLE: If you roll a die two times, what is the probability of rolling two 6's in succession?

The probability of rolling a 6 on the first is 1/6 and on the second roll is 1/6, hence the probability of rolling two 6's is 1/6 × 1/6 = 1/36.

25. If you roll a die two times, what is the probability of rolling two even numbers in succession? _____

    Of rolling an even number and then a 6? _____

26. A jar contains 4 red balls and 4 white balls.

    (a) What is the probability of selecting a white ball? _____

    What is the probability of selecting two white balls in succession if:

    (b) The first ball is replaced before the second is selected? _____

    (c) The first ball is not replaced before the second is selected? _____

Discussion of Exercise 26

Since there are four white balls and a total of eight balls, the probability of selecting a white ball is 4/8 = 1/2. This probability does not change if we replace the ball after the first selection. So for (b), the probability is 1/2 × 1/2 = 1/4. However, if we do not replace the ball before the second selection, then the probability changes. The probability of selecting a white ball in the second draw—assuming we have already picked a white ball—is 3/7. (There are seven balls left and three of them are white.) To find the probability of selecting two white balls in this manner (i.e., without replacement), we multiply the two probabilities: 1/2 × 3/7 = 3/14.

Remember, to find the probability of two events occurring, multiply the probability of the first event by the probability of the second event—assuming the first event has occurred.

27. What is the probability of drawing a diamond from a deck of 52 cards? _____ What is the probability of drawing two diamonds if:

    (a) The first card is replaced before the second draw? _____

    (b) The first card is not replaced before the second draw? _____

28. What is the probability of drawing three aces in succession from a deck of 52 cards, assuming replacement? _____ With no replacement? _____

29. Suppose a pair of dice are rolled twice. What is the probability of

    (a) Rolling a 7 both times? _____

    (b) Rolling a 12 both times? _____

    (c) Rolling a 2 both times? _____

    (d) Rolling a 7 and then an 11? _____

■ We close our discussion of probability with a brief look at the concept of *odds*. Since this term is often used to express probabilities, particularly for gambling events, it is useful to understand it. The term "odds" is defined by the following:

$$\frac{\text{Odds in favor}}{\text{of an event}} = \frac{\text{Number of successful outcomes}}{\text{Number of unsuccessful outcomes}}$$

$$\frac{\text{Odds against}}{\text{an event}} = \frac{\text{Number of unsuccessful outcomes}}{\text{Number of successful outcomes}}$$

For example, the odds in favor of obtaining a 2 in one roll of a die is 1/5—or as is often stated, "one to five," since there are one successful outcome and five unsuccessful outcomes. The odds against getting a 5 are 5/1, or "5 to 1."

Notice that the

$$\frac{\text{Odds in favor}}{\text{of an event}} = \frac{1}{\text{Odds against an event}}$$

30. (a) Find the odds in favor of rolling an even number in one roll of a die. _____

    (b) Find the odds against rolling an even number in one roll of the die. _____

31. A jar contains 30 red marbles, 20 blue marbles, and 50 green marbles. One marble is selected at random.

    (a) What are the odds against selecting a red marble? _____

    (b) What are the odds in favor of selecting a green marble? _____

    (c) What are the odds in favor of selecting a blue marble? _____

32. One card is selected at random from a deck of playing cards.

    (a) What are the odds against it being the seven of hearts? _____

    (b) What are the odds against it being a heart? _____

    (c) What are the odds in favor of it being a black card? _____

33. If two coins are flipped, what are the odds against getting two tails? _____

34. What are the odds against rolling a 7 with a pair of dice? _____ Most gambling casinos pay four to one against rolling a 7. Is this a fair bet? _____

■ Of course, the gambling casino could not exist if it paid off at the same rate as odds of 5 to 1. For example, if the dice were rolled 6,000,000 times, there would be approximately 1,000,000 successes and 5,000,000 failures. If a dollar was wagered at each roll of the die, the casino would pay $5,000,000 for the 1,000,000 sevens that were rolled and collect $5,000,000 for the unsuccessful rolls. So the casino would break even. But then who would pay for the building, the salaries, the electricity, etc.? Think about this before you go into a gambling casino.

35. Roulette wheels in the United States have the numerals 00, 0, 1, 2, 3, . . . , 36. The 0 and 00 are green. Half of the other numerals are red and half are black. What are the odds against the ball landing on a red numeral? _____ The casino pays even money (one to one) for a bet on red. Is that fair? _____

36. Roulette wheels in Europe do not have a double zero. What are the odds against a red appearing on such a roulette wheel? _____ Where is the better place to gamble on roulette? _____

■ Obviously there is a connection between the probability of an event's occurrence and the odds in favor or against it. Let us formulate this. If the number of successful outcomes for a certain experiment is s and the number of unsuccessful outcomes is u, then the total number of outcomes is $t = s + u$. The odds against the event is given by $o = u/s$, and the probability of the event occurring is $p = s/t = s/(s + u)$. We can find an algebraic relation between o and p as follows:

$$\frac{1}{p} = \frac{s + u}{s} = 1 + \frac{u}{s}$$

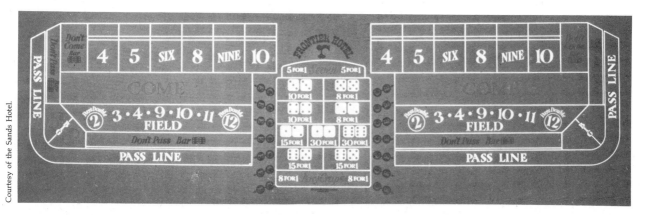

Figure 20-4

Figure 20-5

Hence $u/s = 1/p - 1 = (1 - p)/p$, so that if an event has probability p:

$$\text{Odds against the event} = \frac{1 - p}{p}$$

Letting o denote the odds against, we have $1/p = 1 + o$ or $p = 1/(1 + o)$. Thus if o is the odds against an event:

$$\text{Probability for the event} = \frac{1}{1 + o}$$

37. The probability of rolling a 7 with a pair of dice is 1/6. Use the first rule above to compute the odds against rolling a 7. Compare with Exercise 34.

38. Since there are three ways of rolling a 10 with a pair of dice, the odds against rolling a 10 are 33/3 or 11/1. Use the second rule above to compute the probability of rolling a 10. Compare with the result in Exercise 13.

Important Terms and Concepts

Probability of an event
Tree diagrams
Multiplication principle for probabilities
Odds in favor of an event
Odds against an event

Review Exercises

1. A field is marked as shown below. A parachutist will land in the field. What is the probability of landing inside the figure? _____

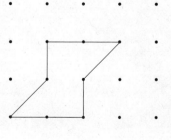

20/Review Exercises

2. You stop two people at random on a street corner. What is the probability they will both have been born on a Saturday? _____

3. A magician has fifteen rabbits in a hat—five white, five brown, and five black. What is the probability the magician will randomly select a white and then a black rabbit? _____ (*Stop!* What must you know before you can answer this question?)

4. A circular board is divided into three equal areas and painted red, yellow, and green. In the center of the board is an arrow that can be spun. If the arrow is spun three times, what is the probability that:
   (a) The arrow will point to red all three times?
   (b) The arrow will point to red first, then green, then yellow?
   (c) The arrow will point to each color exactly once?

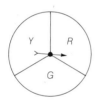

5. Make a tree diagram for the situation described in Review Exercise 4. Use the diagram to check your answers.

6. There are three questions on an exam. The probabilities of answering the questions correctly are 1/2 for the first question, 1/3 for the second question, and 2/3 for the third question.
   (a) What is the probability of answering all three correctly? _____
   (b) What is the probability of answering all three incorrectly? _____
   (c) What is the probability of answering the first question correctly and the third question incorrectly? _____

7. If a tree diagram were to be constructed illustrating three consecutive rolls of a single die, how many entries would be listed in the last column of the tree? _____

8. Instead of taking the sum of a pair of dice, suppose we choose to take the product. As in Exercise 8 in the lab exercises, construct a table that shows all the possible products.

   Calculate the probabilities of:
   (a) Rolling a 9. _____
   (b) Rolling a 2. _____
   (c) Rolling a 7. _____
   (d) Rolling an odd number. _____
   (e) Rolling an even or an odd number. _____
   (f) Rolling an even, then an odd number. _____

9. Refer to the situation in Review Exercise 8.
   (a) What are the odds against rolling an 8? _____
   (b) What are the odds in favor of rolling an 8? _____
   (c) What are the odds against rolling a 36? _____

10. A jar is filled with 30 marbles—ten red, ten purple, and ten white. A marble is selected at random.
    (a) What are the odds against choosing a red marble? ____
    (b) What are the odds in favor of the marble not being white? ____

11. In the above situation, two marbles are selected, with the first marble being replaced before the second marble is selected.
    (a) What are the odds against choosing two red marbles? ____
    (b) What are the odds in favor of choosing two marbles with neither marble being white? ____

12. *Optional.* Let f denote the odds in favor of an event and p denote the probability of the event. Develop a formula that expresses f in terms of p. Develop a formula that expresses p in terms of f.

© 1966 United Feature Syndicate, Inc.

Emmy Noether (1882–1935) was one of the most influential and creative forces in the development of modern abstract algebra. She taught at the University of Gottingen, Germany, during the time when it was the greatest center of mathematical research in the world.

21

Statistics

"It is by the aid of Statistics that law in the social sphere can be ascertained and codified."—Florence Nightingale

Lab Exercises: Set 1

EQUIPMENT: A hand-held calculator would be helpful for some problems but is not required.

In these exercises you will learn concepts commonly used in treating statistical data.

1. Relate an example to your group of how you have heard or seen the word "statistics" used. Discuss what the word "statistics" means.

2. Acme Industries, a small manufacturing firm, has the following salary schedule.

President	$100,000
Vice-President	70,000
Business Manager	35,000
Plant Worker 1	17,000
Plant Worker 2	15,000
Plant Worker 3	15,000
Plant Worker 4	12,000
Secretary 1	12,000
Secretary 2	12,000
Secretary 3	12,000

   The workers are trying to obtain better wages. The firm says that the average salary paid by the company is $30,000. Is this a valid description of the salary situation? Discuss in your group how you might better interpret these data.

3. B-Good Cooperative is formed to compete with Acme Industries. They decide on the following salary schedule.

President	$50,000
Vice-President	40,000
Business Manager	30,000
Plant Worker 1	30,000
Plant Worker 2	25,000
Plant Worker 3	25,000
Plant Worker 4	25,000
Secretary 1	25,000
Secretary 2	25,000
Secretary 3	25,000

   What is the average salary for B-Good Coop?

   _____ Discuss in your group the similarities and differences between the salary schedules of Acme and B-Good. How might you express the differences mathematically?

4. Ann and Martha are avid frisbee players. They decide to have a contest to see who can throw a frisbee the farthest. They agree to throw a frisbee 7 times and compare the distances. The results are as follows.

	Ann	Martha
1st throw	45.3 meters	46.0 meters
2nd throw	50.0 "	50.2 "
3rd throw	42.2 "	42.6 "
4th throw	46.8 "	48.5 "
5th throw	52.3 "	53.1 "
6th throw	47.0 "	48.1 "
7th throw	52.4 "	42.6 "

(a) Ann and Martha both claim that they won the contest. What reason could Ann give for her claim?

(b) What reason could Martha give for her claim?

(c) Discuss in your group the explanation for this situation.

■ In your discussion you may have used the word *average*. Most people think of the "average" of a set of numbers in the same way that Acme Industries used it in Exercise 2, namely the sum of the numbers in a set divided by the number of numbers in that set. The shortcomings of this description of "average" should be obvious from the above exercises. One way to remedy this is to consider other kinds of "average." We use the word "mean" for the concept that Acme Industries was using and introduce two new concepts in the following definition.

Definition: *The* **mean** *of a set of numbers is the sum of the numbers divided by the number of numbers in the set. The* **median** *of a set of numbers is the middle number when the numbers are arranged in order of size. If there are two middle numbers, then the median is the mean of the two middle numbers. The* **mode** *of a set of numbers is the number that occurs most frequently.*

If no number occurs more than once, there is no mode for that set. It is possible to have more than one mode.

Example: Suppose your bowling scores for seven games are 140, 140, 140, 150, 165, 170, and 180.

The mean score is

$$\frac{140 + 140 + 140 + 150 + 165 + 170 + 180}{7}$$

$$= \frac{1{,}085}{7} = 155$$

The median score is 150. The mode is 140. Suppose your next score is 140. Then your scores arranged in order are

140, 140, 140, 140, 150, 165, 170, 180

Since the two middle numbers are 140 and 150, the median of the eight scores is $(140 + 150)/2 = 145$.

5. (a) Find the mean, median, and mode for the salaries paid by Acme Industries and by B-Good Cooperative and record your results below.

	Acme	B-Good
Mean		
Median		
Mode		

(b) Discuss in your group the advantage of having these different concepts to describe the "average."

6. Find the mean and median for the distances in the frisbee contest of Exercise 4 and record your results below.

	Ann	Martha
Mean		
Median		

7. A class is given a 10-point quiz. The scores are as follows.

Score	Number of students with that score
1	1
2	1
3	2
4	3
5	4
6	7
7	5
8	3
9	2
10	2

Discuss in your group how you can find the mean without adding up the thirty individual numbers.

Then compute the mean. _____

8. In the following exercise, we will determine the "average" number of times a die must be rolled so that all six numbers will come up at least once.

   (a) Roll a die and record the numbers that result from each roll on the chart below. When all numbers have occurred, record the total for that trial and start again. For example, the following trial ended when a 3 was rolled.

Trial	1	2	3	4	5	6	Total no. of rolls
EXAMPLE: 0	||	|||	|	ℕ|	|||	||	17
1							
2							
3							
4							
5							
6							
7							
8							
9							
10							
11							
12							
13							
14							
15							
16							
17							
18							
19							
20							
21							
22							
23							
24							
25							

   (b) Determine the median number of rolls needed for all six numbers to appear. _____

   (c) Calculate the mean number of rolls necessary. _____

   (d) Ask for the results from the other groups in your class and record them below.

Medians	Means
_____	_____
_____	_____
_____	_____
_____	_____
_____	_____
_____	_____
_____	_____
_____	_____
_____	_____

   (e) Discuss in your group which of the means or the medians is the "best" prediction of the "average" number of rolls needed.

9. An automobile magazine wishes to test two car models for gas mileage. They rent five cars of each model and obtain the following data. The numbers represent miles per gallon.

   (a) Find the mean and median for each model.

	Model A	Model B
Car 1	18	10
Car 2	19	18
Car 3	20	24
Car 4	16	25
Car 5	17	13

   Mean

   Median

   (b) You should have found that Model A and Model B have the same mean and the same median. Yet if the auto magazine reported only the mean and median, it would be doing its readers a disservice. Explain why.

   (c) Discuss in your group what other information the magazine might provide in order to better clarify the test results. (Of course, it might provide the entire table, but such information might be detailed, especially if a large number of cars were tested.)

■ What is needed in Exercise 9 is some way to describe or measure how the numbers are spread. One way to do this is by specifying the *range*.

Definition: *The* range *of a set of numbers is the difference between the largest and the smallest numbers in the set.*

For example, the range of the set of gas mileages for Model A is 4 and the range for Model B is 15.

10. (a) Find the range of the set of salaries in Exercise 2. ____

   (b) Find the range of the set of salaries in Exercise 3. ____

11. Consider Exercise 4.

   (a) Find the range of the set of distances for Ann. ____

   (b) Find the range of the set of distances for Martha. ____

■ Another measure of how the numbers in a set are spread out is called the *variance*.

Definition: *The* variance *of a set of numbers is calculated by taking the difference between each number in the set and the mean, squaring this difference, then taking the mean of these squares.*

For example, to compute the variance of the set of gas mileages of Model A in Exercise 9, we could compute as shown in the following table.

Numbers	Difference with mean	Squares of differences
18	18 − 18 = 0	0
19	19 − 18 = 1	1
20	20 − 18 = 2	4
16	16 − 18 = −2	4
17	17 − 18 = −1	1
Sum 90		Sum 10
Mean = $\frac{90}{5}$ = 18		Variance = $\frac{10}{5}$ = 2

11. Find the variance of the gas mileages of Model B. ____

12. Find the variance of the set of salaries of Acme Industries (Exercise 2).

13. Find the variance for the set of salaries of B-Good Coop (Exercise 3). ____

■ Related to variance, and more commonly used in statistics, is the *standard deviation*.

Definition: *The* standard deviation *of a set of numbers is the square root of the variance.*

The standard deviation is usually denoted by the Greek letter σ (sigma). Thus the standard deviation of the gas mileages for Model A is given by σ = $\sqrt{2}$.

14. Find the standard deviation of the gas mileages for Model B. ____

15. Suppose Ann did not know statistics and your group has been asked to write a complete set of instructions for her to use in order to compute the standard deviation for her distances. Discuss what you would say and write it below. (It is not necessary to actually compute the standard deviation.)

16. *Optional*—Hand calculator would be helpful. Consider the data your group obtained in Exercise 8. Find the range and the standard deviation for these data. Range = ____ Mean = ____ Standard deviation = ____ Compare with the other groups in your class.

Comment: Basic Statistical Concepts

"Statistical thinking will one day be as necessary for efficient citizenship as the ability to read and write."—H. G. Wells

"There are three kinds of lies: lies, damned lies, and statistics."—Benjamin Disraeli

It might seem as if the above two statements are contradictory. H. G. Wells, a British historian and science-fiction writer who lived between 1866 and 1945, evidently foresaw the role that statistics would play in our modern world. Disraeli, the Prime Minister of Great Britain from 1874 to 1880, was probably concerned about how statistics could be used to misrepresent information. In reality there is no conflict between the two statements. Since statistics deals with the collection, organization, and interpretation of numerical facts—what we refer to as data—and since so many decisions are based on this kind of data, it is useful for a citizen to understand statistics. The fact that some people will use statistics to justify, argue for, or even lie about, their particular point of view makes it even more important that people understand the subject and not be misled.

One of the most often used (and misused) ideas in statistical arguments is the idea of "average." We can read that the "average" family size in the United States is 2.94 persons or the "average" family income (1977) is $18,264.* In order to interpret this information, we should know what "average" is being referred to and something about how the numbers are spread out. Usually, when figures are reported as *averages*, it is the *mean* that is being used. For example, the median U.S. family income in 1977 was $16,060,* considerably lower than the mean. From your work in the exercises above, you should have some idea why the discrepancy of $2,204 occurs. The three "averages" that we studied above (the mean, the median, and the mode) are referred to as *measures of central tendency*. This is a fancy way of saying that in some sense they express the average behavior of a set of numbers. Although the mean is the measure most frequently used, you should now be aware from Exercises 2 and 3 that it does not always give an accurate picture and that sometimes the median is a better concept to use.

The mode also has its uses. For example, suppose a shoe store records the sizes of all the shoes sold during one year and uses these data to place future orders. They would not be interested in the mean or median shoe size, but rather the mode or modes. Can you think of any other situation in which the mode would be useful?

In order to convey a more accurate picture of a set of data, it is helpful to include a measure of the *dispersion*—the "spread"—of the data, as well as a measure of the central tendency. You studied three measures of dispersion in the lab exercises above, namely the range, the variance, and the standard deviation. The range is the simplest. We defined it as the difference between the largest and smallest numbers in the set.

17. Lucia and Roberto have the same mean score on five examinations. Lucia's scores have a range of 10 points. Roberto's scores have a range of 50 points. Write two sets of scores that satisfy these conditions.

18. Can you write scores such that all of Roberto's scores are different?

■ The two other measures of dispersion (the *variance* and the *standard deviation*) are closely related, the standard deviation being defined as the square root of the variance. The standard deviation is the one most commonly used because in many naturally occurring situations it provides additional information about the data. For example, if you were to measure the lifespan of light bulbs and compute the mean m and the standard deviation d, you would find that roughly 68% of all the lifespans lie within $m - d$ and $m + d$, that roughly 96% of the lifespans lie between $m - 2d$ and $m + 2d$, and just about all (99.8%) lie between $m - 3d$ and $m + 3d$. If you measured heights of humans, weights of peas, or densities of stars, similar results would occur. (If you wish to find out more about this subject, find an elementary statistics text and read about *normal distributions*.)

19. Find the mean, median, range, and standard deviation for each of the following sets of numbers.

	Mean	Median	Range	Standard deviation
(a) 7, 8, 9, 10, 11				
(b) 9, 12, 15, 18, 21				
(c) 1, 4, 9, 16, 25				

20. Two families of six people each have the same mean weight. The standard deviation of the weight is small for one family and is large for the other. Give an example of what the weights could be.

■ In most of the examples above, our sets have been small in order to simplify the arithmetic. With the

---
1978 Statistical Abstract of the United States.

availability of electronic computers and hand calculators, large quantities of data can be handled. There are hand calculators, for example, that will automatically compute the mean and standard deviation as you enter a sequence of numbers.

When you deal with a small set of numbers, you must be cautious about using statistical concepts, since the addition of only one or two additional numbers can significantly change the result. The following exercise illustrates this.

21. At a typical college, grade-point averages are calculated by multiplying the number of credit units by the value assigned to each letter grade, then calculating the mean points per credit. Suppose the values assigned are: A = 4 points, B = 3 points, C = 2 points, D = 1 point, and F = 0 point.

    In her first term, Selma's grades are as follows:

Mathematics	3 units	A
Biology	4 units	B
Chemistry	3 units	B
English	4 units	A
Total	14 units	

    Selma's grade-point average is calculated as follows:

Mathematics	$3 \times 4 = 12$
Biology	$4 \times 3 = 12$
Chemistry	$3 \times 3 = 9$
English	$4 \times 4 = 16$
Total	49

    Grade-point average = 49/14 = 3.5.

    (a) Suppose Selma takes a 4-unit history course in summer school and receives a C. Calculate her new grade-point average. ____
    (b) Yvonne has finished 90 units with a GPA of 3.5. Suppose she takes the 4-unit history course and also receives a C. What is her new GPA? ____

22. Write down a set with five numbers such that an additional number will produce a large change in the median. ____

■ There is much more to learn about statistics. One interesting application is the breaking of codes. For an elementary introduction to this subject, see Harold Jacobs, *Mathematics: A Human Endeavor* (San Francisco: W. H. Freeman and Company, 1970), pp. 399–405. For additional information on how statistics can be misused, see Donald Huff, *How to Lie with Statistics* (New York: Norton, 1954).

Important Terms and Concepts

Mean   Range
Median   Variation
Mode   Standard deviation

Review Exercises

1. A student obtains the following scores on examinations: 82, 50, 75, 82, 83, 50, 60, 81, 50, and 81. Find the mean, median, mode, range, and standard deviation. In your opinion, which of the measures of central tendency best reflects the student's performance? Why?

2. On one day in a certain hospital 10 babies are born. Their heights, measured to the nearest centimeter, are 49, 50, 51, 51, 50, 56, 48, 51, 50, and 55. Find the median, mean, mode, range, and standard deviation for these heights.

3. Two students' scores on five examinations have the same mean. One student has done better than the other on four of the five exams. Give an example that shows how this is possible.

4. The test scores of 100 students are given by the frequency distribution table on the following page. (Frequency = number of students obtaining that score.) Find the mean, median, and mode for these scores.

21/Review Exercises

Score	Frequency
100	1
99	0
98	0
97	2
96	2
95	4
94	3
93	4
92	4
91	3
90	5
89	3
88	0
87	6
86	6
85	10
84	7
83	6
82	0
81	8
80	2
79	6
78	4
76	0
75	0
74	3
72	4
71	2
70	0
69	2
68	0
67	1
66	1
65	0
64	1

5. Suppose you know that both the mean and the median of a set of data is 20 and the range is 10. Can you conclude that the data tend to cluster near 20? _____ Justify your answer.

6. A professor states that the mean of 50 test scores is 80 and the range is 20. You received a score of 100. Can the professor's statements be accurate? _____ Justify your answer.

7. Can you write a set of numbers that has more values below the mean than above the mean? ___
   _____

8. Can you write a set of numbers that has more values above the mean than below it? _____
   _____

9. Can you write a set of numbers that has the same number of values above the mean as below the mean, but in which the mean and median are not equal? _____

10. You are told that the standard deviation for a set of numbers is zero. What can you conclude?

© 1960 United Feature Syndicate, Inc.

22

Mathematical Explorations

"The art of teaching is the art of assisting discovery."—Mark Van Doren

This chapter consists of three different and independent explorations in mathematics. The first explores how to develop some equations about numbers. The second develops a formula for the area of a figure on the geoboard. The third explores a subject in topology.

Exploration 1

EQUIPMENT: One container of C-rods for each two people.

1. **(a)** Make a staircase of white rods to illustrate the sum $1 + 2 + 3$.

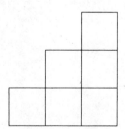

   (b) Construct a similar staircase for $1 + 2 + 3 + 4$ and $1 + 2 + 3 + 4 + 5$.

   (c) Complete the table below.

	Sum
$1 + 2$	3
$1 + 2 + 3$	
$1 + 2 + 3 + 4$	
$1 + 2 + 3 + 4 + 5$	
$1 + 2 + \cdots + 6$	
$1 + 2 + \cdots + 7$	
$1 + 2 + \cdots + 8$	

   (d) Do you see any pattern?

   ■ We want to find a formula to express $1 + 2 + \cdots + n$ in terms of n, where n represents the last number in the sum. Try to guess what it might be before going on.

2. Consider the C-rod staircases you built. Suppose you completed each staircase to a square. Deter-

mine the number of white rods in each square and complete the table below.

	Sum	Total no. of white rods when staircase is completed to a square
1 + 2	3	4
1 + 2 + 3		
1 + 2 + 3 + 4		
1 + 2 + 3 + 4 + 5		
1 + 2 + · · · + 6		
1 + 2 + · · · + 7		
1 + 2 + · · · + 8		

3. If we had a staircase representing $1 + 2 + \cdots + n$, how many white rods would there be in the associated square? (n represents the *last* number in the sum.) ____

4. (a) Make two staircases for $1 + 2 + 3 + 4 + 5$ and complete one to a square. Place the staircase on top of the square.

   (b) Is it approximately one-half of a square? ____

   (c) How many units more or less than half of a square? ____

   (d) Determine the same for each of the staircases and summarize your results in the table at the bottom of the page. Be sure to use the rods to determine at least three of these before figuring it out in your head.

5. Look at a rod staircase. If you had a staircase of n steps where n is some whole number, how many white rods would be in the associated square?

____ In the associated half-square? ____ What would the difference be between the staircase and the half-square? ____ (*Hint:* Consider the diagonal of the square.) Now can you find the formula

$1 + 2 + 3 + \cdots n = $ _____

If not, there are more hints below. If you think you have discovered it, check your answer with some of the sums you calculated previously.

6. If we complete the staircase representing $1 + 2 + 3$ to a square, we get:

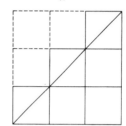

One-half of the square is $3^2/2$. The staircase has some extra half-rods. How many of these does it have? If you write $1 + 2 + 3 = 3^2/2 + \square/2$, what goes in the box?

Do the same for the staircase representing $1 + 2 + 3 + 4$. What goes in the box

$$1 + 2 + 3 + 4 = \frac{\square^2}{2} + \frac{\square}{2}$$

How about

$$1 + 2 + \cdots + n = \frac{\square^2}{2} + \frac{\square}{2}$$

$$= \frac{\square(\square + 1)}{2}$$

7. *Optional.* Can you figure out a formula for sums like $5 + 6 + 7 + 8 + 9 + 10 + 11$ or $56 + 57 + 58 + 59 + 60$? The general form would be $(k + 1) + \cdots + (k + n)$.

	Sum	Square	Half of square	Difference between staircase and half of square
1 + 2				
1 + 2 + 3				
1 + 2 + 3 + 4				
1 + 2 + 3 + 4 + 5				
1 + 2 + · · · + 6				
1 + 2 + · · · + 7				
1 + 2 + · · · + 8				

Exploration 2

EQUIPMENT: One geoboard for each pair of people.

1. (a) With a rubber band, make a figure on the geoboard that touches exactly six pins and has no pins inside the figure, for example:

   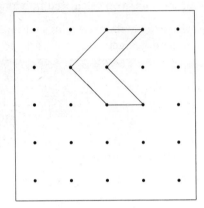

   (b) Use the method of Chapter 14 to determine the area of your figure. It is _____.
   (c) Construct at least four more such figures and determine their areas. What do you notice?

   _____

2. (a) Construct a figure with five pins on the boundary and no pins inside. It has area _____. Construct three more such figures and determine their areas.
   (b) Repeat for figures that have four pins on the boundary and no pins inside.
   (c) The region inside a figure is called the *interior*. If a figure has b pins on the boundary and no pins in the interior, its area is given by $A = $ _____. Check your guess for $b = 3$ and for $b = 7$.

3. Now make a triangle on the geoboard with three pins on the boundary and one pin inside the triangle. What is the area of the triangle? _____ Make another such triangle. Is the area the same? Use the geoboard to fill in the table below, where

b	i	Area of triangle
3	0	
3	1	
3	2	
3	3	

   $b = $ number of pins on the boundary of the triangle and $i = $ number of pins in the interior.

4. (a) Make a triangle with four pins on the boundary and no pins in the interior. What is the area? _____ Make two more such triangles. Now try a nontriangular region and see if it has the same area.
   (b) Repeat (a) with four pins on the boundary and one pin on the interior.
   (c) Repeat (a) with four pins on the boundary and two pins in the interior.
   (d) Fill in the table below.

b	i	Area of region
4	0	
4	1	
4	2	

5. Construct a region with $b = 5$ and $i = 0$. What is its area? _____ Can you construct a region with $b = 5$ and area $2\frac{1}{2}$? Draw it on the diagram of a geoboard below.

6. Can you guess a formula that gives the area of any region in terms of b and i? Be sure to test any of your guesses. Think about this for a few moments before going on.

   $A = $ _____

   The formula we have been attempting to discover is $A = b/2 + i - 1$. It is called Pick's formula.

7. Each person in your group make a complicated figure on the geoboard. Use Pick's formula to find its area.

Exploration 3*

EQUIPMENT: Paper (graph paper will make the cutting easier), scissors, tape.

In these exercises you will explore the properties of various surfaces. There are some surprising discoveries.

Each person cut 10 strips of paper, each 3 centimeters wide and 20 to 30 centimeters long.

1. (a) Take one strip and form it into a cylindrical loop, as shown in Figure 22-1. Before taping the ends, give one end of the strip a half-twist (Figure 22-2). If you label the strip as in Figure 22-3, you will attach corner A to corner D and corner B to corner C. The surface that results is called a *Moebius strip* (Figure 22-4), after the German mathematician who discovered and investigated it in the middle of the nineteenth century.

   (b) Place a pencil midway between the edges of your strip and draw a line along the center until you reach your starting point.

   (c) Describe the result.

   (d) Discuss with your group what you would mean by the *side* of a surface. By the *edge* of a surface. How many sides does the Moebius strip have? _____ How many edges? _____

2. Cut the strip along the line you have drawn. Describe what happens.

3. (a) Make another Moebius strip and cut it parallel to an edge, one-third of the way across (Figure 22-5).

   (b) What happens when you cut around the strip once?

   (c) Continue cutting until you return to your starting point. Be sure to stay the same distance from the edge as you cut. Describe the result.

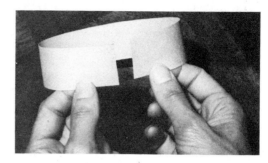

Figure 22-1

Figure 22-2

Figure 22-3

Figure 22-4

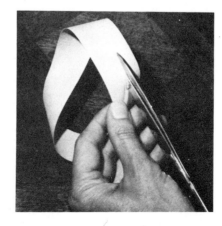

Figure 22-5

---

*This exploration is adapted from H. Jacobs, *Mathematics: A Human Endeavor* (San Francisco, W. H. Freeman and Company, 1970).

4. Make another Moebius strip and cut it parallel to an edge, one-fourth of the way in from the edge. Describe the result. Discuss the similarities and differences between this result and the results in Exercises 2 and 3.

5. (a) Discuss in your group what would result if you cut around the strip one-fifth of the way in from an edge. Write down your prediction.

   (b) Test your prediction in (a) by cutting a Moebius strip.

6. Take a strip of paper and give it two half-twists before taping the ends together. Discuss this surface in your group. How many edges does it have? _____ How many sides? _____ Draw a line down the center to check your answers.

7. (a) Discuss in your group what would result if you cut along the line you drew in Exercise 6. Write down your prediction.

   (b) Cut along the line. Compare the results with your predictions.

8. (a) Make a strip with three half-twists. How many edges does it have? _____ How many sides? _____ Draw a line down the center to check your answers.

   (b) What will happen if you cut the strip along the line? Do it and describe the results.

   (c) Make another strip with three half-twists, and cut it one-third of the way from an edge. Describe the results.

   (d) Discuss what would happen if you cut a strip with three half-twists one-fourth of the way in from an edge. Try it and see what happens.

9. (a) Make a strip with four half-twists. How many edges does it have? _____ How many sides? _____

   (b) Discuss in your group the relationship between the number of half-twists in the strip and the number of sides. Write down your conclusion.

   (c) Cut the strip down the center. What happens?

   (d) If you cut a strip with 80 half-twists down the center, how many loops do you think there would be? _____

10. (a) Take a strip of paper, fold it in half, and cut two slits in both ends, as shown in Figure 22-6. Unfold the strip and number it as shown in Figure 22-7. Put three short pieces of tape on ends 4, 5, and 6 and make a loop, attaching end 4 to end 1. Then pass end 5 *over* end 4, end 2 *under* end 1, and tape end 5 to end 2. Then pass end 6 *between* ends 4 and 5, pass end 3 *over* end 1, and tape end 3 to end 6.

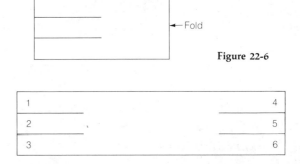

Figure 22-6

Figure 22-7

   (b) Now finish both cuts so that they go completely around the band. Describe the results.

   (c) Prepare another strip as you did in (a). Give end 6 a half-twist and tape it to end 1. Give end 4 a half-twist and tape it to end 2. Give end 5 a half-twist and tape it to end 3. Complete the cuts and describe the results.

22/Exploration 3

■ If you would like to read more about this subject, you will find further information in:

Scientific American, December 1968, pp. 112–115.

The Scientific American Book of Mathematical Puzzles and Diversions by Martin Gardner, (New York: Simon and Schuster, 1959), Chapter 7.

Mathematical Diversions by J. A. Hunter and J. S. Madachy (New York: Dover Publications, 1975), pp. 42–45.

There are also two short stories that involve Moebius strips. They are

"A. Botts and the Moebius Strip," in *Fantasia Mathematica,* ed. Clifton Fadiman (New York: Simon and Schuster, 1958), pp. 155–170.

"Paul Bunyan versus the Conveyor Belt," in *The Mathematical Magpie,* ed. Clifton Fadiman (New York: Simon and Schuster, 1962), pp. 33–35.

"Space is just as crucial a part of nature as matter is, even if (like the air) it is invisible."—Jacob Bronowski

Bibliography

This list is by no means complete. The criterion for a book's inclusion was that I was familiar with it and thought it would be interesting to an elementary school teacher. The books and journals are grouped by category: history of mathematics, anecdotes and aphorisms, teaching and learning, ideas for the classroom, games and puzzles, mathematics and miscellany, journals, and catalogs.

History of Mathematics

Boyer, C. *A History of Mathematics.* New York: Wiley, 1968. [A textbook for a course in the history of mathematics. Clearly written and informative. Some of the mathematical explanations are advanced, but if you are interested only in the history, these can be skipped.]

Groza, V. S. *A Survey of Mathematics: Elementary Concepts and Their Historical Background.* New York: Holt, Rinehart and Winston, 1968. [A good exposition written for elementary teachers.]

Kline, M. *Mathematics in Western Culture.* New York: Oxford University Press, 1953. Paperback. [An interesting and readable book exploring the interplay between mathematics and other aspects of Western culture from the Greeks to modern times.]

Osen, L. *Women in Mathematics.* Cambridge, MA: MIT Press, 1974. [Biographical sketches of women mathematicians.]

Smith, D. E. *History of Mathematics.* New York: Dover Publications, 1958. Paperback. [A classic two-volume work originally published in the 1920s. Volume 1 is a general survey, and Volume 2 covers special topics.]

Smith, D. E. *Number Stories of Long Ago.* Reston, VA: National Council of Teachers of Mathematics, 1969. [Originally published in 1919. Stories about different numeration systems written for young people. My children enjoyed having the stories read aloud to them.]

Zaslavsky, C. *Africa Counts.* Boston: Prindle, Weber and Schmidt, 1968. [A survey of the number systems and geometrical investigations of African civilizations. No mathematical background is necessary.]

Anecdotes and Aphorisms

Eves, H. *In Mathematical Circles.* Boston: Prindle, Weber and Schmidt, 1969. [A three-volume collection of short anecdotes about mathematicians. Fun to read.]

Fadiman, C. *Fantasia Mathematica.* New York: Simon and Schuster, 1958. Paperback. [An entertaining collection of short stories, poems, and limericks, all with some mathematical twist.]

Fadiman, C. *The Mathematical Magpie.* Hamden, CT: Fireside Press, 1964. Paperback. [A sequel to the above.]

Moritz, E. *On Mathematics and Mathematicians.* New York: Dover Publications, 1958. Paperback. [Consists of quotes from mathematicians, historians, philosophers, and others about mathematics.]

Teaching and Learning

Biggs, E., and J. Maclean. *Freedom to Learn.* Reading, MA: Addison-Wesley, 1969. [Of particular interest to elementary school teachers. Explains the "active learning approach."]

Gattegno, C. *For the Teaching of Mathematics.* New York: Educational Solutions, 1963. [Three volumes of essays by a person who worked closely with Cuisenaire on developing Cuisenaire rods.]

Holt, J. *How Children Fail.* New York: Dell, 1964. Paperback. [An observer's penetrating view of what goes on in the classroom. A modern classic.]

Kidd, K. P., et al. *The Laboratory Approach to Mathematics.* Palo Alto, CA: Science Research Associates, 1970. Paperback. [A description of how to set up and use a math lab.]

Kline, M. *Why Johnny Can't Add.* New York: St. Martin's Press, 1973. [A mathematician's view of what is wrong with mathematics education.]

Montessori, M. *The Discovery of the Child.* New York: Ballantine Books, 1967. Paperback. [Maria Montessori's observations about teaching and learning. Written some time ago and still extremely valuable.]

Polya, G. *How to Solve It.* Princeton, NJ: Princeton University Press, 1971. Paperback. [George Polya is well known for his excellent teaching and writing skills. This book explores the teaching of mathematics through problem solving.]

Reys, R. E., and T. R. Post. *A Mathematics Laboratory in the Classroom.* Boston: Prindle, Weber and Schmidt, 1937. [Similar to Kidd et al., but with more emphasis on pedagogy and theory.]

Skemp, R. *Psychology of Learning Mathematics.* New York: Penguin Books, 1971. Paperback. [Readable and reasonably free from jargon.]

Weissglass, J. "Small Groups: An Alternative to the Lecture Method." *Two-Year College Mathematics Journal,* 7 (1976), 15–20. [This article outlines some assumptions about learning and explores one method of organizing a classroom in relation to these assumptions.]

Ideas for the Classroom

Association of Teachers of Mathematics. *Notes on Mathematics for Primary Schools.* Cambridge: Cambridge University Press, 1968. Paperback. [Written by a group of British teachers. Contains some very interesting ideas for getting students involved in doing mathematics.]

Banwell, C. S., K. D. Saunders, and D. G. Tahta. *Starting Points.* New York: Oxford University Press, 1972. Paperback. [Presents ways of providing children with "starting points" for mathematical exploration. Excellent. Every elementary teacher should inspect this excellent book or the *Notes* above.]

Bezuzska, S., M. Kenney, and L. Silvey. *Tessellations: The Geometry of Pattern.* Palo Alto, CA: Creative Publications, 1978. [Contains reproducible worksheets and patterns for student activities.]

Charbonneau, M. *Learning to Think in a Math Lab.* Boston: National Association of Independent Schools, 1971. Available from Creative Publications, Palo Alto, CA. [Contains numerous lab activities.]

Cundy, H. M., and A. P. Rollett. *Mathematical Models.* New York: Oxford University Press, 1968. [Chock-full of information valuable to teachers at every level. Includes instructions on constructing polyhedron models and tessellations of the planes. Also information on dissections, paper-folding, and curve-stitching.]

Gardner, M. *The AHA! Box.* New York: Scientific American, Inc. [The box contains sound filmstrips and a teacher's guide. The focus is on the Aha! that comes when you see the solution to a problem.]

Gardner, M. *The Paradox Box.* New York: Scientific American, Inc. [This box also contains sound filmstrips and a teacher's guide. It is concerned with paradoxes in logic, geometry, time, numbers, probability, and statistics.]

Greenes, C., R. Willcutt, and M. Spikell. *Problem Solving in the Mathematics Laboratory.* Boston: Prindle, Weber and Schmidt, 1972. Paperback. [A thorough investigation of the mathematical uses of four pieces of equipment: attribute blocks, Cuisenaire rods, geoboards, multibase arithmetic blocks.]

Laycock, M., and G. Watson. *The Fabric of Mathematics.* Hayward, CA: Activity Resources, 1975. [A subject-by-subject index of curricular materials and laboratory material available for teaching. Very valuable if you have laboratory equipment available in your school district.]

Mira Math Pub., *Mira Math for Elementary School* and *Mira Activities for Junior High School Geometry.* Palo Alto, CA: Creative Publications, 1973. [Both books provide activities for young people to explore geometry with the Mira.]

Readings in Geometry from the Arithmetic Teacher. Reston, VA: National Council of Teachers of Mathematics, 1970. [A collection of articles describing how some teachers have taught geometrical concepts in the elementary school.]

Games and Puzzles

Burns, M. *The Book of Think.* Boston: Little, Brown, 1976. [Written for young people. Suggests ways to think about various kinds of problems.]

Gardner, M. *Mathematical Puzzles and Diversions.* 2 vols. New York: Simon and Schuster, 1959, 1961. [The author is the editor of the mathematical games section of *Scientific American.* His articles are always fascinating.]

Gardner, M. *New Mathematical Diversions from Scientific American.* New York: Simon and Schuster, 1976. Paperback. [A sequel to the above.]

Gardner, M., ed. *Best Mathematical Puzzles of Sam Lloyd.* New York: Dover Publications, 1959. Paperback. [Sam Lloyd was one of the greatest inventors of puzzles.]

Read, R. *Tangrams.* New York: Dover Publications, 1965. [A short history of the tangram, plus 330 tangram puzzles with solutions.]

Schadler, R., and D. Seymour. *Pic-a-Puzzle.* Palo Alto, CA: Creative Publications, 1970. [A collection of geometric puzzles.]

Bibliography

Smullyan, R. *What Is the Name of This Book?* Englewood Cliffs, N.J.: Prentice-Hall, 1978. [A fascinating collection of riddles, logic puzzles, and paradoxes, this book ends up teaching logic in the most delightful way I can imagine. A rare work.]

Mathematics and Miscellany

Dantzig, T. *Number: The Language of Science.* New York: Free Press, 1967. Paperback. [Very readable.]

Ernst, B. *The Magic Mirror of M. C. Escher.* New York: Ballantine Books, 1976. [A readable and informative survey of Escher's life and work by an artist who knew him. This book contains working sketches and interesting explanations of how Escher created his art.]

Escher, M. C. *The Graphic Work of M. C. Escher.* New York: Ballantine Books, 1969. [A collection of drawings and lithographs with comments by the artist.]

Gamow, G. *One, Two, Three . . . Infinity.* New York: Penguin Books, 1977. Paperback. [A good popularization of mathematical ideas.]

Huff, D. *How to Lie with Statistics.* New York: Norton, 1955. [An entertaining and educational explanation, although the language is somewhat dated.]

Jacobs, H. *Mathematics: A Human Endeavor.* San Francisco: W. H. Freeman and Company, 1970. [A delightful book. Shows that a textbook need not be dull. Attempts to show the beauty of mathematics to nonmathematicians.]

Kline, M., ed. *Mathematics: An Introduction to Its Spirit and Use.* San Francisco: W. H. Freeman and Company, 1978. [A collection of articles from *Scientific American*.]

Messick, D. M., ed. *Mathematical Thinking in Behavioral Sciences.* San Francisco: W. H. Freeman and Company, 1968. [Another collection of articles from *Scientific American*.]

Newman, J. *The World of Mathematics.* 4 vols. New York: Simon and Schuster, 1956–1960. [The title is accurate: a marvelous anthology of articles showing how mathematics relates to just about everything (art, literature, music, humor). Every teacher should at least read the table of contents. A great book.]

Perl, T. *Math Equals.* Menlo Park, CA: Addison-Wesley, 1978. [Contains biographies of nine women mathematicians and their work together with mathematics activities that relate to their work.]

Journals

Arithmetic Teacher. Reston, VA: National Council of Teachers of Mathematics. [For elementary teachers.]

Classroom. Seattle: Rational Island Publishers, P.O. Box 2081, Main Office Station, Seattle, WA 98111. [Contains articles by teachers, parents, and students about making schools more human. Published yearly. Numbers 1–6 are available.]

Mathematics Teacher. Reston, VA: National Council of Teacher of Mathematics. [Primarily for secondary teachers.]

Mathematics Teaching. Cambridge: Association of Teachers of Mathematics. [A British publication that features articles of theoretical interest as well as exciting practical applications.]

Catalogs

Creative Publications, P.O. Box 10328, Palo Alto, CA 94303. [A joy to leaf through.]

Cuisenaire Company of America, 12 Church St., New Rochelle, NY 10805. [Much more than Cuisenaire rods in this catalog.]

Educational Teaching Aids, 159 W. Kinzie St., Chicago, IL 60610. [Sales offices on the East and West Coast. A good selection of material that includes multibase arithmetic blocks.]

Midwest Publications Co., Inc., P.O. Box 129, Troy, MI 48084. [Numerous workbooks and activity cards to accompany laboratory equipment.]

National Council of Teachers of Mathematics Publication List. NCTM, 1906 Association Dr., Reston, VA 22091.

SEE, 43 Bridge St., Newton, MA 02195. [A good selection of laboratory equipment.]

Index of Quotations

Bacon, R., 3
Blake, W., 247
Bronowski, J., 184, 253, 279
Carroll, L., 28
Chinese proverb, 70
Crafte of Nombrynge, The, 53
De Morgan, A., 165
Disraeli, B., 270
Dyson, F., 191
Einstein, A., xix
Escher, M., 281, 220
Fontenelle, B., 112
Fuller, B., 161
Galilei, G., 104, 153
Gauss, C., 85
Hambridge, J., 184
Herbart, J., 94
Herodotus, 142
Hildebrand, 17
Hume, D., 207

Jackins, H., 169
Jefferson, T., 196
Keyser, C., 239
al-Khwarizmi, 43
Kline, M., 98
Kovalévsky, S., 122
Lamb, C., 148
Laplace, P., 260
Lichtenberg, G., 131
Marcke, de la, O., 45
Maxwell, J., 257
Melanchton, P., 55
Menninger, K., 12, 14, 22
Montessori, M., xix
Newman, J., 93
Nightingale, F., 267
Ovid, 25
Pacioli, 64
Philolaus, 76
Plato, xix, 79

Plato's Academy, 136
Plutarch, 234
Poincaré, H., 1, 173
Polya, G., 91, 125
Recorde, R., 62
I Samuel, 203
Sandburg, C., 16, 33
Shakespeare, W., 17
Thompson, D., 199
Thoreau, H., 116
Tillich, P., 59
Van Doren, M., 274
Vinci, da, L., 213
Washington, G., 156
Weissglass, J. 38, 188
Wells, H. G., 270
Weyl, H., 244
Whitehead, A., 47, 100
Wickham, A., 253
Wordsworth, W., 226

Index

Abacus, 39–43, 45–46
A-blocks (attribute blocks), 1
  instructions for making, xvii
  notation for, 5
Absolute value, 131–132
Acute angles, 157
Addition of decimals, 191–192
Addition of rational numbers,
    109–110, 112–113
  associative property of, 113
  commutative property of, 113
Addition of whole numbers,
    28–29, 33
  with abacus, 42–43
  algorithms for, 43–48
  associative property of, 34
  commutative property of, 34
  with Egyptian numerals, 38
  with Mayan numerals, 39
Additive principle, 18, 22
Algebraic expression, 78
Algorithms:
  for addition and subtraction, 38–52
  origin of word, 46
Altitude, 161, 163
  compared with altitude line, 161
Altitude line, 160, 163
  compared with altitude, 161
Angle, 154–155, 156–157
  acute, 157
  obtuse, 157
Angle-side-angle congruence axiom,
    162–163

Area, 169–187
  definition of, 169
  of parallelogram, 170, 178
  of rectangle, 170, 177–178
  of trapezoid, 172, 179
  of triangle, 170, 178–179
Array, 55
Associative property:
  of addition, 34, 113, 126, 127
  of multiplication, 54, 56, 115,
    118, 132
Attribute blocks: *See* A-blocks
Average, 189, 268, 271

Babylonians, 157
Basic principle of measurement, 204
Bisector:
  of angle, 155
  of line segment, 154

Center of gravity. *See* Centroid
Centimeter, 202, 204
Centroid, 160, 163
Changing shapes, 146
Circumcenter, 164
Closed curve, 144
Code, 22
Commutative property:
  of addition, 34, 113, 124, 126, 127
  of multiplication, 54, 55–56, 115,
    118, 130, 132
Complement, 6–7
Composite number, 77, 79

Congruence axioms, 158, 161–163
Congruent figures, 148
Congruent regions, 100
Connected network, 94
Connected region, 149
Convex polygon, 147
Counting, 12–17
Counting boards, 45, 46, 47
Counting numbers, 16
  as subset of rational numbers, 108
Cubit, 200
Curve, 144, 148–150
  closed, 144, 148–149
  simple closed, 144, 149

Decimal point, 190
Decimals, 188–198
  addition of, 191–192
  division of, 193–194
  multiplication of, 192–193
  notation for, 190, 191
  repeating, 195, 196
  terminating, 191, 195, 196
Decimeter, 202, 204
Degree (of vertex), 92
Degree (angle measure), 157
Dekameter, 202, 204
Difference:
  of trains, 32, 35
  of sets, 34
Digit, 200
Distance-preserving property, 154

Distance-preserving transformation, 227, 234
Distributive property:
  for division, 72
  of multiplication over addition, 54, 56, 115, 118, 130, 132
  of multiplication over subtraction, 57
Division of decimals, 193–194
Division of rational numbers, 115, 119, 133
Division of whole numbers, 67–75
  definition of, 71
  long, 73
  with MBA-blocks, 68–70
  right-hand distributive property of, 72
Divisor, 68, 77, 79
Duplation, 59–60, 62

Egyptian numerals, 17
  addition and subtraction of, 38–39
Empty set, 4
Equiangular polygon, 215
Equilateral polygon, 215
Equivalent fraction, 102–103, 106–107
Eratosthenes, sieve of, 80
Errors in measurement, 206–209
ESCHER, M. C., 213
Euclidean algorithm, 86
EULER, L., 90
Euler's formula, 94–98
Exponents, 81, 192
Exterior (of curve), 149
Exterior angles, 214

Factor, 77, 79
  common, 83
  greatest common, 83
Factoring:
  with C-rods, 77
  by factor trees, 81
Factory game, 23
FERMAT, P., 79
First law of exponents, 192
Fraction(s), 101, 106
  denominator of, 106
  equivalent, 102–103, 106–107
  numerator of, 106
  reduced form of, 107
Fundamental theorem of arithmetic, 81

GARFIELD, J., proof of Pythagorean theorem, 183–184
GCF: See Greatest common factor
Geoblocks, 136
Geometry, meaning of, 142
Goldbach's conjecture, 86–87
Gram, 209
Greater than relationship:
  for rational numbers, 108
  for whole numbers, 35
Greatest common factor (GCF), 83
  finding by Chinese method, 84–85
  finding by Euclidean algorithm, 86
Greenland Eskimo number words, 16
Grouping, 22
Group of symmetries, 253

Hectometer, 202, 204
Hindu-Arabic numerals, 22–23
  computation with, 43–44, 47–49
Hypotenuse of right triangle, 179

Identity element for multiplication, 54, 56, 118
  of symmetry transformations, 252
Identity transformation, 153
Image, 153–226
Incenter, 165
Integers, 123, 125
  addition of, 125
  addition of with red and black counters, 127
  multiplication of, 128–133
  subtraction of, 126
Interior (of curve), 149
Interior angles, 214
Intersection, 4
Inverse of symmetry transformation, 252, 253–254
Irrational number, 185
  decimal representation of, 197
Isometry, 239
Isosceles triangle, 159

Kilometer, 202, 204
Königsberg bridge problem, 90–94

Lattice method (for multiplication), 61–62, 64
Least common multiple (LCM), 83–84
Legs (of right triangle), 179
Less than relationship:
  for rational numbers, 168
  for whole numbers, 36
Lines, 141, 143
  parallel, 141, 144
  skew, 141, 144
Line segment, 143
Line of symmetry, 244
Liter, 209

Match, matching, 10
Mayan numerals, 20–21
  addition and subtraction of, 39
Mean, 189, 268, 271
Measurement, 199–212
Measures of central tendency, 271
Median:
  of set of numbers, 268, 271
  of triangle, 160, 163
Meter, 201, 202, 203–204
Metric system, 201–205
Milliliter, 209
Millimeter, 202, 204

Mira, 153–154
Mira line, 154
Mode, 268, 271
Moebius strip, 277–279
Multiple, 83
Multiplication of decimals, 192–193
Multiplication of rational numbers, 113–119
  definition of, 117, 133
  properties of, 115, 118
Multiplication of symmetries, 250–254
Multiplication of whole numbers, 53–66
  associative property of, 54, 56
  commutative property of, 54, 55–56
  definition of, 55
  distributive property of, 54, 56, 57
  duplication of, 59–60
  grating or lattice method of, 61–62, 64
  with MBA-blocks, 60
  partial-products algorithm for, 63–64
  of trains of C-rods, 53
Multiplication principle for probabilities, 261–262
Multiplicative principle, 19, 22

Negative integers, 123
Negative numbers, 122–134
  addition of, 122–125
  subtraction of, 124, 126
Negative rational numbers, 125
Network, 91
n-gon, 215
Number, 14, 22
  odd and even, 87
Number line, 108, 122–123
Number names, 14, 22
Number theory, 76–89
Numerals, 17, 22
  Chinese, 19–20
  Egyptian, 18
  Hindu-Arabic, 22–23
  Mayan, 20–21
  Roman, 17–18
Numeration systems, 17–25

Obtuse angles, 157
Odds, 262–264
One-to-one correspondence, 15
Opposites, 125
Orthocenter, 161, 163

Parallel lines, 141, 144
  construction of, 156
Parallelogram, area of, 170
Parallel planes, 143
Partial-products algorithm:
  for division, 73
  for multiplication, 63–64
Path, 94

Index

Pattern blocks, 216
Pentominos, 220
Percent, 194
Percentage error, 207
Perpendicular, 155
Perpendicular bisector, 155
Pick's formula, 276
Plane, 141–144
Point, 143
Polygon, 146–148
   equiangular, 215
   equilateral, 215
   regular, 215
   sum of angles of, 214–215
Polyhedron, 97
Positional numeration system, 20, 25
Positive integers, 123
Prime numbers, 77–79
Probability, 257–266
   of event (definition), 260
   multiplication principle for, 261–262
   with or without replacement, 262
PYTHAGORAS, 79, 184
Pythagorean theorem, 179–186
   Bhaskara's proof of, 185
   Chinese method of proof of, 184
   Garfield's proof of, 183–184

Quotient, 68, 119
   and remainder, 72

Range, 270
Rational numbers, 100–121, 125
   addition of, 109–110, 112–113, 125
   division of, 115, 119, 133
   multiplication of, 113–115, 116–118
   positive and negative, 125
   subtraction of, 110, 113
   trichotomy law for, 108
Ray, 143
Reciprocal, 116
Reciprocal property, 118
Rectangle, area of, 170
Reduced form of fraction, 107
Reflectional symmetry, 246
Reflections:
   with Mira, 153–154, 226
   products of, 230–232, 237–240
   rotations and, 239
   translations and, 232–235
Regions, 94, 144–145
Regular polygon, 215
   measure of angle of, 215, 216, 218–219
Regular tessellation, 216–219
Repeating decimal, 195, 196
Reptiles (Escher), 213
Right angles, 157
Roman numerals, 18–19
   subtractive principle for, 19, 22
Rotational symmetry, 244
Rotations, 237–239

Second law of exponents, 192
Semiregular tessellation, 217, 222–223
Set(s), 3
   empty, 4
   equal, 3
Set theory, 1–11
Side-angle-side congruence axiom, 158, 161
Simple closed curve, 144
Skew lines, 141, 144
Soroban (Japanese abacus), 45–46
Space, 141, 143
Span, 200
Sphere, 145
Square root, 181–182
Standard deviation, 270, 271
Statistics, 267–273
Suan-pan (Chinese abacus), 45
Subset, 3
Subtraction:
   with abacus, 42–43
   additive (or Austrian) method of, 44
   with Egyptian numerals, 39
   with Mayan numerals, 45
   of rational numbers, 110, 113
   of whole numbers, 29, 34–35
Sum of angles of polygon, 214–215, 218
Surface, 141
Symmetries: *See* Symmetry transformations
Symmetry, 243–256
   lines of, 244, 246
   reflectional (bilateral), 246
   rotational, 244
   translational, 247, 249
Symmetry transformations (symmetries), 246
   algebra of, 253–254
   of equilateral triangle, 250–252
   product of, 250–252, 253–254
   of rectangle, 252

Tally sticks, 17
Tangrams, 136–141
Terminating decimal, 191, 195, 196
Tessellation, 213–225
   definition of, 216
   with parallelograms, 219
   with pentominos, 220
   with quadrilaterals, 219–220
   regular, 216
   semiregular, 217
   with triangles, 219
Tiling (of plane), 216
Topology, 90–99
Torus, 145
Trains (of C-rods), 29
   difference of, 32, 35
   sum of, 32
Transformations, 226–242
   distance-preserving, 227, 234
   product of, 230
Translational symmetry, 246
Translations, 227–232, 234–235
Trapezoid, 172
Tree diagram, 261
Triangles, 158–166
   area of, 170
Trichotomy law:
   for rational numbers, 168
   for whole numbers, 36
Twin primes, 86–87

Union, 5
Universe, 6

Value, 2
Variance, 270
Venn diagram, 5
Vertex:
   of angle, 156
   of network, 91
   of tessellation, 216

WEISSGLASS, J., 112
Well-defined operation, 112
Whole number, 16
Win the Block (game), 13